T0319512

**Deterministic and Stochastic Modeling
in Computational Electromagnetics**

Deterministic and Stochastic Modeling in Computational Electromagnetics

Integral and Differential Equation Approaches

Dragan Poljak and Anna Šušnjara
University of Split
Croatia

IEEE Antennas and
Propagation Society

IEEE Press Series on Electromagnetic Wave Theory

IEEE PRESS
WILEY

Published by John Wiley & Sons, Inc., Hoboken, New Jersey.
Published simultaneously in Canada.

For general information on our other products and services or for technical support, please contact our Customer Care Department within the United States at (800) 762-2974, outside the United States at (317) 572-3993 or fax (317) 572-4002.

Wiley also publishes its books in a variety of electronic formats. Some content that appears in print may not be available in electronic formats. For more information about Wiley products, visit our web site at www.wiley.com.

Library of Congress Cataloging-in-Publication Data
Names: Poljak, D. (Dragan), author. | Šušnjara, Anna, author.
Title: Deterministic and stochastic modeling in computational
   electromagnetics : integral and differential equation approaches /
   Dragan Poljak, Anna Šušnjara.
Description: Hoboken, New Jersey : Wiley, [2024] | Includes index.
Identifiers: LCCN 2023037537 (print) | LCCN 2023037538 (ebook) | ISBN
   9781119989240 (hardback) | ISBN 9781119989257 (adobe pdf) | ISBN
   9781119989264 (epub)
Subjects: LCSH: Electromagnetism–Mathematical models. | Stochastic models.
Classification: LCC QC760 .P625 2024 (print) | LCC QC760 (ebook) | DDC
   537.01/515–dc23/eng/20231017
LC record available at https://lccn.loc.gov/2023037537
LC ebook record available at https://lccn.loc.gov/2023037538

Cover Image and Design: Wiley

Set in 9.5/12.5pt STIXTwoText by Straive, Pondicherry, India

To our beloved ones...

Contents

About the Authors

Dragan Poljak (Senior Member, IEEE) received the Ph.D. degree in electrical engineering from the University of Split, Croatia, in 1996. He is currently a full professor with the Department of Electronics and Computing, University of Split. He is also involved in ITER Physics EUROfusion Collaboration and in the Croatian Center for Excellence in Research for Technology Sciences. He has published more than 160 journals and 250 conference papers and authored some books, e.g. two by Wiley, Hoboken, NJ, USA, and one by Elsevier, St. Louis, MO, USA. His research interests include computational electromagnetics (electromagnetic compatibility, bioelectromagnetics, and plasma physics). From May 2013 to June 2021, he was a member of the Board of the Croatian Science Foundation. He is a member of the Editorial Board of Engineering Analysis with Boundary Elements, Mathematical Problems in Engineering, and IET Science, Measurement and Technology. He was awarded several prizes for his achievements, such as the URSI Young Scientists Award in 1999, the National Prize for Science in 2004, the Croatian Section of IEEE Annual Award in 2016, the Technical Achievement Award of the IEEE EMC Society in 2019, and the George Green Medal from the University of Mississippi in 2021. He is active in a few working groups of the IEEE/International Committee on Electromagnetic Safety (ICES) Technology Committee 95 SC6 EMF Dosimetry Modeling.

Anna Šušnjara received her PhD degree in electrical engineering from the University of Split, Croatia, in 2021. She is currently a postdoc researcher at the Faculty of Electrical Engineering, Mechanical Engineering and Naval Architecture, University of Split. Her research interests include numerical modeling, uncertainty quantification, and sensitivity analysis in computational electromagnetics. Dr. Šušnjara is involved in ITER physics EUROfusion collaboration. From 2015 to 2021, she was a member of EUROfusion work package for code development in European Transport Solver (ETS), while from 2021, she has been a member of IFMIF-DONES project. Dr. Šušnjara is a member of IEEE and BIOEM societies. She currently serves as Vice President of IEEE EMC Croatian chapter. To date, Dr. Šušnjara has (co)authored 19 journal and more than 40 conference papers. She serves as a reviewer for seven journals and two conferences. Dr. Šušnjara gave lectures about computational electromagnetics at several European academic institutions and tutorials at international scientific conferences. She was awarded several prizes for her achievements. In 2023, she received the URSI Young Scientist Award at the 35th URSI GASS in Sapporo, Japan. She was also awarded the National Prize for Science and the University of Split Prize for Science in 2021 and 2022, respectively. In 2016, she received the best poster paper award at BioEM conference in Ghent, Belgium, and spent one month at Politecnico di Torino as SPI2016 Young Investigator Training Program awardee.

Preface

Most of the computational models used in engineering electromagnetics are deterministic in nature, i.e. one deals with an exact set of input data in a sense of either material properties or geometry. However, in many scenarios, there are problems with uncertainty in the input data set as some system properties are partly or entirely unknown. Therefore, a stochastic approach is required to determine the relevant statistics about the given responses, thus providing the assessment of the related confidence intervals in the set of numerical results obtained as an output of a given deterministic model. Of particular interest are nonintrusive stochastic approaches that could be easily coupled with widely used well-established deterministic models, by efficiently postprocessing numerical results arising from deterministic models.

The goal of this book is to demonstrate the efficiency of parallel use of deterministic and stochastic models featuring combination of well-established analytical/ numerical methods with stochastic analysis techniques. The nonintrusive stochastic approach presented in the book can be readily incorporated into majority of computational electromagnetics (CEM) models with little effort aiming to quickly provide a more detailed insight into the relationship between the input parameters and the output of interest.

A variety of examples throughout the book are presented to clearly demonstrate the efficiency of deterministic-stochastic approaches in CEM models, and a reference list is given at the end of each chapter. The book provides computational examples illustrating successful application of stochastic collocation (SC) technique in the areas of ground-penetrating radars (GPRs), grounding systems, radiation from 5G systems, human exposure to electromagnetic fields, transcranial magnetic stimulation (TMS), transcranial electric stimulation (TES), transient analysis of buried wires, and design of instrumental landing system (ILS).

The book is divided into three parts. Part I outlines the fundamentals of classical electromagnetics and basics of numerical modeling. Part II deals with deterministic models pertaining to analysis of thin wires in both frequency and time domains, human exposure to electromagnetic fields in GHz frequency range, and multiphysics phenomena such as plasma confinement in tokamak. Finally, Part III is entirely devoted to stochastic modeling covering a detailed description of SC method and sensitivity analysis. Part III also contains a number of applications arising from electromagnetic-thermal dosimetry, biomedical applications, electromagnetic compatibility (EMC), GPRs, grounding systems, air traffic control systems, and transport phenomena in tokamak.

The material given in this book is dominantly based on papers previously published by the authors, suitably modified to present the results in a uniform design and format. Additional related material is planned to be prepared aiming to complete some details and extensions of the published work. An extensive reference list of other related work is also included. The goal is to provide a reference on the deterministic-stochastic modeling in different areas of CEM covering frequency and time domain analyses of wire antennas and their various applications, EMC, computational models of lines and cables, lightning, grounding systems, GPRs, magnetohydrodynamics, bioelectromagnetics, and biomedical applications of electromagnetic fields.

As the book covers multidisciplinary phenomena, such as electromagnetic-thermal dosimetry, magnetohydrodynamics, and plasma physics, the authors hope it could be of interest to multidisciplinary researchers, engineers, physicists, and mathematicians.

The Authors
Split, Croatia, 2023

Part I

Some Fundamental Principles in Field Theory

1

Least Action Principle in Electromagnetics

Laws of nature are governed by following fundamental principles – the action principle, locality, Lorentz invariance, and gauge invariance [1]. Hamilton's principle, or the least action principle, is originally developed for classical mechanics stating that a particle, among all of the trajectories between fixed time instants t_1 and t_2, follows the path which minimizes the *action*. Action is defined as time integral of the difference between the kinetic energy and potential energy, respectively. Thus, Hamilton's principle somehow requires the time averages of the kinetic energy and potential energy to distribute as equally as possible (equipartition) [2]. In classical mechanics, Hamilton's principle and Newton's second law represent equivalent formulations.

An extension of Hamilton's principle from classical mechanics to classical electromagnetics can be undertaken starting with the analysis of the motion of single charged particle [3]. Next step is to construct a Lagrangian for the electromagnetic field by extending the Lagrangian pertaining to classical mechanics. From the corresponding Lagrangians, featuring Noether's theorem and gauge invariance, it is possible to derive equation of continuity for the charge, Lorentz force, and Maxwell's equations, which can be found elsewhere, e.g. [2–5].

Generally, when a functional is extremal, Noether's theorem yields the conservation law. Thus, invariance of the system under a time translation results in the energy conservation. It is also worth noting that space translation invariance corresponds to the conservation of linear momentum, rotation invariance corresponds to the conservation of angular momentum, while gauge invariance yields the charge conservation [1, 2].

These derivations are recently reviewed in [6–8].

This chapter first deals with derivation of continuity equation and Lorentz force, and a derivation of Maxwell's equations from the electromagnetic field

Deterministic and Stochastic Modeling in Computational Electromagnetics: Integral and Differential Equation Approaches, First Edition. Dragan Poljak and Anna Šušnjara.
© 2024 The Institute of Electrical and Electronics Engineers, Inc.
Published 2024 by John Wiley & Sons, Inc.

functional is carried out. Finally, a variational basis of numerical solution methods in electromagnetics is discussed.

1.1 Hamilton Principle

Hamilton variational principle represents not only the basis of modern analytical dynamics but also of universal physical laws, i.e. fundamental laws of classical physics can be understood in terms of action.

This section first deals with Hamilton's variational principle in mechanics, Newton's equation of motion, and Noether's theorem. Then the variational principle in electromagnetics is discussed.

For simplicity, a system with one degree of freedom represented by a generalized coordinate q is considered together with related function of position, velocity, and time $L(q, \dot{q}, t)$ where $\dot{q}$ denotes the time derivative.

The task is to determine how a point particle should move in this one-dimensional space so that the time integral of L is minimized compared with the integral over the conceivable paths between the same starting and end points, as depicted in Fig. 1.1. The solution is given by stating q as a function of time $q = q(t)$.

To compare all paths having the same starting and end points, the variation of function q is zero at both ends [1–4]

$$\delta q = 0 \tag{1.1}$$

i.e. all alternatives start at instant t_1 and arrive together at instant t_2.

The minimum condition is then given by a functional F expressed in terms of the integral [1–4]

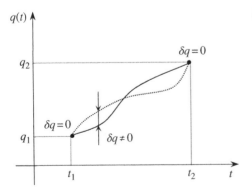

Figure 1.1 The varied function $q(t)$.

$$F = \int_{t_1}^{t_2} L(q, \dot{q}, t)dt = \text{min} \tag{1.2}$$

In classical mechanics, function L is referred to as Lagrangian and is expressed as

$$L = W_{kin} - W_{pot} \tag{1.3}$$

where W_{kin} and W_{pot} are the kinetic energy and potential energy, respectively.

According to the calculus of variation, the functional approaches minimum value

$$F = \int_{t_1}^{t_2} \left(W_{kin} - W_{pot}\right)dt = \text{min} \tag{1.4}$$

when its variation vanishes, i.e.:

$$\delta F = 0 \tag{1.5}$$

which can also be written as

$$\delta \int_{t_1}^{t_2} Ldt = 0 \tag{1.6}$$

Therefore, function $q(t)$ minimizes functional (1.2) or (1.4), respectively, and it follows

$$\delta F = \int_{t_1}^{t_2} \delta Ldt \tag{1.7}$$

For the simplest case given by $L(q, \dot{q}, t)$ the variation of function L is given by

$$\delta L = \frac{\partial L}{\partial q}\delta q + \frac{\partial L}{\partial \dot{q}}\delta \dot{q} \tag{1.8}$$

And by performing some further mathematical manipulation, one readily obtains

$$\delta F = \int_{t_1}^{t_2} \left[\frac{\partial L}{\partial q} - \frac{d}{dt}\left(\frac{\partial L}{\partial \dot{q}}\right)\right]\delta qdt + \frac{\partial L}{\partial \dot{q}}\delta q\Big|_{t_1}^{t_2} = 0 \tag{1.9}$$

As $\delta q = 0$ at the ends of the path, the second term at the right-hand side automatically vanishes.

Furthermore, according to the *fundamental lemma* of variational calculus [4] the first integral term at the right-hand side of (1.9) vanishes if the following condition is satisfied

$$\frac{\partial L}{\partial q} - \frac{d}{dt}\left(\frac{\partial L}{\partial \dot{q}}\right) = 0 \tag{1.10}$$

It is worth noting that the second order differential Eq. (1.10) relates the position q for the time t, and also determines the true path of the system when two end positions and times are given. This equation is known as Lagrange–Euler equation of motion [1, 2].

In the case of a single function of multiple variables, $L(q_k, \dot{q}_k, \xi, t)$ (1.10) becomes

$$\frac{\partial L}{\partial q} - \frac{\partial L}{\partial \xi}\left(\frac{\partial L}{\partial q_\xi}\right) - \frac{\partial}{\partial t}\left(\frac{\partial L}{\partial q_t}\right) = 0 \tag{1.11}$$

Finally, if few functions of multiple independent variables $L(q_k, \dot{q}_k, x, y, z, t)$ are considered, it follows

$$\sum_{k=1}^{N}\left\{\frac{\partial L}{\partial q_k} - \frac{\partial}{\partial x}\left[\frac{\partial L}{\partial(q_k)_x}\right] - \frac{\partial}{\partial y}\left[\frac{\partial L}{\partial(q_k)_y}\right] - \frac{\partial}{\partial z}\left[\frac{\partial L}{\partial(q_k)_z}\right]\frac{\partial}{\partial t}\left(\frac{\partial L}{\partial(q_k)_t}\right)\right\} = 0 \tag{1.12}$$

which, for example, pertains to three-dimensional problems in electromagnetics.

Hamilton variational principle can be considered a general law not only for particle dynamics but also for the dynamics of continuous materials.

An extension to three-dimensional problems in continuous materials, i.e. for physical fields, the variational principle corresponding to Eq. (1.6) is given by

$$\delta \int_{t_1}^{t_2} \int_{V} \bar{L}_d \, dV dt = 0 \tag{1.13}$$

where $\bar{L}_d$ is the so-called Lagrange density defined as

$$L = \int_{V} \bar{L}_d \, dV \tag{1.14}$$

and has a unit of energy per volume.

It is worth noting that the variational principle is an invariant for coordinate transformations [1].

1.2 Newton's Equation of Motion from Lagrangian

Lagrangian in classical mechanics for a particle with mass m with displacement $\vec{r}$ at time t $L_0\left(\vec{r}, \dfrac{d\vec{r}}{dt}, t\right)$ is of the form

$$L_0\left(\vec{r}, \frac{d\vec{r}}{dt}, t\right) = \frac{1}{2}m\left(\frac{d\vec{r}}{dt}\right)^2 - W_{pot}\left(\vec{r}\right) \tag{1.15}$$

where $\dfrac{d\vec{r}}{dt}$ is the particle velocity and $W_{pot}\left(\vec{r}\right)$ stands for potential energy (not due to electromagnetic field).

The corresponding action F_0 related to Lagrangian (1.15) is given by integral

$$F_0 = \int_{t_1}^{t_2} L_0 dt \tag{1.16}$$

Now varying the action (1.16)

$$\delta F_0 = 0 \tag{1.17}$$

It can be written as:

$$\frac{\partial L}{\partial x} = -\frac{\partial W_{pot}}{\partial x}, \frac{\partial L}{\partial y} = -\frac{\partial W_{pot}}{\partial y}, \frac{\partial L}{\partial z} = -\frac{\partial W_{pot}}{\partial z} \tag{1.18}$$

$$\frac{\partial L}{\partial \dot{x}} = m\dot{x}, \quad \frac{\partial L}{\partial \dot{y}} = m\dot{y}, \quad \frac{\partial L}{\partial \dot{z}} = m\dot{z} \tag{1.19}$$

and it follows:

$$
\begin{aligned}
-\frac{\partial W_{pot}}{\partial x} - \frac{d}{dt}(m\dot{x}) &= 0 \\
-\frac{\partial W_{pot}}{\partial y} - \frac{d}{dt}(m\dot{y}) &= 0 \\
-\frac{\partial W_{pot}}{\partial z} - \frac{d}{dt}(m\dot{z}) &= 0
\end{aligned} \tag{1.20}
$$

which simply gives:

$$
\begin{aligned}
\frac{d^2(mx)}{dt^2} &= -\frac{\partial W_{pot}}{\partial x} \\
\frac{d^2(my)}{dt^2} &= -\frac{\partial W_{pot}}{\partial y} \\
\frac{d^2(mz)}{dt^2} &= -\frac{\partial W_{pot}}{\partial z}
\end{aligned} \tag{1.21}
$$

If mass m is regarded as constant quantity, right-hand side of (1.21) represents the components of mechanical force:

$$F_{mech,x} = m\frac{d^2x}{dt^2}$$

$$F_{mech,y} = m\frac{d^2y}{dt^2} \qquad (1.22)$$

$$F_{mech,z} = m\frac{d^2z}{dt^2}$$

As the mechanical force is defined as a negative gradient of the potential energy

$$\vec{F}_{mech} = -\nabla W_{pot} \qquad (1.23)$$

one simply obtains a set of Newton's equations of motion

$$m\frac{d^2\vec{r}}{dt^2} = -\nabla W_{pot}\left(\vec{r}\right) \qquad (1.24)$$

where the right-hand side of (1.24) represents a force acting on the particle.

1.3   Noether's Theorem and Conservation Laws

There are physical quantities that do not change throughout the time development of physical systems. These quantities are stated to be conserved under certain conditions which are governed by conservation laws. It can be shown that conservation laws are a consequence of the symmetry properties of a physical system (invariance properties of a system under a group of transformations [2, 9]). The symmetry properties of the system and conservation laws are connected with Noether's theorem, e.g. [2, 9].

Let $u_k(x)$ $(k = 1,2,...n)$ be a set of differentiable functions of the independent variable x and let $v_k(x)$ be the first derivatives, i.e. it can be written as

$$v_k(x) = \frac{du_k}{dx} \qquad (1.25)$$

Now Lagrangian L is defined as a function of x, and n functions of u_k and n functions of v_k.

$$L = L[x, u_k, v_k] \qquad (1.26)$$

If one considers an infinitesimal transformation T

$$x \rightarrow x' = x + \delta x \qquad (1.27)$$

Furthermore, under T, one consequently has:

$$u_k(x) \rightarrow u'_k(x') = u_k(x) + \delta u_k(x) \tag{1.28}$$

$$v_k(x) \rightarrow v'_k(x') = v_k(x) + \delta v_k(x) \tag{1.29}$$

Now assuming the functional

$$S = \int_\Omega L dx \tag{1.30}$$

to be invariant under T so that the transformation (1.30) which maps the interval Ω into Ω' does not change, it follows

$$\delta S = \int_{\Omega'} L' dx - \int_\Omega L dx = 0 \tag{1.31}$$

Performing some mathematical manipulations, one obtains the following expression

$$\left[\frac{d}{dx}\left(\frac{\partial L}{\partial v_k}\right) - \frac{\partial L}{\partial u_k} \right](\delta u_k - v_k \delta x) = \frac{d}{dx}\left[L\delta x + \frac{\partial L}{\partial v_k}(\delta u_k - v_k \delta x) \right] \tag{1.32}$$

Finally, Noether's theorem states that if functional S is invariant under the infinitesimal one-parameter group of transformations T, then the set of n equations

$$\left[\frac{d}{dx}\left(\frac{\partial L}{\partial v_k}\right) - \frac{\partial L}{\partial u_k} \right] = 0 \tag{1.33}$$

simply gives

$$\frac{d}{dx}\left[L\delta x + \frac{\partial L}{\partial v_k}(\delta u_k - v_k \delta x) \right] = 0 \tag{1.34}$$

i.e. it can be written as

$$L\delta x + \frac{\partial L}{\partial v_k}(\delta u_k - v_k \delta x) = const \tag{1.35}$$

namely, the expression is independent of x variable.

On the other hand, Newton's equation of motion can be expressed in terms of Lagrangian being a function of the generalized coordinates q_k ($k = 1, 2, ..., n$) and their derivatives, i.e.

$$\frac{d}{dt}\left(\frac{\partial L}{\partial \dot{q}_k}\right) - \frac{\partial L}{\partial q} = 0 \tag{1.36}$$

which is equivalent to the request

$$\delta \int_{t_1}^{t_2} L \, dt = 0 \tag{1.37}$$

Furthermore, Hamilton's principle (the least action principle), which states that solutions q_k of the system of mass points for which the integral in (C6) is an extremum contrary to all functions q'_k

$$q'_k(t) = q_k(t) + \delta q_k(t) \tag{1.38}$$

If one compares (1.34) with (1.35) with $\delta x = 0$ and $\delta u_k = 0$, it follows that these statements are identical Namely, it can be written as

$$L\delta t + \frac{\partial L}{\partial \dot{q}_k}(\delta q_k - \dot{q}_k \delta t) = const \tag{1.39}$$

i.e. the left-hand side is a constant of motion.

As any time translation is expressed as $\delta t \neq 0$ and $\delta q_k = 0$, the following expression is obtained

$$H = \frac{\partial L}{\partial \dot{q}_k} \dot{q}_k - L = const \tag{1.40}$$

stating that Hamiltonian H which is equal to the total energy of the system (sum of kinetic energy and potential energy) is constant of the motion.

In other words, time translation symmetry corresponds to the conservation of energy. Therefore, Noether's theorem states that for a system with time translation symmetry, the energy of the system is conserved.

The energy conservation can be demonstrated by using Lagrangian and Hamiltonian in classical mechanics for one-dimensional case.

Thus, Lagrangian is of the form

$$L = \frac{1}{2}m\dot{x}^2 - W_{pot}(x) \tag{1.41}$$

where

$$\dot{x} = \frac{dx}{dt} \tag{1.42}$$

while Hamiltonian of the system, according to (1.40), is then

$$H = \frac{\partial L}{\partial \dot{x}}\dot{x} - L \tag{1.43}$$

As from (1.42), one simply obtains

$$\frac{\partial L}{\partial \dot{x}} \dot{x} = m\dot{x} \tag{1.44}$$

and the Hamiltonian now can be written as

$$H = \frac{1}{2} m\dot{x}^2 + W_{pot}(x) \tag{1.45}$$

For a system with a time translation symmetry

$$t \rightarrow t' = t + \delta t \tag{1.46}$$

the Hamiltonian of the system remains invariant.

To check if the Hamiltonian is a conserved quantity under a time translation, it should be differentiated with respect to time, i.e. it follows

$$\frac{dH}{dt} = \frac{d}{dt}\left(\frac{\partial L}{\partial \dot{x}} \dot{x} - L\right) = \frac{d}{dt}\left(\frac{\partial L}{\partial \dot{x}} \dot{x}\right) - \frac{dL}{dt} \tag{1.47}$$

and, one has

$$\frac{d}{dt}\left(\frac{\partial L}{\partial \dot{x}} \dot{x}\right) = \frac{d}{dt}\left(\frac{\partial L}{\partial \dot{x}}\right)\dot{x} + \frac{\partial L}{\partial \dot{x}}\frac{d\dot{x}}{dt} = \frac{d}{dt}\left(\frac{\partial L}{\partial \dot{x}}\right)\dot{x} + \frac{\partial L}{\partial \dot{x}}\ddot{x} \tag{1.48}$$

$$\frac{dL}{dt} = \frac{\partial L}{\partial x}\frac{dx}{dt} + \frac{\partial L}{\partial \dot{x}}\frac{d\dot{x}}{dt} + \frac{\partial L}{\partial t} = \frac{\partial L}{\partial x}\dot{x} + \frac{\partial L}{\partial \dot{x}}\ddot{x} + \frac{\partial L}{\partial t} \tag{1.49}$$

Now, inserting (1.48) and (1.49) into (1.47) yields

$$\frac{dH}{dt} = \frac{d}{dt}\left(\frac{\partial L}{\partial \dot{x}}\right)\dot{x} + \frac{\partial L}{\partial \dot{x}}\ddot{x} - \left(\frac{\partial L}{\partial x}\dot{x} + \frac{\partial L}{\partial \dot{x}}\ddot{x} + \frac{\partial L}{\partial t}\right) \tag{1.50}$$

and one obtains

$$\frac{dH}{dt} = \frac{d}{dt}\left(\frac{\partial L}{\partial \dot{x}}\right)\dot{x} - \frac{\partial L}{\partial x}\dot{x} - \frac{\partial L}{\partial t} \tag{1.51}$$

which can be written:

$$\frac{dH}{dt} = \left(\frac{d}{dt}\frac{\partial L}{\partial \dot{x}} - \frac{\partial L}{\partial x}\right)\dot{x} - \frac{\partial L}{\partial t} \tag{1.52}$$

Now, as Euler Lagrange equation vanishes

$$\frac{d}{dt}\frac{\partial L}{\partial \dot{x}} - \frac{\partial L}{\partial x} = 0 \tag{1.53}$$

and (1.52) becomes

$$\frac{dH}{dt} = -\frac{\partial L}{\partial t} \tag{1.54}$$

Now, considering Lagrangian (1.41) it is evident that its time derivative is zero, as there is no explicit time dependence. Consequently, the total time derivative of the Hamiltonian is also zero, and it can be concluded that Hamiltonian is conserved under the time translation (1.46).

It is worth noting that space-time invariance corresponds to the conservation of linear momentum, and rotation invariance corresponds to the conservation of angular momentum.

1.4 Equation of Continuity from Lagrangian

A change in system which does not affect the action integral, or the equation of motions, is referred to as invariance or symmetry [1].

Lagrangian L in classical mechanics is defined as difference between kinetic energy (W_{kin}) and potential energy (W_{pot}) of the system

$$L = W_{kin} - W_{pot} \tag{1.55}$$

while the corresponding Euler–Lagrange equation of motion governing the considered trajectory of a particle is given by

$$\frac{\partial L}{\partial q} - \frac{d}{dt}\left(\frac{\partial L}{\partial \dot{q}}\right) = 0 \tag{1.56}$$

where q pertains to the position of the particle and $\dot{q}$ denotes its time derivative. Furthermore, Lagrangian for particle of mass m with displacement $\vec{r}$ at time t $L\left(\vec{r}, \dfrac{d\vec{r}}{dt}, t\right)$ is given by [3, 4]

$$L\left(\vec{r}, \frac{d\vec{r}}{dt}, t\right) = \frac{1}{2}m\left|\frac{d\vec{r}}{dt}\right|^2 - W_{pot}\left(\vec{r}\right) = \frac{1}{2}m\dot{r} - W_{pot}\left(\vec{r}\right) \tag{1.57}$$

where $\left|\dfrac{d\vec{r}}{dt}\right| = \dot{r}$ is the particle velocity and $W_{pot}\left(\vec{r}\right)$ stands for potential energy (in the absence of electromagnetic field).

The corresponding action F related to Lagrangian (1.57) is given by integral

$$F = \int_{t_1}^{t_2} L\, dt \tag{1.58}$$

Now varying action (1.58)

$$\delta F = 0 \tag{1.59}$$

yields Newton's second law:

$$m\frac{d^2\vec{r}}{dt^2} = -\nabla W_{pot}\left(\vec{r}\right) \tag{1.60}$$

where the right-hand side represents a force acting on the particle.

Now, according to the symmetry of the Lagrangian stemming from Noether's theorem [2, 3, 9], a total time derivative may be added to Lagrangian (1.57) without changing the equation of motion [3].

Thus, choosing an arbitrary differentiable scalar function $q\Lambda\left(\vec{r}, t\right)$, it can be written as

$$L' = L + q\frac{d\Lambda}{dt} \tag{1.61}$$

where q stands for the electric charge.

Now, a new action F' can be defined as follows

$$F' = \int_{t_1}^{t_2} L'dt \tag{1.62}$$

and using (1.61), it follows

$$F' = \int_{t_1}^{t_2}\left(L + q\frac{d\Lambda}{dt}\right)dt = \int_{t_1}^{t_2} Ldt + \int_{t_1}^{t_2} q\frac{d\Lambda}{dt}dt \tag{1.63}$$

Furthermore, one obtains

$$F' = F + q\left\{\Lambda\left[\vec{r}(t_2), t_2\right] - \Lambda\left[\vec{r}(t_1), t_1\right]\right\} \tag{1.64}$$

As the end points of the interval are fixed

$$\delta\vec{r} = 0 \tag{1.65}$$

performing the variation of functional F' simply yields

$$\delta F' = \delta F \tag{1.66}$$

Therefore, the Lagrangian L and L' are equivalent.

For the point particle, the charge density can be written as

$$\rho\left(\vec{R}, t\right) = q\delta\left[\vec{R} - \vec{r}(t)\right] \tag{1.67}$$

where δ [ ] is the Dirac delta function.

In the next step, the current density can be expressed as charge in motion, i.e.

$$\vec{J}\left(\vec{R},t\right) = q\frac{d\vec{r}(t)}{dt}\delta\left[\vec{R} - \vec{r}(t)\right] \tag{1.68}$$

Now, taking into account the total differential of scalar function $\Lambda\left(\vec{r},t\right)$

$$d\Lambda = \frac{\partial\Lambda}{\partial x}dx + \frac{\partial\Lambda}{\partial y}dy + \frac{\partial\Lambda}{\partial z}dz + \frac{\partial\Lambda}{\partial t}dt = \nabla\Lambda \cdot d\vec{r} + \frac{\partial\Lambda}{\partial t}dt \tag{1.69}$$

total time derivative is obtained as

$$\frac{d\Lambda}{dt} = \nabla \cdot \Lambda\frac{d\vec{r}}{dt} + \frac{\partial\Lambda}{\partial t} = \frac{\partial\Lambda}{\partial n}\vec{r}_0\frac{d\vec{r}}{dt} + \frac{\partial\Lambda}{\partial t} \tag{1.70}$$

and the Lagrangian can be written as follows

$$L' = L + q\left[\frac{\partial\Lambda}{\partial n}\frac{dr}{dt} + \frac{\partial\Lambda}{\partial t}\right] \tag{1.71}$$

Integrating over volume V yields

$$\int_V \rho\left(\vec{R},t\right)dV = \int_V q\delta\left[\vec{R} - \vec{r}(t)\right]dV \tag{1.72}$$

and one simply has

$$q = \int_V \rho\left(\vec{R},t\right)dV \tag{1.73}$$

Finally, Lagrangian (1.61) takes the form

$$L' = L + \int_V \rho\left(\vec{R},t\right)\frac{d\vec{r}}{dt} \cdot \nabla\Lambda\left(\vec{R},t\right)dV + \int_V \rho\left(\vec{R},t\right)\frac{\partial\Lambda\left(\vec{R},t\right)}{\partial t}dV \tag{1.74}$$

which, taking into account (1.68), becomes

$$L' = L + \int_V \vec{J}\left(\vec{R},t\right) \cdot \nabla\Lambda\left(\vec{R},t\right)dV + \int_V \rho\left(\vec{R},t\right)\frac{\partial\Lambda\left(\vec{R},t\right)}{\partial t}dV \tag{1.75}$$

If V tends to infinity and current is assumed to vanish at infinity after performing integration by parts, it follows

$$L' = L - \int_V \nabla \cdot \vec{J}\left(\vec{R},t\right) \Lambda\left(\vec{R},t\right) dV + \frac{d}{dt} \int_V \rho\left(\vec{R},t\right) \Lambda\left(\vec{R},t\right) dV$$

$$- \int_V \frac{\partial \rho\left(\vec{R},t\right)}{\partial t} \Lambda\left(\vec{R},t\right) dV \tag{1.76}$$

which now can be written as

$$L' = L - \int_V \left[\nabla \cdot \vec{J}\left(\vec{R},t\right) - \frac{\partial \rho\left(\vec{R},t\right)}{\partial t} \right] \Lambda\left(\vec{R},t\right) dV + \frac{d}{dt} \int_V \rho\left(\vec{R},t\right) \Lambda\left(\vec{R},t\right) dV \tag{1.77}$$

Now Lagrangian L' can be equivalent to (1.57) if the second term of the right-hand side is equal to zero.

According to the fundamental lemma of variational calculus, as $\Lambda\left(\vec{R},t\right)$ is an arbitrary space-dependent function, the second integral term from the right-hand side vanishes identically if the following condition is satisfied

$$\nabla \cdot \vec{J}\left(\vec{R},t\right) - \frac{\partial \rho\left(\vec{R},t\right)}{\partial t} = 0 \tag{1.78}$$

Expression (1.78) is a well-known equation of continuity relating to charge and current density, respectively.

Therefore, one of the basic equations of classical electromagnetics is derived from the gauge invariance of classical mechanics, i.e. from the symmetry property of the Lagrangian.

Namely, the gauge invariance of the Lagrangian implies conservation of electric charge.

It is worth noting that most of standard textbooks used for various courses in electromagnetics start from Maxwell's equations and the continuity equation is usually derived from Maxwell's equations (which are considered rigorous mathematical expressions of previously discovered laws as a consequence of experiments) via certain mathematical manipulations.

This chapter exploits the approach that goes the other way around by exploiting the symmetry properties of the Lagrangian, similar to the approach presented in [3, 6–8].

It is rather worth stressing that within the framework of such an approach, electromagnetic potentials are regarded as more fundamental entities than in the case in which one considers corresponding equation of motions instead of Lagrangian. Namely, in latter approach, starting from the electric and magnetic fields being gauge invariant, potentials are no more than auxiliary functions, i.e. pure mathematical constructs with no physical meaning.

1.5 Lorentz Force from Gauge Invariance

In the next step, Lagrangian (1.75) is rewritten by introducing vector function

$$\vec{A} = \nabla \Lambda \left(\vec{R}, t \right) \tag{1.79}$$

and scalar function

$$\varphi = -\frac{\partial \Lambda \left(\vec{R}, t \right)}{\partial t} \tag{1.80}$$

which now gives new Lagrangian

$$L_i = L + \int_V \vec{J} \left(\vec{R}, t \right) \cdot \vec{A} \left(\vec{R}, t \right) dV - \int_V \rho \left(\vec{R}, t \right) \varphi \left(\vec{R}, t \right) dV \tag{1.81}$$

Therefore, current density $\vec{J} \left(\vec{R}, t \right)$ is coupled to the vector potential $\vec{A} \left(\vec{R}, t \right)$, while charge density is coupled to the scalar potential $\varphi \left(\vec{R}, t \right)$ [3, 4].

Therefore, integral terms in (1.81) describe interaction of the charged particle with the electromagnetic field potentials.

Again, to obtain an equivalent Lagrangian, the total time derivative is added

$$L_i' = L_i + q \frac{d\Lambda}{dt} \tag{1.82}$$

and the obtained Lagrangian is

$$L_i' = L + \int_V \vec{J} \cdot \vec{A} dV - \int_V \rho \varphi dV + q \frac{d\Lambda}{dt} \tag{1.83}$$

which can be also written as

$$L' = L + \int_V \vec{J} \cdot \vec{A}' dV - \int_V \rho \varphi' dV \tag{1.84}$$

If the potentials are defined as follows [2, 3]

$$\vec{A}' = \vec{A} + \nabla \Lambda \left(\vec{R}, t \right) \tag{1.85}$$

$$\varphi' = \varphi - \frac{\partial \Lambda\left(\vec{R}, t\right)}{\partial t} \tag{1.86}$$

The vector and scalar potentials in (1.83) correspond to the additional terms in (1.75).

Expressions (1.85, 1.86) are referred to as gauge transformations added to ensure invariance of the Lagrangian under the addition of a total time derivative.

The corresponding equation of motions of the particle interacting with electromagnetic field can now be obtained from the least action principle.

The total action for the charged particle F_i can be written as

$$F_i = \int_{t_1}^{t_2} L_i dt \tag{1.87}$$

Varying functional (1.87) and taking into account

$$\frac{d\vec{A}}{dt} = \frac{\partial \vec{A}}{\partial x}\frac{dx}{dt} + \frac{\partial \vec{A}}{\partial y}\frac{dy}{dt} + \frac{\partial \vec{A}}{\partial z}\frac{dz}{dt} + \frac{\partial \vec{A}}{\partial t} = \left(\vec{v} \cdot \nabla\right)\vec{A} + \frac{\partial \vec{A}}{\partial t} \tag{1.88}$$

leads to the following equation of motion [3]

$$m\frac{d^2\vec{r}}{dt^2} = -\nabla W_{pot}\left(\vec{r}\right) + q\left(-\nabla\varphi - \frac{\partial \vec{A}}{\partial t}\right) + q\frac{d\vec{r}}{dt} \times \left(\nabla \times \vec{A}\right) \tag{1.89}$$

representing Newton's second law with additional forces due to the existence of the electromagnetic field.

Note that (1.89) is equivalent to expression

$$m\frac{d^2\vec{r}}{dt^2} = -\nabla W_{pot}\left(\vec{r}\right) + q\left(-\nabla\varphi' - \frac{\partial \vec{A}'}{\partial t}\right) + q\frac{d\vec{r}}{dt} \times \left(\nabla \times \vec{A}'\right) \tag{1.90}$$

The vector and scalar potentials are not unique and consequently not measurable physical quantities aiming as they are mathematical constructs and do not represent physical fields [1, 3, 4].

However, the electric field and magnetic field can be defined in terms of potentials A and φ as they are invariant under gauge transformations (1.85) and (1.86) as follows [3]:

$$\vec{B} = \nabla \times \vec{A} \tag{1.91}$$

$$\vec{E} = -\nabla\varphi - \frac{\partial \vec{A}}{\partial t} \tag{1.92}$$

where B denotes the magnetic field and E stands for the electric field.

It can be concluded that electric and magnetic fields are obtained from the equations of motion for charge particles and they are gauge invariant.

The total equation of motion can be finally written as follows

$$m\frac{d^2\vec{r}}{dt^2} = -\nabla W_{pot}\left(\vec{r}\right) + q\vec{E} + q\frac{d\vec{r}}{dt} \times \vec{B} \tag{1.93}$$

Equation (1.93) represents Newton's second law of motion with the Lorentz force included, i.e. it can be written as

$$m\frac{d^2\vec{r}}{dt^2} = \vec{F}_{mech} + \vec{F}_{EM} \tag{1.94}$$

where $\vec{F}_{mech}$ is the force acting to the particle due to the potential energy

$$\vec{F}_{mech} = -\nabla W_{pot} \tag{1.95}$$

while the Lorentz force due to the interaction of particles with the electromagnetic field is

$$\vec{F}_{EM} = q\left(\vec{E} + \vec{v} \times \vec{B}\right) \tag{1.96}$$

Therefore, (1.96) is the second fundamental equation of electromagnetics derived from the gauge invariance starting from action in classical mechanics.

Namely, according to the gauge invariance, vector potential can be changed without any effect on the charged particle. This can be referred to as gauge transformation.

It is also worth noting that Lorentz force (1.96), being a basis of classical particle electromagnetics, cannot be obtained from Maxwell's equations for stationary media, while (1.78, 1.91, 1.92) can be readily derived from Maxwell's equation, which is available elsewhere, e.g. in [4].

References

1 L. Suskind and A. Friedman, Special Relativity and Classical Field Theory, New York, UK: Penguin, 2017.

2 D. E. Neuenschwander, Emmy Noether's Wonderful Theorem, Baltimore, USA: The Johns Hopkins University Press, 2011.

3 D. H. Cobe, "Derivation of Maxwell's equations from the gauge invariance of classical mechanics," *American Journal of Physics*, vol. 48, no. 348, pp. 348–353, 1980.

4 D. Poljak, Advanced Modeling in Computational Electromagnetic Compatibility, New Jersey, USA: John Wiley & Sons, Inc., 2007.

5 C. Civelek and T. F. Bechteler, Lagrangian formulation of electromagnetic fields in nondispersive medium by means of the extended Euler–Lagrange differential equation, *International Journal of Engineering Science*, vol. 46, no. 12, pp. 1218–1227, 2008.

6 D. Poljak, "Review of Least Action Principle in Electromagnetics, Part II: Derivation of Continuity Equation and Lorentz Force, Maxwell's Equations", in *2022 International Conference on Software, Telecommunications and Computer Networks (SoftCOM)*, Split, Croatia, SoftCOM, 2022.

7 D Poljak, "Review of Least Action Principle in Electromagnetics, Part II: Derivation of Maxwell's Equations," SoftCOM, 2022.

8 D. Poljak, "Review of Least Action Principle in Electromagnetics, Part III: Variational Basis for Numerical methods," SoftCOM, 2022.

9 T. B. Mieling, "Noether's Theorem Applied to Classical Electrodynamics," Lecture Notes, November 2017.

2

Fundamental Equations of Engineering Electromagnetics

This chapter deals with a derivation of Maxwell's equations from Hamilton's principle in electromagnetics. It is well known that electromagnetic fields can essentially be treated as mechanical systems [1, 2]. As in Chapter 1, Hamilton's principle has been used to derive continuity equation and Lorentz force law, this variational principle should also provide a single equation from which Maxwell's equation can be deduced. Namely, as Hamilton's principle is applicable to continuous media, it can also be used to describe the behavior of fields, i.e. it should be possible to set up a suitable Lagrangian function of the field variables, such that Hamilton's principle yields the equations of motion of the field – Maxwell's equations.

Though there is no general rule for the derivation of the Lagrangian function, the electromagnetic energy density is defined as a function of position in space in terms of the field variables.

An intuitive approach would be to correlate the kinetic field energy with magnetic field energy and the potential energy with the electric field energy [1, 2].

Kinematical Maxwell's equations are derived from gauge symmetry, while two dynamical Maxwell's equations are derived from functional of electromagnetic energy.

2.1 Derivation of Two-Canonical Maxwell's Equation

It has been shown in Section 1.5, featuring the gauge symmetry, that electric and magnetic fields can be expressed in terms of the vector and scalar potentials regardless of the fact that these potentials are not unique and thus not measurable. Namely, these potentials do not represent physical quantities [3].

Deterministic and Stochastic Modeling in Computational Electromagnetics: Integral and Differential Equation Approaches, First Edition. Dragan Poljak and Anna Šušnjara.
© 2024 The Institute of Electrical and Electronics Engineers, Inc.
Published 2024 by John Wiley & Sons, Inc.

On the other hand, the electric and magnetic fields can be defined in terms of these mathematical constructs which are invariant under gauge transformations [3], i.e. it can be written as

$$\vec{B} = \nabla \times \vec{A} \tag{2.1}$$

$$\vec{E} = -\nabla \varphi - \frac{\partial \vec{A}}{\partial t} \tag{2.2}$$

where A and φ stand for magnetic vector and electric scalar potential, respectively.

Now the two kinematical equations that electric and magnetic fields satisfy can be obtained from Eqs (2.1) and (2.2) in which electric and magnetic fields are expressed in terms of potentials.

Thus, taking the curl of (2.1), the right-hand side vanishes

$$\nabla \cdot \vec{B} = \nabla \cdot \left(\nabla \times \vec{A} \right) = 0 \tag{2.3}$$

and one obtains the Gauss's law for the magnetic field in a differential form

$$\nabla \cdot \vec{B} = 0 \tag{2.4}$$

i.e. the fourth Maxwell's equation states that no magnetic monopoles exist.

Furthermore, taking the curl of (2.2), the first term of the right-hand side vanishes

$$\nabla \times \vec{E} = \nabla \times \left(-\nabla \varphi - \frac{\partial \vec{A}}{\partial t} \right) = -\frac{\partial}{\partial t} \left(\nabla \times \vec{A} \right) \tag{2.5}$$

One obtains Faraday law in the differential form

$$\nabla \times \vec{E} = -\frac{\partial \vec{B}}{\partial t} \tag{2.6}$$

i.e. the first curl Maxwell's equation.

So, the kinematical Maxwell's equations are obtained from the electric and magnetic fields expressed in terms of vector and scalar potentials featuring the use of gauge symmetry.

2.2 Derivation of Two-Dynamical Maxwell's Equation

It has been shown in [3] that Lagrangian for the charged particle is given in the form

$$L = \frac{1}{2} m \dot{r}^2 - W_{pot} + \int_V \left(\vec{J} \cdot \vec{A} - \rho \cdot \phi \right) dV \tag{2.7}$$

Now this difference between kinetic energy and potential energy can be written by relating kinetic energy with magnetic field energy and potential energy with the electric field energy, i.e. the new Lagrangian is given by [1, 2]

$$L = \int_V \left[\frac{1}{2} \left(\vec{B} \cdot \vec{H} - \vec{D} \cdot \vec{E} \right) + \vec{J} \cdot \vec{A} - \phi \cdot \rho \right] dV = \int_V L_d \cdot dV \tag{2.8}$$

where L_d stands for Lagrangian density of the form

$$L_d = \frac{1}{2} \left(\vec{B} \cdot \vec{H} - \vec{E} \cdot \vec{D} \right) + \vec{J} \cdot \vec{A} - \rho \cdot \phi \tag{2.9}$$

which is quantity introduced in mathematical physics when fields are considered, rather than particles [1, 2].

Therefore, total Lagrangian density is composed of a sum of Lagrangian for the fields and sources, respectively.

Functional of electromagnetic energy containing source densities in the volume of interest is of the form [4]

$$F = \int_{t_1}^{t_2} \int_V \left(\frac{1}{2} \cdot \vec{B} \cdot \vec{H} - \frac{1}{2} \cdot \vec{D} \cdot \vec{E} + \vec{J} \cdot \vec{A} - \phi \cdot \rho \right) dVdt \tag{2.10}$$

Now, taking into account (2.1) and (2.2), the electric and magnetic fields are expressed in terms of their potentials resulting in the following functional

$$F = \int_{t_1}^{t_2} \int_V \left[\frac{1}{2\mu} \cdot \left(\nabla \times \vec{A} \right)^2 - \frac{1}{2} \cdot \varepsilon \left(\nabla \varphi + \frac{\partial \vec{A}}{\partial t} \right)^2 + \vec{J} \cdot \vec{A} - \varphi \cdot \rho \right] dVdt \tag{2.11}$$

And related Lagrangian density is given by

$$L_d = \frac{1}{2\mu} \cdot \left(\nabla \times \vec{A} \right)^2 - \frac{1}{2} \cdot \varepsilon \left(\nabla \varphi + \frac{\partial \vec{A}}{\partial t} \right)^2 + \vec{J} \cdot \vec{A} - \varphi \cdot \rho \tag{2.12}$$

Furthermore, Lagrangian density (2.12) is requested to satisfy Lagrange–Euler equations [4]

$$\sum_{k=1}^{N} \left\{ \frac{\partial L}{\partial q_k} - \frac{\partial}{\partial x} \left[\frac{\partial L}{\partial (q_k)_x} \right] - \frac{\partial}{\partial y} \left[\frac{\partial L}{\partial (q_k)_y} \right] - \frac{\partial}{\partial z} \left[\frac{\partial L}{\partial (q_k)_z} \right] - \frac{\partial}{\partial t} \left(\frac{\partial L}{\partial (q_k)_t} \right) \right\} = 0 \tag{2.13}$$

Furthermore, the following expressions are obtained:

$$\frac{\partial L}{\partial \varphi} = -\rho \tag{2.14}$$

$$\frac{\partial}{\partial x} \frac{\partial L}{\partial \left(\frac{\partial \varphi}{\partial x}\right)} = \frac{\partial}{\partial x}\left(-\varepsilon \frac{\partial \varphi}{\partial x}\right) = \frac{\partial}{\partial x}(\varepsilon E_x) = \frac{\partial D_x}{\partial x} \tag{2.15}$$

$$\frac{\partial}{\partial y} \frac{\partial L}{\partial \left(\frac{\partial \varphi}{\partial y}\right)} = \frac{\partial}{\partial y}\left(-\varepsilon \frac{\partial \varphi}{\partial y}\right) = \frac{\partial}{\partial y}(\varepsilon E_y) = \frac{\partial D_y}{\partial y} \tag{2.16}$$

$$\frac{\partial}{\partial z} \frac{\partial L}{\partial \left(\frac{\partial \varphi}{\partial z}\right)} = \frac{\partial}{\partial z}\left(-\varepsilon \frac{\partial \varphi}{\partial z}\right) = \frac{\partial}{\partial z}(\varepsilon E_z) = \frac{\partial D_z}{\partial z} \tag{2.17}$$

which simply gives

$$\frac{\partial D_x}{\partial x} + \frac{\partial D_y}{\partial y} - \frac{\partial D_z}{\partial z} = \rho \tag{2.18}$$

i.e. one obtains third divergence from Maxwell's equation

$$\nabla \cdot \vec{D} = \rho \tag{2.19}$$

Proceeding with the application of Lagrange–Euler Eq. (2.13), one has:

$$\frac{\partial L}{\partial A_z} = J_z \tag{2.20}$$

$$\frac{\partial}{\partial t} \frac{\partial L}{\partial \left(\frac{\partial A_z}{\partial t}\right)} = \frac{\partial}{\partial t}\left[-\varepsilon\left(\frac{\partial \varphi}{\partial z} + \frac{\partial A_z}{\partial t}\right)\right] = -\frac{\partial D_z}{\partial t} \tag{2.21}$$

$$\frac{\partial}{\partial x} \frac{\partial L}{\partial \left(\frac{\partial A_z}{\partial x}\right)} = -\frac{1}{\mu}\frac{\partial}{\partial x}\left(\frac{\partial A_z}{\partial x} - \frac{\partial A_x}{\partial z}\right) = -\frac{\partial H_y}{\partial x} \tag{2.22}$$

$$\frac{\partial}{\partial y} \frac{\partial L}{\partial \left(\frac{\partial A_z}{\partial y}\right)} = -\frac{1}{\mu}\frac{\partial}{\partial y}\left(\frac{\partial A_z}{\partial y} - \frac{\partial A_y}{\partial z}\right) = \frac{\partial H_x}{\partial y} \tag{2.23}$$

$$\frac{\partial}{\partial z} \frac{\partial L}{\partial \left(\frac{\partial A_z}{\partial z}\right)} = 0 \tag{2.24}$$

which can be written

$$J_z + \frac{\partial D_z}{\partial t} + \frac{\partial H_x}{\partial y} - \frac{\partial H_y}{\partial x} = 0 \tag{2.25}$$

Performing similar mathematical operations for other vector potential components yields, the following set of partial differential equations [4]

$$\frac{\partial H_y}{\partial x} - \frac{\partial H_x}{\partial y} = J_z + \frac{\partial D_z}{\partial t} \tag{2.26}$$

$$\frac{\partial H_z}{\partial y} - \frac{\partial H_y}{\partial z} = J_x + \frac{\partial D_x}{\partial t} \tag{2.27}$$

$$\frac{\partial H_x}{\partial z} - \frac{\partial H_z}{\partial x} = J_y + \frac{\partial D_y}{\partial t} \tag{2.28}$$

which can be written in a vector form as second curl Maxwell's equation

$$\nabla x \vec{H} = \vec{J} + \frac{\partial \vec{D}}{\partial t} \tag{2.29}$$

representing generalized Ampere's law in the differential form.

Thus, Maxwell's equations implicitly contain minimum energy principle. Consequently, any equation derived from Maxwell's equations satisfies this principle. Similar analysis could be found elsewhere, e.g. in [4, 5].

2.3 Integral Form of Maxwell's Equations, Continuity Equations, and Lorentz Force

Integral form of Maxwell's equations can be obtained from the differential form of Maxwell's equations by applying Stokes' theorem and Gauss's divergence theorem.

Integral form of the Faraday law states that any change of magnetic flux density B through any closed loop induces an electromotive force around the loop. Taking the surface integration over (2.6) and applying the Stokes' theorem yields:

$$\oint_c \vec{E} d\vec{s} = -\int_S \frac{\partial \vec{B}}{\partial t} \cdot d\vec{S} \tag{2.30}$$

where the line integral is taken around the loop and with $d\vec{S} = \vec{n} dS$.

The voltage induced by a varying flux has a polarity such that the induced current in a closed path gives rise to a secondary magnetic flux opposing the change in original time-varying magnetic flux.

The integral form of the Ampere law is obtained by integrating (2.29) and applying the Stokes' theorem

$$\oint_c \vec{H} d\vec{s} = \int_S \vec{J} d\vec{S} + \int_S \frac{\partial \vec{D}}{\partial t} \cdot d\vec{S} \tag{2.31}$$

The generalized Ampere law states that either an electric current or a time-varying electric flux gives rise to magnetic field.

Taking the volume integral over (2.19) and applying the Gauss's divergence theorem yields:

$$\oint_S \vec{D} d\vec{S} = \int_V \rho dV \tag{2.32}$$

where the right-hand side represents the total charge within the volume V.

Equation (2.32) is the Gauss's flux law for the electric field stating that the electric flux density corresponds to the total electric charge within the domain.

The Gauss's flux law for the magnetic field can be derived by taking the volume integral of (2.4) and applying the Gauss's divergence theorem, i.e.:

$$\oint_S \vec{B} d\vec{S} = 0 \tag{2.33}$$

stating that the flux of B vector over any closed surface S is identically zero.

The integral form of the continuity equation, stating that the rate of charge moving out of a region is equal to the time rate of charge density decrease, is obtained by performing the volume integration of differential Eq. (1.78)

$$\int_V \nabla \cdot \vec{J} dV = -\frac{\partial}{\partial t} \int_V \rho dV \tag{2.34}$$

and applying the Gauss's divergence theorem

$$\int_V \nabla \cdot \vec{J} dV = \oint_S \vec{J} d\vec{S} \tag{2.35}$$

The integral form of the continuity equation is then given by

$$\oint_S \vec{J} d\vec{S} = -\frac{\partial Q}{\partial t} \tag{2.36}$$

where the unit normal in $d\vec{S}$ is the outward-directed normal and Q is the total charge within the volume:

$$Q = \int_V \rho dV \tag{2.37}$$

Equation (2.36) represents the Kirchhoff conservation law widely used in the circuit theory.

Finally, integral form of Lorentz force (1.94) is obtained if an electromagnetic system is subjected to an external force, where force density to charge density ρ and current density $J = \rho v$ is given by:

$$\vec{f} = \rho\vec{E} + \vec{J} \times \vec{B} \tag{2.38}$$

The total force on the matter contained within volume V, i.e. to charges and currents is defined by expression:

$$\vec{F} = \int_V \vec{f}\, dV = \int_V \left(\rho\vec{E} + \vec{J} \times \vec{B}\right) dV \tag{2.39}$$

Note that expression

$$\int_V \vec{J} \times \vec{B}\, dV \tag{2.40}$$

is usually simplified for the case of current carrying conductor c, as follows

$$\int_c I d\vec{s} \times \vec{B} \tag{2.41}$$

where $I d\vec{s}$ stands for a current element.

2.4 Phasor Form of Maxwell's Equations

In many applications, one deals with a continuous wave (CW) excitation and a time-harmonic variation of electromagnetic fields can be assumed. In addition, transient excitations can be analyzed by transforming the governing equations in the frequency domain. In such cases, it is convenient to represent the variables of interest in a complex phasor form in which an arbitrary time-dependent vector field $F(r,t)$ can be expressed as follows [4]:

$$\vec{F}\left(\vec{r},t\right) = \mathrm{Re}\left[\vec{F}_s\left(\vec{r}\right)e^{j\omega t}\right] \tag{2.42}$$

where $F_s(r)$ is the phasor form of $F(r,t)$, and $F_s(r)$ is in general complex with an amplitude and a phase changing with position. Then Re [ ] implies taking the real part of quantity in brackets, and ω is the angular frequency of the sinusoidal signal.

When using the phasor representation, the differentiation and integration with respect to time results in:

$$\frac{\partial}{\partial t}\left[\vec{F}_s\left(\vec{r}\right)e^{j\omega t}\right] = j\omega \vec{F}_s\left(\vec{r}\right)e^{j\omega t} \tag{2.43a}$$

$$\int \left[\vec{F}_s\left(\vec{r}\right)e^{j\omega t}\right]dt = \frac{1}{j\omega}\vec{F}_s\left(\vec{r}\right)e^{j\omega t} \tag{2.43b}$$

As in linear, homogeneous, and isotropic mediums, one deals with the following constitutive equations:

$$\vec{D} = \varepsilon \vec{E} \tag{2.44}$$

$$\vec{J} = \sigma \vec{E} \tag{2.45}$$

$$\vec{B} = \mu \vec{H} \tag{2.46}$$

assuming the time-harmonic variation of fields, curl Maxwell's equations become:

$$\nabla \times \vec{E} = -j\omega\mu\vec{H} \tag{2.47}$$

$$\nabla \times \vec{H} = \vec{J} + j\omega\varepsilon\vec{E} \tag{2.48}$$

while the integral form of these equations is given by:

$$\oint_c \vec{E}d\vec{s} = -j\omega\mu \int_S \vec{H}d\vec{S} \tag{2.49}$$

$$\oint_c \vec{H}d\vec{s} = (\sigma + j\omega\varepsilon) \int_S \vec{E}d\vec{S} \tag{2.50}$$

Furthermore, the differential form of continuity equation for time-harmonic excitation is

$$\nabla \cdot \vec{J} = -j\omega\rho \tag{2.51}$$

while the corresponding integral form is given by

$$\oint_S \vec{J}d\vec{S} = -j\omega \int_V \rho dV = -j\omega Q \tag{2.52}$$

Therefore, assuming the time-harmonic variation of fields eliminates the time dependence from Maxwell's equations, thereby reducing the space-time dependence to space dependence only.

2.5 Continuity (Interface) Conditions

To solve Maxwell's equations for a given problem, the continuity conditions at the interface of two media with different electrical properties must be specified [4]:

$$\vec{n} \times \left(\vec{E}_1 - \vec{E}_2 \right) = 0 \tag{2.53}$$

$$\vec{n} \times \left(\vec{H}_1 - \vec{H}_2 \right) = \vec{J}_s \tag{2.54}$$

$$\vec{n} \left(\vec{D}_1 - \vec{D}_2 \right) = \rho_s \tag{2.55}$$

$$\vec{n} \left(\vec{B}_1 - \vec{B}_2 \right) = 0 \tag{2.56}$$

where $\vec{n}$ is a unit normal vector directed from medium 1 to medium 2, subscripts 1 and 2 denote fields in regions 1 and 2.

Derivation of continuity conditions (2.53)–(2.56) is available elsewhere, e.g. in (Wiley 2007). Equations (2.53) and (2.56) state that the tangential components of E and the normal components of B are continuous across the boundary. Equation (2.54) implies that the tangential component of H is discontinuous by the surface current density J_s on the boundary, while (2.55) means that the discontinuity in the normal component of D is the same as the surface charge density ρ_s on the boundary.

Note that the electric field E and magnetic field H vanish within the perfectly conducting (PEC) medium. These fields are replaced by the surface charge density ρ_s and surface current density J_s at the interface. At higher frequencies, there exists skin effect which confines current largely to surface regions.

As no time-varying field exists in PEC conductor, the electric flux density is entirely normal to the conductor and supported by a surface charge density at the interface

$$D_n = \rho_s \tag{2.57}$$

The magnetic field is entirely tangential to PEC conductor and is equilibrated by a surface current density:

$$H_s = J_s \tag{2.58}$$

Finally, for the current density from the continuity equation, one obtains the following continuity condition for current density

$$\vec{n} \left(\vec{J}_1 - \vec{J}_2 \right) = - \frac{\partial \rho_s}{\partial t} \tag{2.59}$$

For time-harmonic case, one has

$$\vec{n}\left(\vec{J}_1 - \vec{J}_2\right) = -j\omega\rho_s \tag{2.60}$$

More details on the derivation can be found elsewhere, e.g. in [4].

2.6 Poynting Theorem

The general conservation law of energy in the macroscopic electromagnetic field can be readily derived from curl Maxwell's equations.

It is instructive to discuss the concept of differential form of Poynting theorem. Namely, while continuity equation for sources in terms of charge and current density, respectively, represents a conservation of electric charge

$$\nabla \cdot \vec{J} = -\frac{\partial\rho}{\partial t} \tag{2.61}$$

Poynting theorem for source-free region (continuity of fields in source-free and lossless medium) addresses a conservation of electromagnetic energy and can be written as follows

$$\nabla \cdot \vec{S} = -\frac{\partial w}{\partial t} \tag{2.62}$$

where $\vec{S}$ is space-time dependent Poynting vector (power density)

$$\vec{S} = \vec{E} \times \vec{H} \tag{2.63}$$

and w is energy density defined as

$$w = \frac{1}{2} \cdot \left(\vec{E}\cdot\vec{D} + \vec{H}\cdot\vec{B}\right) \tag{2.64}$$

When sources are included, (2.63) is to be extended as follows [4]

$$\nabla \cdot \left(\vec{E} \times \vec{H}\right) = -\frac{\partial}{\partial t}\left(\vec{E}\cdot\vec{D} + \vec{H}\cdot\vec{B}\right) + \vec{E}'\cdot\vec{J} - \vec{E}\cdot\vec{J} \tag{2.65}$$

where $\vec{E}'$ represents the field due to sources, while the last term on the right-hand side accounts for conductive losses.

Taking the volume integral over (2.65) yields

$$\int_V \nabla \cdot \left(\vec{E} \times \vec{H}\right) dV = -\frac{\partial}{\partial t}\int_V \left(\frac{\vec{E}\cdot\vec{D} + \vec{H}\cdot\vec{B}}{2}\right) dV - \int_V \vec{E}\cdot\vec{J}\, dV \tag{2.66}$$

And applying the Gauss's integral theorem to the left-hand side term, the volume integral transforms to the surface integral over the boundary, where $d\vec{S}$ is the outward-drawn normal vector surface element, (2.67) becomes

$$\int_V \vec{E}' \cdot \vec{J} dV = \frac{\partial}{\partial t} \int_V \frac{1}{2} \left(\vec{E} \cdot \vec{D} + \vec{H} \cdot \vec{B} \right) dV + \int_V \frac{\left|\vec{J}\right|}{\sigma} \vec{J} dV' + \oint_S \left(\vec{E} \times \vec{H} \right) \cdot d\vec{S}$$

(2.67)

The sources within the volume of interest are balanced with the rate of increase of electromagnetic energy in the domain, the rate of flow of energy through the domain surface and the Joule heat production in the domain.

For the time-harmonic quantities, the complex Poynting vector counterpart of (2.63) is given by

$$\underline{\vec{S}} = \frac{1}{2} \left(\vec{E} \times \vec{H}^* \right)$$

(2.68)

Taking the divergence of Poynting vector yields

$$\nabla \cdot \underline{\vec{S}} = \frac{1}{2} \left(\vec{H}^* \nabla \times \vec{E} - \vec{E} \nabla \times \vec{H}^* \right)$$

(2.69)

The divergence of complex power density can be expressed in terms of rate of stored energy density, power density losses, and source density

$$\nabla \cdot \underline{\vec{S}} = -j\frac{\omega}{2} \left(\mu \left|\vec{H}\right|^2 - \varepsilon \left|\vec{E}\right|^2 \right) - \frac{\left|\vec{J}\right|}{\sigma} + \frac{1}{2}\sigma \left|\vec{E}\right|^2$$

(2.70)

which can be also written as

$$\nabla \cdot \underline{\vec{S}} = -j\frac{\omega}{2} \left(\mu \left|\vec{H}\right|^2 - \varepsilon \left|\vec{E}\right|^2 \right) - \frac{1}{2}\sigma \left| \vec{E} \right| + \frac{1}{2}\sigma \left|\vec{E}'\right|^2$$

(2.71)

Now the integration over a volume of interest yields

$$\int_V \nabla \cdot \vec{S} dV = -j\frac{\omega}{2} \int_V \left(\mu \left|\vec{H}\right|^2 - \varepsilon \left|\vec{E}\right|^2 \right) dV - \frac{1}{2} \int_V \sigma \left|\vec{E}\right|^2 dV + \frac{1}{2} \int_V \sigma \left|\vec{E}'\right|^2 dV$$

(2.72)

And applying the Gauss's theorem, one obtains

$$\frac{1}{2} \int_V \vec{E} \times \vec{H}^* d\vec{S} = -j\frac{\omega}{2} \int_V \left(\mu \left|\vec{H}\right|^2 - \varepsilon \left|\vec{E}\right|^2 \right) dV - \frac{1}{2} \int_V \sigma \left|\vec{E}\right|^2 dV + \frac{1}{2} \int_V \sigma \left|\vec{E}'\right|^2 dV$$

(2.73)

Finally, the real and imaginary parts, respectively, of the Poynting flow can be written

$$\frac{1}{2} \operatorname{Re} \int_V \vec{E} \times \vec{H}^* \, d\vec{S} = -\frac{1}{2} \int_V \sigma \left| \vec{E} \right|^2 dV + \frac{1}{2} \int_V \sigma \left| \vec{E}' \right|^2 dV \tag{2.74}$$

$$\frac{1}{2} \operatorname{Im} \int_V \vec{E} \times \vec{H}^* \, d\vec{S} = -\frac{\omega}{2} \int_V \left(\mu \left| \vec{H} \right|^2 - \varepsilon \left| \vec{E} \right|^2 \right) dV \tag{2.75}$$

The real part of the integral over Poynting vector represents the total average power while the imaginary part of the integral over Poynting vector is proportional to the difference between average stored magnetic energy in the volume and average stored energy in the electric field.

The ½ factor appears because E and H fields represent peak values, and it should be omitted for root-mean-square (*rms*) values. The total average power can, for example, represent the radiated power by an antenna. In addition, the first volume integral on the right-hand side of (2.74) represents power loss in the conduction currents.

2.7 Electromagnetic Wave Equations

The wave equations can be readily derived from Maxwell's equations. Taking curl on both sides of Eq. (2.29) yields

$$\nabla \times \nabla \times \vec{H} = \nabla \times \vec{J} + \frac{\partial}{\partial t} \left(\nabla \times \vec{D} \right) \tag{2.76}$$

Using constitutive Eqs. (2.44)–(2.45) and assuming homogeneous medium, one has

$$\nabla \times \nabla \times \vec{H} = \sigma \nabla \times \vec{E} + \varepsilon \frac{\partial}{\partial t} \left(\nabla \times \vec{E} \right) \tag{2.77}$$

According to Maxwell's Eq. (2.6), curl of E is replaced by the rate of change of magnetic flux density

$$\nabla \times \nabla \times \vec{H} = -\mu \sigma \frac{\partial \vec{H}}{\partial t} - \mu \varepsilon \frac{\partial^2 \vec{H}}{\partial t^2} \tag{2.78}$$

Using the standard vector identity valid for any vector E:

$$\nabla \times \nabla \times \vec{H} = \nabla \cdot \left(\nabla \cdot \vec{H} \right) - \nabla^2 \vec{H} \tag{2.79}$$

And taking into account solenoidal nature of the magnetic field (2.4), the wave equation is obtained

$$\nabla^2 \vec{H} - \mu\sigma \frac{\partial \vec{H}}{\partial t} - \mu\varepsilon \frac{\partial^2 \vec{H}}{\partial t^2} = 0 \tag{2.80}$$

If source-free medium is considered, (2.80) simplifies into

$$\nabla^2 \vec{H} - \frac{1}{v^2} \frac{\partial^2 \vec{H}}{\partial t^2} = 0 \tag{2.81}$$

where v denotes the wave propagation velocity in lossless homogeneous medium

$$v = \frac{1}{\sqrt{\mu\varepsilon}} \tag{2.82}$$

The velocity of wave propagation in free space is the velocity of light:

$$c = \frac{1}{\sqrt{\mu_0 \varepsilon_0}} \tag{2.83}$$

where $c = 3 \times 10^8$ m/s, approximately.

For time-harmonic quantities, the wave Eq. (2.81) results in the following equation of the Helmholtz type:

$$\nabla^2 \vec{H} - \gamma^2 \vec{H} = 0 \tag{2.84}$$

where γ is the complex constant of propagation given by

$$\gamma = \sqrt{j\omega\mu\sigma - \omega^2 \mu\varepsilon} \tag{2.85}$$

For a lossless medium, the Helmholtz Eq. (2.84) simplifies into:

$$\nabla^2 \vec{H} + k^2 \vec{H} = 0 \tag{2.86}$$

where k is a wave number of a lossless medium:

$$k = \omega\sqrt{\mu\varepsilon} \tag{2.87}$$

The complex form of potential wave equations could be derived similarly. Thus, taking the divergence (2.2), one obtains:

$$\nabla \cdot \vec{E} = -\frac{\partial \left(\nabla \cdot \vec{A} \right)}{\partial t} - \nabla(\nabla\varphi) \tag{2.88}$$

Now, utilizing Gauss's law, (2.19) yields:

$$\nabla^2 \varphi + \frac{\partial \left(\nabla \cdot \vec{A} \right)}{\partial t} = -\frac{\rho}{\varepsilon} \tag{2.89}$$

Furthermore, taking the curl of (2.29) and taking into account (2.1), it follows

$$\nabla \times \nabla \times \vec{A} = \mu \vec{J} + \mu\varepsilon \left[-\nabla \left(\frac{\partial \varphi}{\partial t} \right) - \frac{\partial \vec{A}}{\partial t} \right] = 0 \tag{2.90}$$

Using the vector identity

$$\nabla \times \nabla \times \vec{A} = \nabla \left(\nabla \cdot \vec{A} \right) - \nabla^2 \vec{A} \tag{2.91}$$

expression (2.91) can be written as

$$\nabla \left(\nabla \cdot \vec{A} \right) - \nabla^2 \vec{A} = \mu \vec{J} - \mu\varepsilon \nabla \left(\frac{\partial \varphi}{\partial t} \right) - \mu\varepsilon \frac{\partial^2 \vec{A}}{\partial t^2} \tag{2.92}$$

At this stage, some reference points for potentials are required. As it is discussed in the chapter, through gauge transformations some new potentials can be obtained by adding new functions to existing potentials for which E and B fields remain the same. Particular choice of such functions is referred to as a gauge condition.

Such a gauge is obtained by choosing the divergence of A as follows

$$\nabla \cdot \vec{A} + \mu\varepsilon \frac{\partial \varphi}{\partial t} = 0 \tag{2.93}$$

Expression (2.93) is usually referred to as Lorentz gauge.

Combining (2.92) and (2.93), potential wave functions are obtained

$$\nabla^2 \varphi - \mu\varepsilon \frac{\partial^2 \varphi}{\partial t^2} = -\frac{\rho}{\varepsilon} \tag{2.94}$$

$$\nabla^2 \vec{A} - \mu\varepsilon \frac{\partial^2 \vec{A}}{\partial t^2} = -\mu \vec{J} \tag{2.95}$$

Set of Eqs. (2.94)–(2.95) can be regarded as a set of inhomogeneous potential wave equations for lossless media.

Provided all sources are known within the domain of interest, solutions to potential wave Eqs. (2.94) and (2.95) are given in the form of particular integrals often regarded as retarded potentials [4]:

$$\varphi \left(\vec{r}, t \right) = \frac{1}{4\pi\varepsilon} \int_{V'} \frac{\rho \left(\vec{r}', t - R/c \right)}{R} dV' \tag{2.96}$$

$$\vec{A}(r, t) = \frac{\mu}{4\pi} \int_{V'} \frac{\vec{J} \left(\vec{r}', t - R/c \right)}{R} dV' \tag{2.97}$$

where R is a distance from the source point to the observation point.

If all quantities of interest change harmonically in time, the potential equations are given by:

$$\nabla^2 \varphi + k^2 \varphi = -\frac{\rho}{\varepsilon} \tag{2.98}$$

$$\nabla^2 \vec{A} + k^2 \vec{A} = -\mu \vec{J} \tag{2.99}$$

and the particular integrals for the retarded potentials for time-harmonic dependence are given by [4]:

$$\varphi\left(\vec{r}\right) = \frac{1}{4\pi\varepsilon} \int_{V'} \frac{\rho\left(\vec{r}'\right) e^{-jkR}}{R} dV' \tag{2.100}$$

$$\vec{A}(r) = \frac{\mu}{4\pi} \int_{V'} \frac{\vec{J}\left(\vec{r}'\right) e^{-jkR}}{R} dV' \tag{2.101}$$

while factor $e^{j\omega t}$ is understood and omitted.

2.8 Plane Wave Propagation

Plane waves represent sufficient approximation for many realistic scenarios, such as radio waves at large distances from the transmitter, or from scattering obstacles. Also, complicated wave patterns can be regarded as a superposition of plane waves. Finally, the basic ideas of propagation, reflection, and refraction in the plane wave approach are useful in understanding more complex wave problems.

Considering the case of source-free homogeneous medium and y-directed component of the electric field, the corresponding Helmholtz equation is given by:

$$\frac{\partial^2 E_y}{\partial x^2} + k^2 E_y = 0 \tag{2.102}$$

and the general solution can be written in the form

$$E_y = Ce^{-jkx} + De^{+jkx} \tag{2.103}$$

where k is the wavenumber of a lossless medium, while C and D are unknown constants.

As a medium is unbounded, only the forward wave exists:

$$E_y = E_0 e^{-jkx} \tag{2.104}$$

where E_0 denotes the magnitude of the electric field.

The corresponding magnetic field can be obtained from the time-harmonic curl Maxwell's equation

$$\nabla \times \vec{E} = -j\omega\mu\vec{H} \tag{2.105}$$

Furthermore, in cartesian coordinates, it follows:

$$\vec{H} = -\frac{1}{j\omega\mu} \begin{vmatrix} \vec{e_x} & \vec{e_y} & \vec{e_z} \\ \frac{\partial}{\partial x} & 0 & 0 \\ 0 & E_y & 0 \end{vmatrix} = -\vec{e_z}\frac{1}{j\omega\mu}\frac{\partial E_y}{\partial x} \tag{2.106}$$

and the magnetic field is

$$H_z = \frac{E_0}{Z_0}e^{-jkx} = H_0 e^{-jkx} \tag{2.107}$$

where

$$Z_0 = \frac{E_0}{H_0} = \sqrt{\frac{\mu}{\varepsilon}} \tag{2.108}$$

is the wave impedance of the medium, and for free space, it is approximately $Z_0 = 120\pi$.

Now the space-time notation can be used

$$E_y(x, t) = \mathrm{Re}\left[E_0 e^{j(\omega t - kx)}\right] \tag{2.109}$$

and the electric and magnetic fields can be written as follows

$$E_y(x, t) = E_0 \cos(\omega t - kx) \tag{2.110}$$

$$H_z(x, t) = \frac{E_0}{Z_0} \cos(\omega t - kx) \tag{2.111}$$

The corresponding plane wave is shown in Fig. 2.1.

Note that in circuit theory, contrary to electromagnetism, a sinusoidal dependence, rather than co-sinusoidal, is chosen.

As there is no variation of the field quantities in the plane perpendicular to the propagation direction, Fig. 2.1 is regarded as a *plane wave*.

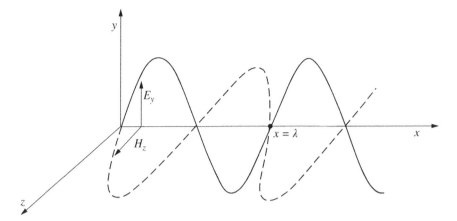

Figure 2.1 The electric and magnetic fields of an *x*-directed plane wave.

2.9 Hertz Dipole as a Simple Radiation Source

In classical electrodynamics, radiation is a phenomenon caused due to the acceleration of charged particles. Fields generated by accelerated charges can be analyzed by using microscopic approach featuring average fields over the charge distributions are considered [4].

The macroscopic approach is based on Maxwell's equations from which the radiated fields due to their sources in terms of charge and current densities are presented.

Radiation of electromagnetic energy is an undesired leakage phenomenon or a desired process for exciting waves in space. In the case of desired radiation, the goal is to excite waves from the given source in the required direction, as efficiently as possible. The matching unit between the source and waves in space is known as the *radiator, antenna*, or *areal*. The results developed for radiating or transmitting antennas can be applied to the same antenna when used for receiving applications if antenna does not contain active components (the principle of reciprocity). The relation of the radiating to the receiving case is rigorously expressed through the principle of reciprocity [4].

The simplest radiating system is that of an ideal short linear element (Hertz dipole) with current distribution uniform over its entire length. More complex antenna structures can be regarded as a superposition of such elementary antennas with the corresponding magnitudes and phases of their currents.

Fig. 2.2 shows Hertz dipole located in *z*-axis direction, with its location at the origin of a set of spherical coordinates.

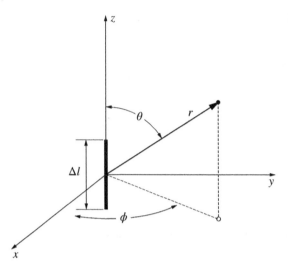

Figure 2.2 Hertzian dipole.

The length Δl of this electrically short antenna is very small compared with wavelength. As the wire radius a is small compared to the wavelength, the particular integral for retarded potential can be written in the form

$$\vec{A}(r) = \frac{\mu}{4\pi} \int_{S'} \frac{\vec{J}\left(\vec{r'}\right)\vec{S}e^{-jkR}}{R} d\vec{s'} \tag{2.112}$$

which results in

$$\vec{A}(z) = \frac{\mu}{4\pi} \int_L \frac{I(z')e^{-jkR}}{R} \vec{e} \, dz' \tag{2.113}$$

and, finally, if the current along the short antenna is assumed to be constant and expressed in the following phasor form:

$$I(z') = I_0 \tag{2.114}$$

and following expression for the retarded potential is obtained:

$$A_z(z) = \frac{\mu}{4\pi} \int_L \frac{I(z')e^{-jkR}}{R} dz' = \frac{\mu I_0 e^{-jkr}}{4\pi r} \int_{-h}^{h} dz' = \frac{\mu h I_0}{4\pi r} e^{-jkr} \tag{2.115}$$

If the system of spherical coordinates is considered, it follows

$$A_r = A_z \cos\theta = \mu \frac{h I_0}{4\pi r} e^{-jkr} \cos\theta \tag{2.116}$$

$$A_\theta = -A_z \sin\theta = -\mu \frac{hI_0}{4\pi r} e^{-jkr} \sin\theta \qquad (2.117)$$

Combining the complex phasor form of Eqs (2.2) and (2.94)

$$\vec{E} = -j\omega\vec{A} - \nabla\phi \qquad (2.118)$$

$$\nabla\vec{A} = -j\omega\mu\varepsilon\phi \qquad (2.119)$$

the electric field can be expressed as follows:

$$\vec{E} = -j\omega\vec{A} + \frac{1}{j\omega\mu\varepsilon}\nabla\left(\nabla\vec{A}\right) \qquad (2.120)$$

Finally, the field components are:

$$H_\varphi = \frac{hI_0}{4\pi} e^{-jkr}\left(\frac{jk}{r} + \frac{1}{r^2}\right)\sin\theta \qquad (2.121)$$

$$E_r = \frac{hI_0}{4\pi} e^{-jkr}\left(\frac{2Z_0}{r^2} + \frac{2}{j\omega\varepsilon r^3}\right)\cos\theta \qquad (2.122)$$

$$E_\theta = \frac{hI_0}{4\pi} e^{-jkr}\left(\frac{j\omega\mu}{r} + \frac{1}{j\omega\varepsilon r^3} + \frac{Z_0}{r^2}\right)\sin\theta \qquad (2.123)$$

At large distances from the source ($r \gg \lambda$), the only significant terms for E and H are those varying as $1/r$. This is the far-field region and the corresponding components are

$$H_\varphi = \frac{jkhI_0}{4\pi r} e^{-jkr}\sin\theta \qquad (2.124)$$

$$E_\theta = \frac{j\omega\mu hI_0}{4\pi r} e^{-jkr}\sin\theta = Z_0 H_\varphi \qquad (2.125)$$

The total power radiated in the far field can be obtained by using the expression (2.41). Namely, the total power is the integral of the time average Poynting vector over any surrounding surface. For simplicity, this surface is a sphere of radius r:

$$P = \oint_S \frac{1}{2}\text{Re}\left(\vec{E}\text{x}\vec{H}^*\right)d\vec{S} = \int_0^{2\pi}\int_0^\pi \frac{1}{2}\text{Re}\left(E_\theta H_\phi^*\right)r^2\sin\theta d\theta d\phi$$

$$= \frac{Z_0 k^2 I_0^2 h^2}{16\pi}\int_0^\pi \sin^3\theta d\theta = \frac{Z_0\pi I_0^2}{3}\left(\frac{h}{\lambda}\right)^2 \qquad (2.126)$$

As the power radiated by the electrically short antenna is proportional to the squared value of the ratio h/λ, the Hertz dipole with the entire length h is rather small compared to wavelength λ and represents a radiator with a very small efficiency.

2.9.1 Determination of the Q-Factor

The complex (apparent) power P_s, according to the notation of circuit theory, is given by left-hand side of (2.74)

$$P_s = \frac{1}{2} \int_A \vec{E} \times \vec{H}^* \, d\vec{A} \tag{2.127}$$

and can be also written as follows:

$$P_s = P_{rad} + j2\omega(W_E - W_M) \tag{2.128}$$

where W_E and W_M stand for the energies stored in electric and magnetic fields:

$$W_E = \frac{1}{4} \int_V \varepsilon |\vec{E}|^2 \, dV \tag{2.129}$$

$$W_M = \frac{1}{4} \int_V \mu |\vec{H}|^2 \, dV \tag{2.130}$$

The quality factor (Q-factor) is defined as follows [6]

$$Q = \left| \frac{\text{Im}(P_s)}{\text{Re}(P_s)} \right| = \frac{2\omega(W_E - W_M)}{P_{rad}} \tag{2.131}$$

Therefore, the Q-factor directly stems from the Poynting theorem and is simply obtained by the ratio of imaginary and real parts of complex power (surface integral over complex Poynting vector). Thus, numerator represents active power in the notation of circuit theory, or the average radiated power in the notation of the electromagnetic field theory, while denominator pertains to the reactive power or measure of the energy stored in the electric and magnetic fields, respectively.

Note that electrically small antenna (ESA) implies an antenna inside a sphere of radius $a = 1/k$. The minimum radius of a sphere, a, which encloses a lossless antenna, is related to the maximum quality factor of the antenna, Q. As it is well-known, ESA positioned within a given volume has relatively small value of Q which corresponds to a limit of its impedance bandwidth.

The complex Poynting vector (2.127) for Hertz dipole is:

$$\vec{S} = \frac{1}{2} \left(\vec{E} \times \vec{H}^* \right)$$
$$= \frac{I^2 (\Delta l)^2}{32\pi^2} \omega\mu \frac{(kr)^3 - j}{r^5 k^2} \sin^2\theta \cdot \vec{e}_r - j\frac{I^2 (\Delta l)^2}{32\pi^2} \frac{1 + r^2 k^2}{r^5 k} \sin(2\theta) \cdot \vec{e}_\theta \tag{2.132}$$

and, consequently, by integrating (2.132), the apparent power for $r = a$ is obtained

$$P_s = \int\limits_0^{2\pi} \int\limits_0^\pi \frac{I^2(\Delta l)^2}{32\pi^2} \omega\mu \frac{(kr)^3 - j}{r^5 k^2} \sin^2\theta \cdot r^2 \sin\theta d\theta d\phi = \frac{I^2(\Delta l)^2}{12\pi} \omega\mu \left(k - j\frac{1}{k^2 a^3}\right)$$

$$(2.133)$$

The Q-factor is then [1]:

$$Q = \left|\frac{\text{Im}(P_s)}{\text{Re}(P_s)}\right| = \frac{2\omega(W_E - W_M)}{P_{rad}} = \frac{\dfrac{I^2(\Delta l)^2}{12\pi k^2 a^3}\omega\mu}{\dfrac{I^2(\Delta l)^2}{12\pi}k\omega\mu} = \frac{1}{k^3 a^3} \qquad (2.134)$$

Note that the minimum value radius of a sphere a is related to the maximum value of quality factor of the antenna Q.

2.10 Wire Antennas of Finite Length

The finite-length wire antennas (linear antennas) can be analyzed as a super-position of infinitesimal radiation sources. Antenna problem can be considered in two different modes: radiation (transmitting) mode and scattering (receiving) mode.

The key to understanding the behavior of radiated/scattered fields is the knowledge of the current distribution along the antenna. Contrary to the case of ESAs where the length is small compared to the wavelength, the current can be assumed to be uniform if the wire length increases and the current distribution changes significantly. There are two approaches to determining the current distribution along finite-length wires; to assume, or to calculate the current waveform. Assuming the current distribution may provide satisfactory results when calculating radiated power or radiation pattern which deals with far-field expressions. Nevertheless, there are scenarios where an accurate current distribution calculation is necessary. Such an approach is necessary when computing quantities using near-field expressions (assessment of input or mutual impedance, respectively). A more rigorous approach to this problem is to obtain the current distribution by solving the corresponding integral equation which is discussed in detail in Subsection 2.10.2. In particular, radiation from thin wires can be rigorously analyzed by a corresponding type of either space-frequency or space-time integral equation, of Pocklington or Hallen type respectively.

This subsection deals with the approximate current distributions.

2.10.1 Dipole Antennas

One of the simplest antennas most commonly used in practice and also as an EMC model is the center-fed dipole antenna of length L, shown in Fig. 2.3.

Most commonly used approximation for the antenna current at high frequencies is the sinusoidal distribution defined as:

$$I(z) = I_0 \sin\left[k\left(\frac{L}{2} - |z|\right)\right] \tag{2.135}$$

At sufficiently low frequencies where $\lambda > \dfrac{L}{2}$, i.e. $k\left(\dfrac{L}{2} - |z|\right) \ll 1$, Eq. (2.135) becomes

$$I(z) = I_0\left(1 - \frac{2|z|}{L}\right) \tag{2.136}$$

The radiated far field E_θ, where the following condition is satisfied

$$kr = 2\pi\frac{r}{\lambda} \gg 1 \tag{2.137}$$

can be determined from the magnetic vector potential

$$E_\theta = -j\omega A_\theta = j\omega A_z \sin\theta \tag{2.138}$$

Approximating the distance from the source to the observation point

$$R = r - z' \cos\theta \tag{2.139}$$

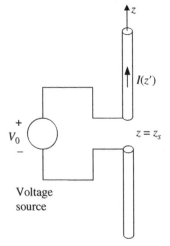

Figure 2.3 Center-fed dipole antenna.

where the magnetic vector potential for the case of sinusoidal current distribution is given by

$$A_z(z) = \frac{\mu}{4\pi} \int_L \frac{I(z')e^{-jkR}}{R}dz' = \frac{\mu I_0}{4\pi r}e^{-jkr} \int_{-L/2}^{L/2} \sin\left[k\left(\frac{L}{2} - |z'|\right)\right]e^{jkz'\cos\theta}dz'$$

(2.140)

Inserting (2.140) and (2.138) yields

$$E_\theta = jk\frac{Z_0}{4\pi r}e^{-jkr}\sin\theta \int_{-L/2}^{L/2} \sin\left[k\left(\frac{L}{2} - |z'|\right)\right]e^{jkz'\cos\theta}dz'$$

(2.141)

$$= jZ_0\frac{I_0}{2\pi r}e^{-jkr}\frac{\cos\left(k\frac{L}{2}\cos\theta\right) - \cos\left(k\frac{L}{2}\right)}{\sin\theta}$$

and the magnetic field is simply given by

$$H_\varphi = \frac{E}{Z_0} = j\frac{I_0}{2\pi r}e^{-jkr}\frac{\cos\left(k\frac{L}{2}\cos\theta\right) - \cos\left(k\frac{L}{2}\right)}{\sin\theta}$$

(2.142)

Finally, the total power radiated by the dipole antenna is given by

$$P_{rad} = \int_0^{2\pi}\int_0^\pi \frac{|E_\theta|^2}{2Z_0}r^2\sin\theta d\theta d\varphi = Z_0\frac{I_0^2}{4\pi}\int_0^\pi \frac{\left[\cos\left(k\frac{L}{2}\cos\theta\right) - \cos\left(k\frac{L}{2}\right)\right]}{\sin\theta}d\theta$$

(2.143)

In the case of half-length dipole, the total power is

$$P_{rad} = Z_0\frac{I_0^2}{4\pi}\int_0^\pi \frac{\cos^2\left(k\frac{L}{2}\cos\theta\right)}{\sin\theta}d\theta$$

(2.144)

Definition of other antenna parameters can be found elsewhere, e.g. in (Poljak Wiley 2007, 2018). The radiation resistance of half-length dipole is approximately equal to 73 Ω.

2.10.2 Pocklington Integro-Differential Equation for Straight Thin Wire

Consider a dipole antenna insulated in free space having length L and of radius a, as shown in Fig. 2.4, is considered.

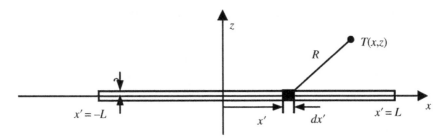

Figure 2.4 Single wire antenna insulated in free space.

The current distribution along the wire is unknown and has to be determined for a given excitation. This current flowing along the thin wire antenna is governed by the frequency domain Pocklington integro-differential equation. The Pocklington equation can be derived starting from Maxwell's equation for time-harmonic fields.

The thin wire approximation (TWA) requires wire dimensions to satisfy the conditions

$$a \ll \lambda_0 \text{ and } a \ll L \tag{2.145}$$

where λ_0 is the wavelength of a plane wave in free space.

According to TWA, only axial component of $\vec{A}$ exists along the wire, and (2.92) can be written as

$$E_x = \frac{1}{j\omega\mu\varepsilon_0} \left[\frac{\partial^2 A_x}{\partial x^2} + k^2 A_x \right] \tag{2.146}$$

The magnetic vector potential A_x is expressed by a particular integral over the unknown axial current $I(x)$ along the wire

$$A_x = \frac{\mu}{4\pi} \int_{-L}^{L} I(x') \frac{e^{-jkR}}{R} dx' \tag{2.147}$$

where R is a distance from the source point to the observation point and k is the wave number of free space.

Assuming the wire to be PEC, and according to the continuity conditions for the tangential electric field components, the total tangential electric field vanishes on the PEC wire surface, it can be written as

$$E_x^{inc} + E_x^{sct} = 0 \tag{2.148}$$

where E_x^{inc} is the incident field and E_x^{sct} is the scattered field due to the presence of the PEC surface.

Combining (2.146) and (2.147), the electric field scattered from the antenna surface is given by

$$E_x^{sct} = \frac{1}{j4\pi\omega\varepsilon_0} \int\limits_{-L}^{L} \left(\frac{\partial^2}{\partial x^2} + k^2\right) g_0(x,x')I(x')dx' \tag{2.149}$$

Inserting (2.149) into (2.148) leads to the *Pocklington integro-differential equation* for the unknown current distribution along the single straight wire antenna insulated in free space

$$E_x^{inc} = -\frac{1}{j4\pi\omega\varepsilon_0} \int\limits_{-L}^{L} \left(\frac{\partial^2}{\partial x^2} + k^2\right) g_0(x,x')I(x')dx' \tag{2.150}$$

where $g_0(x, x')$ is the free space Green function

$$g_0(x,x') = \frac{e^{-jkR}}{R} \tag{2.151}$$

while the distance R from the source to the observation point, respectively, is

$$R = \sqrt{(x-x\prime)^2 + a^2} \tag{2.152}$$

Once the axial current on the antenna is determined by solving (2.150), other important antenna parameters, such as radiated field, radiation pattern, or input impedance can be calculated. The details can be found elsewhere, e.g. in [4].

Very similar derivation could be undertaken for imperfectly conducting wires. Analyses for thin wires in the presence of inhomogeneous media could be found elsewhere, e.g. in [4] and [7]. Numerical solution of integro-differential Eq. (2.131) is discussed in Subsection 2.2.

References

1 L. Suskind and A. Friedman, Special Relativity and Classical Field Theory, New York, UK: Penguin, 2017.

2 D. E. Neuenschwander, Emmy Noether's Wonderful Theorem, Baltimore, USA: The Johns Hopkins University Press, 2011.

3 D. H. Cobe, "Derivation of Maxwell's equations from the gauge invariance of classical mechanics," *American Journal of Physics*, vol. 48, no. 348, pp. 348–353, 1980.

4 D. Poljak, Advanced Modeling in Computational Electromagnetic Compatibility, New Jersey, USA: John Wiley & Sons, Inc., 2007.

5 C. Civelek and T. F. Bechteler, Lagrangian formulation of electromagnetic fields in nondispersive medium by means of the extended Euler–Lagrange differential equation, *International Journal of Engineering Science*, vol. 46, no. 12, pp. 1218–1227, 2008.

6 M. Manteghi, "Fundamental limits, bandwidth, and information rate of electrically small antennas," *IEEE AP Magazine*, vol. 61, no. 3, pp. 14–26, 2019.

7 D. Poljak and K. El Khamlichi Drissi, Computational Methods in Electromagnetic Compatibility: Antenna Theory Approach versus Transmission line Models, Hoboken, New Jersey, USA: John Wiley & Sons, 2018.

3

Variational Methods in Electromagnetics

Applying Hamilton's variational principle, one assigns an integral form to differential equations. Therefore, a differential equation solution can be carried out the other way around by indirect approach, i.e. by determining a function which minimizes a corresponding functional.

Variational approach can be also applied to the solution of integral equations of Fredholm type [1, 2]. Finally, there is a correlation between variational approach to solving operator equations and widely used weighted residual approach (projective approach) which is regarded as a keystone for majority of numerical methods used in physics and engineering [3, 4].

This chapter features the variational approach to handle simple problems with spherical geometry. It is followed by derivation of functionals for Poisson's equation and scalar potential integral equation (SPIE). It is followed by a derivation of a correlation between variational approach and weighted residual procedure. Finally, the Ritz method is outlined.

3.1 Analytical Methods

Variational approach can be used to handle some canonical problems. This section deals with the determination of capacity of metallic sphere insulated in free space and resistance of a simple grounding system (metallic sphere immersed in an unbounded soil).

3.1.1 Capacity of Insulated Charged Sphere

The capacitance of the metallic sphere of radius can be obtained by minimizing the functional of the electrostatic energy in an unbounded dielectric medium, as it is shown in Fig. 3.1.

Deterministic and Stochastic Modeling in Computational Electromagnetics: Integral and Differential Equation Approaches, First Edition. Dragan Poljak and Anna Šušnjara.
© 2024 The Institute of Electrical and Electronics Engineers, Inc.
Published 2024 by John Wiley & Sons, Inc.

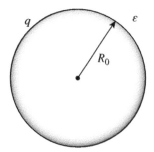

Figure 3.1 Metallic sphere in free space.

Distribution of charge over the sphere occurs in accordance with the principle of minimum energy.

According to the Poynting theorem the functional of electrostatic energy is given by

$$\psi(D) = \int_V \frac{D^2}{\varepsilon} dV - 2\varphi_0 \cdot q \tag{3.1}$$

where D is the electric flux density, ε is the permittivity of the medium, φ_0 is the potential of the metallic sphere, and q is the charge.

Taking into account spherical symmetry of the geometry, (3.1) can be written as follows

$$\psi(q) = \int_V \frac{q^2}{(4r^2\pi)^2\varepsilon} dV - 2\varphi_0 \cdot q \tag{3.2}$$

So, one is to seek the charge distribution for which functional (3.2) is stationary. Now, taking the variation of the functional

$$\delta\psi = \int_V \frac{1}{(4r^2\pi)^2\varepsilon} 2q \cdot \delta q \cdot dV - 2\varphi_0 \cdot \delta q \tag{3.3}$$

and satisfying the stationary condition

$$\frac{\delta\psi}{\delta q} = 0 \tag{3.4}$$

it follows

$$\int_V \frac{1}{(4r^2\pi)^2\varepsilon} 2q \cdot dV - 2\varphi_0 = 0 \tag{3.5}$$

In spherical coordinates, it can be written as

$$\frac{2q}{\varepsilon} \int_{R_0}^{\infty} \int_{0}^{\pi} \int_{0}^{2\pi} \frac{1}{(4r^2\pi)^2} r^2 \sin\theta \, dr \, d\theta \, d\phi - 2\varphi_0 = 0 \tag{3.6}$$

and taking the straightforward integration yields

$$\frac{q}{4\pi\varepsilon R_0} - \varphi_0 = 0 \tag{3.7}$$

Therefore, the charge over the sphere is

$$q = 4\pi\varepsilon R_0 \cdot \varphi_0 \tag{3.8}$$

and the sphere capacitance is given by

$$c = 4\pi\varepsilon R_0 \tag{3.9}$$

The same results can be obtained by applying the Gauss's law for electrostatics.

3.1.2 Spherical Grounding Resistance

The resistance of the grounding sphere, representing a simple grounding system, Fig. 3.2 can be obtained by minimizing the functional of the power losses in a conducting medium.

Figure 3.2 Grounding sphere.

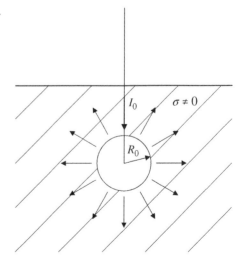

Distribution of current in a lossy medium occurs in accordance with the principle of minimum of energy, i.e. the corresponding losses are minimal.

According to the Poynting theorem, the functional of losses (dissipated power) is given by

$$\psi(J) = \int_V \frac{J^2}{\sigma} dV - 2\varphi_0 \cdot I \tag{3.10}$$

where J is the volume current density, σ is the conductivity of the soil, φ_0 is the potential of the metallic sphere, and I is the lightning current illuminating the spherical grounding system. Taking into account spherical symmetry, (3.10) can be written as

$$\psi(I) = \int_V \frac{I^2}{(4r^2\pi)^2\sigma} dV - 2\varphi_0 \cdot I \tag{3.11}$$

Now, the current for which functional (3.11) is stationary needs to be determined.

Taking the variation of the functional

$$\delta\psi = \int_V \frac{1}{(4r^2\pi)^2\sigma} 2I \cdot \delta I \cdot dV - 2\varphi_0 \cdot \delta I \tag{3.12}$$

and satisfying the stationary condition

$$\frac{\delta\psi}{\delta I} = 0 \tag{3.13}$$

one obtains

$$\int_V \frac{1}{(4r^2\pi)^2\sigma} 2I \cdot dV - 2\varphi_0 = 0 \tag{3.14}$$

In spherical coordinates, it follows

$$\frac{2I}{\sigma} \int_{R_0}^{\infty} \int_0^{\pi} \int_0^{2\pi} \frac{1}{(4r^2\pi)^2} r^2 \sin\theta \, dr d\theta d\phi - 2\varphi_0 = 0 \tag{3.15}$$

Now the straightforward integration yields

$$\frac{I}{4\pi\sigma R_0} - \varphi_0 = 0 \tag{3.16}$$

and the current through the soil is given by

$$I = \frac{\varphi_0}{\frac{1}{4\pi\sigma R_0}} \qquad (3.17)$$

i.e. the grounding resistance is

$$R = \frac{1}{4\pi\sigma R_0} \qquad (3.18)$$

The same result can be obtained by using the continuity equation [5].

3.2 Variational Basis for Numerical Methods

As already stated, an important fact in numerical solution of differential equations is that they can be formulated by an integral expression. Variational approach to numerical modeling of operator equations is illustrated on functionals for Poisson's equation and SPIE.

3.2.1 Poisson's Equation

Consider the Poisson's equation of type

$$\nabla\varphi = -\frac{\rho}{\varepsilon} \qquad (3.19)$$

where φ stands for scalar potential and ρ is the volume charge density.

It is also assumed that some boundary conditions are prescribed. Next step is to insert Poisson's equation into the integral

$$\delta F = \int_\Omega \left(\nabla\varphi + \frac{\rho}{\varepsilon}\right)\delta\varphi d\Omega \qquad (3.20)$$

where Ω is a domain of interest.

Taking into account

$$\nabla(\nabla\varphi \cdot \delta\varphi) = \nabla^2\varphi\delta\varphi + \nabla\varphi \cdot \nabla(\delta\varphi) \qquad (3.21)$$

the variation of the functional becomes

$$\delta F = \int_\Omega \nabla(\nabla\varphi \cdot \delta\varphi)d\Omega - \int_\Omega \nabla\varphi \cdot \delta\nabla\varphi d\Omega + \int_\Omega \frac{\rho}{\varepsilon}\delta\varphi d\Omega \qquad (3.22)$$

Applying the divergence theorem, it follows

$$\delta F = \int_{\Gamma} \frac{\partial \varphi}{\partial n} \cdot \delta \varphi d\Gamma - \int_{\Omega} \nabla \varphi \cdot \delta \nabla \varphi d\Omega + \int_{\Omega} \frac{\rho}{\varepsilon} \delta \varphi d\Omega \tag{3.23}$$

and the desired functional is obtained

$$F = \int_{\Omega} \left[\frac{1}{2} (\nabla \varphi)^2 - \frac{\rho}{\varepsilon} \varphi \right] d\Omega - \int_{\Gamma} \frac{\partial \varphi}{\partial n} \varphi d\Gamma \tag{3.24}$$

where Γ stands for the domain boundary and $\dfrac{\partial \varphi}{\partial n}$ is the Neumann boundary condition.

Thus, in problems arising in physics and engineering, it is in principle possible to replace an integration of differential equations by finding a function minimizing a given integral (functional).

3.2.2 Scalar Potential Integral Equation (SPIE)

A similar procedure can be used for the solution of integral equations [1, 2].

Consider an integral operator

$$\int_{L} \rho_l(s')G(s,s')ds' = \phi(s) \tag{3.25}$$

where $\phi(s)$ is a known potential function, $\rho_l(s')$ is a linear charge density distribution, and the Green function of an unbounded medium is given by

$$G(s,s') = \frac{1}{4\pi\varepsilon R(s,s')} \tag{3.26}$$

where $R(s, s')$ is the distance from the source to the observation point, respectively.

Integral Eq. (3.25) is usually referred to as SPIE [5].

Performing a similar procedure as in the case of Poisson's equation, the following functional is obtained

$$F = \int_{l} \rho_l(s) \int_{l} \rho_l(s')G(s,s')ds' - 2 \int_{l} \rho_l(s)\phi(s)ds \tag{3.27}$$

Minimization of the functional (3.27) yields [6]

$$\int_{l} \rho_l(s) \int_{l} \rho_l(s')G(s,s')ds' = \int_{l} \rho_l(s)\phi(s)ds \tag{3.28}$$

Equation (3.28) is commonly used for the numerical solution procedures of integral Eq. (3.25) [6, 7].

3.2.3 Correlation Between Variational Principle and Weighted Residual (Galerkin) Approach

Assume one seeks a solution of an operator differential equation

$$L(q, t) = 0 \tag{3.29}$$

where L is corresponding differential operator and q is a time-dependent quantity of interest. Differential (operator) equation is now regarded as Lagrange equation. Thus, it can be written as

$$\delta F = \int_{t_1}^{t_2} L(q, t) \delta q \, dt \tag{3.30}$$

Furthermore, in the framework of weighted residual procedure, construction of a corresponding functional is not required.

What is of interest is to write the variation of the functional in a way that the differential equation is multiplied by an arbitrary (weight) function.

Provided that the variation of the functional is zero, it can be written as

$$\int_{t_1}^{t_2} L(q, t) \psi(t) \, dt = 0 \tag{3.31}$$

Integral $\int_{t_1}^{t_2} L(q, t) \psi(t) \, dt$ is scalar (inner) product of functions $L(q, t)$ and $\psi(t)$, while request that integral (3.31) vanishes is equivalent to the orthogonality of these functions [6].

3.2.4 Ritz Method

As the use of analytical methods is limited to a very narrow class of problems pertaining to canonical geometries, realistic problems arising from physics and engineering require an application of approximate methods of solution. In particular, from mid-60s, due to the aid of digital computers, there has been a rapid development of numerical methods. It can be stated that most of these methods stem from variational methods to minimize a given functional. One such method is the Ritz procedure [7].

Provided the function φ, minimizing the functional:

$$\psi = \int_{\Omega} F\left(x, \varphi, \frac{\partial \varphi}{\partial x}\right) d\Omega \tag{3.32}$$

is known, one also has the solution of the corresponding differential equation.

Contrary to the Fourier series expansion, the Ritz method is based on the expansion of function φ into finite series with the convergence rate being improved by adding more terms of series.

First step is to approximate unknown function φ by linear combination of basic functions f_i

$$\varphi = \sum_{i=1}^{n} C_i f_i \tag{3.33}$$

where C_i denotes unknown coefficients to be determined from the request that the functional is minimal

$$\psi = \int_{\Omega} F\left(x, \varphi, \frac{\partial \varphi}{\partial x}\right) d\Omega = \min \tag{3.34}$$

Varying the functional over unknown coefficients, it can be written as

$$\delta\psi = \int_{\Omega} \delta F(C_i) d\Omega = \int_{\Omega} \sum_{i=1}^{n} \frac{\partial F(C_i)}{\partial C_i} \delta C_i d\Omega = 0 \tag{3.35}$$

Now, the condition for the extrema of function of multiple variables is given by

$$\frac{\partial \psi(C_i)}{\partial C_i} = 0 \tag{3.36}$$

thus enabling one to determine unknown coefficients C_i and construct the desired approximate solution.

References

1 J. G. Van Bladel, Electromagnetic Fields, Second Edition, New York: Wiley, 2007.
2 B. H. McDonald, M. Friedman, A. Wexler, "Variational solution of integral equations," *IEEE Transactions on MTT*, vol. 22, no 3, pp. 237–248, 1974.
3 V. Jovic, Introduction to Engineering Numerical Modeling (in Croatian), Split: Acquarius Engineering, 1993.
4 D. Poljak, Electromagnetic Field Theory with Engineering Applications (in Croatian), Skolska knjiga, Zagreb, 2014.

5 D. Poljak, Advanced Modeling in Computational Electromagnetic Compatibility, New Jersey, USA: John Wiley & Sons, Inc., 2007.

6 D. E. Neuenschwander, Emmy Noether's Wonderful Theorem, Baltimore, USA: The Johns Hopkins University Press, 2011.

7 D. Poljak, Electromagnetic Field Theory with Engineering Applications (in Croatian), Skolska knjiga, Zagreb, 2014.

4

Outline of Numerical Methods

Phenomena in classical electromagnetics can be studied by solving operator equations (differential, integral, or variational equations). Generally, there are two basic approaches to solve problems in electromagnetics; the differential or the field approach, and the integral or the source approach.

The field approach implies a solution of a corresponding differential equation with associated boundary conditions, specified at a boundary of a computational domain. Such an approach is rather useful to treat the interior field problems (Fig. 4.1).

A standard boundary-value problem can be formulated in terms of the operator equation

$$L(u) = p \tag{4.1}$$

on the domain Ω with conditions $F(u) = q|_\Gamma$ prescribed on the boundary Γ (Fig. 4.1) where L is linear differential operator, u solution of the problem, and p is the excitation function representing the known sources inside the domain.

Methods for the solution of the interior field problem are generally referred to as differential methods or field methods.

The source, or the integral approach, is based on the solution of a corresponding integral equation, and it is particularly convenient for the treatment of exterior field problems (Fig. 4.2).

Integral formulation can be generally written as follows:

$$g(u) = h \tag{4.2}$$

where g represents an integral operator, unknowns u are related to field sources, i.e. charge densities or current densities distributed along the boundary Γ', while h is excitation function, e.g. the electric field illuminating the metallic object, thus inducing the charge density.

Deterministic and Stochastic Modeling in Computational Electromagnetics: Integral and Differential Equation Approaches, First Edition. Dragan Poljak and Anna Šušnjara.
© 2024 The Institute of Electrical and Electronics Engineers, Inc.
Published 2024 by John Wiley & Sons, Inc.

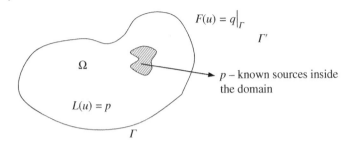

Figure 4.1 Differential approach.

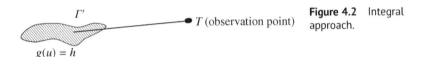

Figure 4.2 Integral approach.

Once the sources are determined, field at an arbitrary point, inside or outside the domain, can be obtained by integrating the sources.

Methods of solutions for the exterior field problems are referred to as the integral methods or method of sources.

Operator equations can be handled analytically and/or numerically, respectively. Analytical solution methods yield exact solutions but are limited to a narrow range of applications, mostly related to canonical problems. There are not many practical engineering problems that can be solved in the closed form. Numerical methods are applicable to almost all scientific engineering problems, with the main drawbacks pertaining to the approximation limit in model itself, space, and time discretization. Moreover, the criteria for accuracy, stability, and convergence are not always straightforward. The most commonly used methods in computational electromagnetics (CEM), among others, are: Finite Difference Method (FDM), Finite Element Method (FEM), Boundary Element Method (BEM), and Method of Moments (MoM).

The problems being analyzed can be regarded as steady state or transient, and the solution methods are usually classified as frequency or time domain. The frequency and time domain techniques for solving transient electromagnetic phenomena have been reported elsewhere, e.g. in [1].

According to the differential and integral formulation, respectively, numerical methods per se can be classified as domain, boundary, or source simulation methods [2, 3].

There are some basic differences between domain methods (e.g. FEM), boundary methods (e.g. BEM), and source simulation methods (charge or current simulation method).

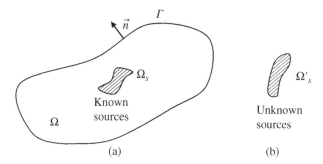

Figure 4.3 Method of fields (a) versus method of sources (b).

Modeling of partial differential equations (PDEs) via domain discretization methods is undertaken by discretizing the entire calculation domain Ω, and integrating any known sources Ω_s (if any) within the domain (Fig. 4.3a). Dirichlet and Neumann boundary conditions are specified along the boundary Γ for both FDM and FEM. Finally, an algebraic equation system provides a sparse matrix, usually banded and, in many cases, symmetric.

Modeling via the Boundary Integral Equation Method involves direct and indirect approaches, respectively.

Integral formulations of PDEs along the boundary are carried out using Green's integral theorem. The resulting equations are modeled by discretizing only the boundary and by integrating any known sources Ω_s within a given subdomain. Modeling of PDEs in terms of related boundary integral formulations results in less unknowns but dense matrices [1]. The boundary element approach method using field and potential quantities rather than field sources is usually referred to as the *direct BEM formulation*. However, there are many applications in CEM of boundary integral formulation for which it is more convenient to deal with sources. Such an approach is known as *direct BEM* and related integral equations are posed in terms of sources distributed over the boundary. This method can be considered a special case of the *direct BEM approach* involving integration over unknown charge density or current density. Integral equations over unknown sources Ω'_s (Fig. 4.3b) can be also derived from Green's integral theorem and the solution method can be referred to as a special variant of the BEM – indirect BEM.

However, as classic BEM uses potential or field on the domain boundary and this integral equation approach deals with unknown sources, some authors use the term *finite elements for integral operators* [4]. On the other hand, in order to stress the integration over sources, the term *source element method* (SEM) or *source integration method* (SIM) is suggested [2].

4.1 Variational Basis for Numerical Methods

Presented formalism of approximation of functions can be readily applied to the approximate solution of differential equations.

Any differential equation can be considered in the general implicit form:

$$A(u) = 0 \tag{4.3}$$

where:

$$A(u) \equiv L(u) - p \tag{4.4}$$

Therefore, an operator differential equation can be written as:

$$L(u) - p = 0 \tag{4.5}$$

where L is a given differential operator and p stands for an excitation.

The boundary conditions are:

$$M(u) - r = 0 \tag{4.6}$$

Therefore, the corresponding residuals are given by:

$$\xi_\Omega = L(\bar{u}) - p \tag{4.7}$$

$$\xi_\Gamma = M(\bar{u}) - r \tag{4.8}$$

And the error due to approximate solution is taken into account by satisfying the fundamental lemma of the variational calculus.

Thus, *weighted residual integrals* over the domain Ω and boundary Γ vanish, i.e.

$$\int_\Omega [L(\bar{u}) - p] W_j d\Omega + \int_\Gamma [M(\bar{u}) - r] W_j d\Gamma = 0 \tag{4.9}$$

In principle, without loss of generality, it is possible to choose base functions to automatically satisfy

$$M(\bar{u}) - r = 0 \tag{4.10}$$

Thus, one has

$$\int_\Omega [L(\bar{u}) - p] W_j d\Omega = 0 \tag{4.11}$$

and writing an approximate solution in the form

$$\bar{u}(x) = \sum_{i=1}^{n} \alpha_i N_i(x) \tag{4.12}$$

it follows

$$\int_{\Omega} \left[L\left(\sum_{i=1}^{n} \alpha_i N_i(x) \right) - p \right] W_j d\Omega = 0 \tag{4.13}$$

Finally, an original operator (differential equation) is transformed into linear equation system:

$$\sum_{i=1}^{n} \alpha_i \int_{\Omega} L(N_i(x)) W_j d\Omega = \int_{\Omega} p W_j d\Omega \quad j = 1, 2, ..., n \tag{4.14}$$

which can be written as

$$\sum_{i=1}^{n} a_{ji} \alpha_i = b_j \quad j = 1, 2, ..., n \tag{4.15}$$

or, in the matrix notation

$$[a]\{\alpha\} = \{b\} \tag{4.16}$$

where the general matrix term and right side vector are:

$$a_{ji} = \int_{\Omega} L(N_i) W_j d\Omega \tag{4.17}$$

$$b_j = \int_{\Omega} p W_j d\Omega \tag{4.18}$$

Choosing certain types of base and test functions, one deals with different numerical techniques for the solution of PDEs.

4.2 The Finite Element Method

The FEM is one of the most commonly used numerical methods in science and engineering. The method is highly automatized and convenient for computer implementation based on step-by-step algorithms. The special features of FEM are related to efficient modeling of complex shape geometries and inhomogeneous domains, providing highly banded and symmetric matrix, same as accuracy refinement by higher order approximation. The method also provides automatic inclusion of natural (Neumann) boundary conditions. This method generally gives more accurate results than highly robust FDM approaches [2].

4.2.1 Basic Concepts of FEM – One-Dimensional FEM

The calculation domain is discretized to sufficiently small segments – finite elements. The unknown solution over a finite element is expressed in terms of linear combination of local interpolation functions (shape functions), for 1D problem, as it is shown in Fig. 4.4.

The global base functions assigned to nodes are assembled from local shape functions, assigned to elements, as it is shown in Fig. 4.5 for the case of linear approximation.

The approximate solution of a problem of interest can be written as:

$$\bar{f} = \sum_{i=1}^{n} \alpha_i N_i \tag{4.19}$$

where coefficient α_i represents the solution at the global nodes, while n denotes the total number of nodes.

The approximate solution along two finite elements is shown in Fig. 4.6.

Elements of local and global nodes and linear approximation of the unknown solution over a domain of interest are shown in Fig. 4.7.

The accuracy can be improved by finer discretization, or by implementation of higher order approximation.

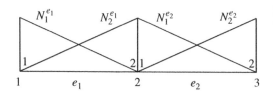

Figure 4.4 Linear shape functions over two elements.

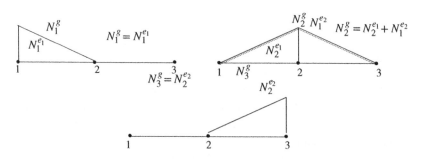

Figure 4.5 Assembling of global functions from local shape functions.

Figure 4.6 Approximate solution above two finite elements.

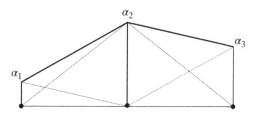

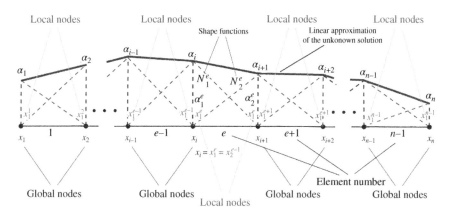

Figure 4.7 Local and global quantities.

Many problems in science and engineering can be formulated in terms of second-order differential equations which can be generally written via one-dimensional Helmholtz equation

$$\frac{d}{dx}\left[\lambda\frac{du}{dx}\right] + k^2 u - p = 0 \tag{4.20}$$

where λ and k are the properties of a medium and p represents the excitation function, i.e. the sources within the domain of interest. For one-dimensional case, the domain is related to interval $[a, b]$.

Substituting the approximate solution $\bar{u}$ into the differential Eq. (4.20) and integrating over the calculation domain according to the weighted residual approach [2], it follows

$$\int_a^b \left[\frac{d}{dx}\left(\lambda\frac{d\bar{u}}{dx}\right) + k^2\bar{u} - p\right] W_j dx = 0, \quad j = 1, 2, \cdots, n \tag{4.21}$$

which can also be written as

$$\int_a^b \frac{d}{dx}\left(\lambda\frac{d\overline{u}}{dx}\right)W_j dx + \int_a^b k^2\overline{u}W_j dx = \int_a^b pW_j dx \tag{4.22}$$

Equation (4.22) represents the *strong formulation* of the problem. Within strong formulation, base functions must be in the domain of the differential operator, and automatically satisfy the prescribed boundary conditions. The strong requirements can be avoided moving to the *weak formulation* of the problem.

The order of differentiation can be decreased by carefully performing integration by parts. Differentiation of product of two functions can be written as

$$\frac{d}{dx}\left(\lambda\frac{d\overline{u}}{dx}W_j\right) = \frac{d}{dx}\left(\lambda\frac{d\overline{u}}{dx}\right)W_j + \lambda\frac{d\overline{u}}{dx}\frac{dW_j}{dx}; \qquad \lambda = \lambda(x) \tag{4.23}$$

Somewhat rearranging Eq. (4.23) and integrating along the interval

$$\frac{d}{dx}\left(\lambda\frac{d\overline{u}}{dx}\right)W_j = \frac{d}{dx}\left(\lambda\frac{d\overline{u}}{dx}W_j\right) - \lambda\frac{d\overline{u}}{dx}\frac{dW_j}{dx} \bigg/ \int_a^b dx \tag{4.24}$$

one obtains

$$\int_a^b \frac{d}{dx}\left(\lambda\frac{d\overline{u}}{dx}\right)W_j dx = \lambda\frac{d\overline{u}}{dx}W_j\bigg|_a^b - \int_a^b \lambda\frac{d\overline{u}}{dx}\frac{dW_j}{dx}dx \tag{4.25}$$

Substituting expression (4.25) into (4.22), the weak formulation is obtained

$$\int_a^b k^2\overline{u}W_j dx + \lambda\frac{d\overline{u}}{dx}W_j\bigg|_a^b - \int_a^b \lambda\frac{d\overline{u}}{dx}\frac{dW_j}{dx}dx = \int_a^b pW_j dx \tag{4.26}$$

Second term on the right-hand side of Eq. (4.26) is the natural boundary condition (Neumann condition), thus being directly included in the weak formulation and representing the flux density at the ends of interval.

Applying the finite element algorithm, the unknown solution $\overline{u}$ is expanded into linear combination of base functions. Implementing the Bubnov–Galerkin procedure (the same choice of base and test functions) yields the following matrix equation:

$$[a]\{\alpha\} = \{b\} + \{Q\} \tag{4.27}$$

where $[a]$ is the global matrix of the system, $\{\alpha\}$ is the solution vector, $\{b\}$ represents the excitation vector, and $\{Q\}$ denotes the flux density.

The local approximation for the unknown function over a finite element is given by

$$\bar{u}^e = \alpha_1^e N_1^e(x) + \alpha_2^e N_2^e(x) \tag{4.28}$$

where $N_1^e(x), N_2^e(x)$ are the linear shape functions.

Finite element matrix and vector, respectively, are given by integrals

$$a_{ji}^e = \int_{x_1^e}^{x_2^e} k^2 N_i^e N_j^e dx - \int_{x_1^e}^{x_2^e} \lambda \frac{dN_i^e}{dx} \frac{dN_j^e}{dx} dx \tag{4.29}$$

$$b_j^e = \int_{x_1^e}^{x_2^e} p N_j^e dx \tag{4.30}$$

and the form of the local matrix equation is then:

$$\begin{bmatrix} a_{11}^e & a_{12}^e \\ a_{21}^e & a_{22}^e \end{bmatrix} \cdot \begin{bmatrix} \alpha_1^e \\ \alpha_2^e \end{bmatrix} = \begin{bmatrix} b_1^e \\ b_2^e \end{bmatrix} + \begin{bmatrix} -\lambda \dfrac{du}{dx}\Big|_{x=x_1^e} \\ \lambda \dfrac{du}{dx}\Big|_{x=x_2^e} \end{bmatrix} \tag{4.31}$$

The resulting global matrix system is assembled from the local ones and it is given by

$$\begin{bmatrix} a_{11}^{e_1} & a_{12}^{e_1} & 0 & \cdots & 0 \\ a_{21}^{e_1} & a_{22}^{e_1}+a_{11}^{e_2} & a_{12}^{e_2} & & 0 \\ 0 & a_{21}^{e_2} & a_{22}^{e_2}+a_{11}^{e_3} & \vdots & \\ \vdots & & & \ddots & a_{12}^{e_n} \\ 0 & 0 & \cdots & a_{21}^{e_n} & a_{22}^{e_n} \end{bmatrix} \cdot \begin{bmatrix} \alpha_1 \\ \alpha_2 \\ \alpha_3 \\ \vdots \\ \alpha_n \end{bmatrix} = \begin{bmatrix} b_1^{e_1} \\ b_2^{e_1}+b_1^{e_2} \\ b_2^{e_2}+b_1^{e_3} \\ \vdots \\ b_2^{e_n} \end{bmatrix} + \begin{bmatrix} -\lambda \dfrac{du}{dx}\Big|_{x=a} \\ 0 \\ 0 \\ \vdots \\ \lambda \dfrac{du}{dx}\Big|_{x=b} \end{bmatrix} \tag{4.32}$$

Note that the flux densities vanish at internal nodes, and only the values related to the domain boundary (in 1D case interval ends) are non-zero terms.

Contrary to the Neumann boundary conditions, which are automatically included in the weak formulation of the FEM, Dirichlet (forced) boundary conditions are incorporated into matrix system subsequently, i.e. it follows

$$\alpha_1 = u(a), \quad \alpha_n = u(b) \tag{4.33}$$

thus decreasing the number of unknowns, i.e. the first and last rows of the matrix equations are omitted, and the global matrix equation results in matrix system of $(n-2)$ unknowns

$$
\begin{bmatrix}
a_{22}^{e_1} + a_{11}^{e_2} & a_{12}^{e_2} & 0 & \cdots & 0 \\
a_{21}^{e_2} & a_{22}^{e_2} + a_{11}^{e_3} & a_{12}^{e_3} & & 0 \\
0 & a_{21}^{e_3} & a_{22}^{e_3} + a_{11}^{e_4} & & \vdots \\
\vdots & & & \ddots & a_{12}^{e_{n-1}} \\
0 & 0 & \cdots & a_{21}^{e_{n-1}} & a_{22}^{e_{n-1}} + a_{11}^{e_n}
\end{bmatrix}
\cdot
\begin{bmatrix}
\alpha_2 \\
\alpha_3 \\
\alpha_4 \\
\vdots \\
\alpha_{n-1}
\end{bmatrix}
$$

$$
=
\begin{bmatrix}
b_2^{e_1} + b_1^{e_2} \\
b_2^{e_2} + b_1^{e_3} \\
b_2^{e_3} + b_1^{e_4} \\
\vdots \\
b_2^{e_{n-1}} + b_1^{e_n}
\end{bmatrix}
+
\begin{bmatrix}
0 \\
0 \\
0 \\
\vdots \\
0
\end{bmatrix}
-
\begin{bmatrix}
\alpha_1 a_{21}^{e_1} \\
0 \\
0 \\
\vdots \\
0
\end{bmatrix}
-
\begin{bmatrix}
0 \\
0 \\
0 \\
\vdots \\
\alpha_n a_{12}^{e_n}
\end{bmatrix}
$$

$$(4.34)$$

Once the unknown coefficients α_2 to α_{n-1} are determined, it is possible from first and last rows (equation) to obtain the flux densities at the ends of the interval

$$
\begin{aligned}
\lambda \frac{du}{dx}\bigg|_{x=a} &= b_1^{e_1} - \alpha_1 a_{11}^{e_1} - \alpha_2 a_{12}^{e_1} \\
\lambda \frac{du}{dx}\bigg|_{x=b} &= -b_2^{e_n} + \alpha_{n-1} a_{21}^{e_n} + \alpha_n a_{22}^{e_n}
\end{aligned}
$$

$$(4.35)$$

Note that if the Dirichlet condition is prescribed at the one end of the interval, and the Neumann condition at the other, then the global system consists of $n-1$.

4.2.2 Two-Dimensional FEM

The simplest discretization of a 2D domain can be carried out using the so-called triangular elements (Fig. 4.8). The shape functions are given via plane equations in 3D space. The approximate solution is shown in Fig. 4.9, while the corresponding shape functions over a triangle are shown in Fig. 4.10. Finally, Fig. 4.11 depicts the global functions assigned to i-th node, assembled from neighboring shape functions. (Note that in 1D case, global bases always consist of two neighboring shape functions only).

Figure 4.8 Discretization of a 2D domain via triangular elements.

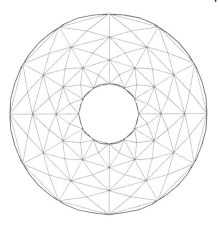

Figure 4.9 Approximate solution for 2D problem.

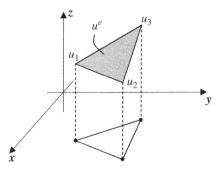

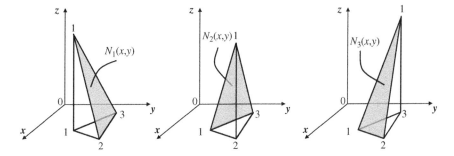

Figure 4.10 Shape functions over a triangle.

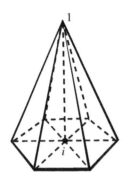

Figure 4.11 Global base function assigned to *i*-th node.

The solution on a triangular element is given by

$$u^e = \alpha_1^e N_1^e(x,y) + \alpha_2^e N_2^e(x,y) + \alpha_3^e N_3^e(x,y)$$
(4.36)

where N_1, N_2, N_3 are 2D shape functions determined by

$$N_1(x,y) = \frac{1}{2A}(2A_1 + b_1 x + a_1 y)$$
(4.37a)

$$N_2(x,y) = \frac{1}{2A}(2A_2 + b_2 x + a_2 y)$$
(4.37b)

$$N_3(x,y) = \frac{1}{2A}(2A_3 + b_3 x + a_3 y)$$
(4.37c)

Thus, an *i*-th shape function can be written as:

$$N_i(x,y) = \frac{1}{2A}(2A_i + b_i x + a_i y); \quad i = 1, 2, 3$$
(4.38)

where A denotes the area of a triangle:

$$2A = 2(A_1 + A_2 + A_3)$$
(4.39)

and A_1–A_3, a_1–a_3 and b_1–b_3 are auxiliary variables:

$$2A_1 = x_2 y_3 - x_3 y_2 \quad a_1 = x_3 - x_2 \quad b_1 = y_2 - y_3$$
(4.40a)

$$2A_2 = x_3 y_1 - x_1 y_3 \quad a_2 = x_1 - x_3 \quad b_2 = y_3 - y_1$$
(4.40b)

$$2A_3 = x_1 y_2 - x_2 y_1 \quad a_3 = x_2 - x_1 \quad b_3 = y_1 - y_2$$
(4.40c)

or, it can be simply written as

$$2A_i = x_j y_k - x_k \quad y_j a_i = x_k - x_j \quad b_i = y_j - y_k$$
(4.41)
$$i = 1, 2, 3 \quad j = 2, 3, 1 \quad k = 3, 1, 2$$

Now, the solution on a triangle can be written in the following form

$$u^e = \frac{1}{2A} \sum_{i=1}^{3} (2A_i + b_i x + a_i y)\alpha_i$$
(4.42)

where $n_e = 3$

Generalized Helmholtz equation, as in the 1D case of FEM solution, is implemented through the weak formulation of the problem and for 2D and 3D problems can be written in the following form

$$\nabla(k\nabla u) + ru = p$$
(4.43)

where u is the unknown solution, k and r are the material properties, while p represents the sources inside the domain of interest.

Applying the weighted residual approach yields

$$\int_\Omega [\nabla(k\nabla u) + ru - p]W_j d\Omega = 0 \tag{4.44}$$

i.e. one obtains:

$$\int_\Omega \nabla(k\nabla u)W_j d\Omega + \int_\Omega ruW_j d\Omega - \int_\Omega pW_j d\Omega = 0 \tag{4.45}$$

Applying the straightforward differentiation rule

$$\nabla(k\nabla u)W_j = \nabla\big[(k\nabla u)W_j\big] - (k\nabla u)\nabla W_j \tag{4.46}$$

and generalized Gauss's integral theorem

$$\int_\Omega \nabla \vec{A} d\Omega = \oint_\Gamma \vec{A}\vec{n} d\Gamma \tag{4.47}$$

the weak formulation of the Helmholtz Eq. (4.43) is obtained

$$\int_\Omega [(k\nabla u)\nabla W_j - ruW_j]d\Omega = \int_\Gamma k\frac{\partial u}{\partial n}W_j d\Gamma - \int_\Omega pW_j d\Omega \tag{4.48}$$

Integral expression (4.48) is sometimes regarded as *the variational equation* [2, 5].

The term on the left-hand side gives rise to the finite element matrix while the first term on the right-hand side is the flux through the part of the domain boundary in which the Neumann boundary condition (flux density) is prescribed. Second term on the right-hand side contains the known sources in the domain if such sources exist.

Applying the finite element algorithm, the unknown solution over an element is expressed in terms of linear combination of shape functions.

In the matrix form, the approximate solution can be written as

$$u^e = \{N\}^T\{\alpha\} = [N_1 \ N_2 \ N_3]\begin{bmatrix} \alpha_1 \\ \alpha_2 \\ \alpha_3 \end{bmatrix} \tag{4.49}$$

where $\{\alpha\}$ denotes the unknown solution coefficients.

The gradient of scalar function u in 2D is simply determined by relation

$$\nabla u = \frac{\partial u}{\partial x}\vec{e}_x + \frac{\partial u}{\partial y}\vec{e}_y \tag{4.50}$$

Inserting (4.49) into (4.50) yields

$$\nabla u = \begin{bmatrix} \dfrac{\partial u}{\partial x} \\ \dfrac{\partial u}{\partial y} \end{bmatrix} = \begin{bmatrix} \dfrac{\partial N_1}{\partial x} & \dfrac{\partial N_2}{\partial x} & \dfrac{\partial N_3}{\partial x} \\ \dfrac{\partial N_1}{\partial y} & \dfrac{\partial N_2}{\partial y} & \dfrac{\partial N_3}{\partial y} \end{bmatrix} \begin{bmatrix} \alpha_1 \\ \alpha_2 \\ \alpha_3 \end{bmatrix} \tag{4.51}$$

Implementation of the Bubnov–Galerkin procedure ($W_j = N_j$) results in the following finite element matrix

$$[a]^e = k_e \int_\Omega \begin{bmatrix} \dfrac{\partial N_1}{\partial x} & \dfrac{\partial N_1}{\partial y} \\ \dfrac{\partial N_2}{\partial x} & \dfrac{\partial N_2}{\partial y} \\ \dfrac{\partial N_3}{\partial x} & \dfrac{\partial N_3}{\partial y} \end{bmatrix} \begin{bmatrix} \dfrac{\partial N_1}{\partial x} & \dfrac{\partial N_2}{\partial x} & \dfrac{\partial N_3}{\partial x} \\ \dfrac{\partial N_1}{\partial y} & \dfrac{\partial N_2}{\partial y} & \dfrac{\partial N_3}{\partial y} \end{bmatrix}$$

$$d\Omega - r_e \int_\Omega \begin{bmatrix} N_1 \\ N_2 \\ N_3 \end{bmatrix} \begin{bmatrix} N_3 & N_2 & N_3 \end{bmatrix} d\Omega \tag{4.52}$$

Note that it is plausible to discretize the domain into sufficiently small elements, thus ensuring constant quantities over an element. Otherwise, parameters k and r become spatially dependent, which increases the complexity of integration.

Derivatives of shape functions are simply given by

$$\frac{\partial N_i}{\partial x} = \frac{b_i}{2A}; \quad \frac{\partial N_i}{\partial y} = \frac{a_i}{2A} \tag{4.53}$$

Performing certain mathematical manipulations yields:

$$[a]^e = \frac{k_e}{4A} \begin{bmatrix} a_1^2 + b_1^2 & a_1a_2 + b_1b_2 & a_1a_3 + b_1b_3 \\ a_2a_1 + b_2b_1 & a_2^2 + b_2^2 & a_2a_3 + b_2b_3 \\ a_3a_1 + b_3b_1 & a_3a_2 + b_3b_2 & a_3^2 + b_3^2 \end{bmatrix} - \frac{r_eA}{12} \begin{bmatrix} 2 & 1 & 1 \\ 1 & 2 & 1 \\ 1 & 1 & 2 \end{bmatrix} \tag{4.54}$$

The related global matrix is obtained by assembling the contributions from the local ones.

Once obtaining the solution for scalar potential (all coefficients α_i are known), it is possible to compute Q_i which are consequence of the potentials. Total flux Q on the part of the domain boundary is given by integral

$$Q = \int_\Gamma k \frac{\partial u}{\partial n} N_j d\Gamma \tag{4.55}$$

Quantity $q = k \frac{\partial u}{\partial n}$ represents the flux density, i.e. the Neumann (natural) boundary condition.

On the finite element located on a part of the boundary, the flux can be expressed by integral

$$Q_e = \int_{\Gamma_e} q_e N_j d\Gamma \tag{4.56}$$

The flux density q is generally variable but assumed constant over an element, provided the element is sufficiently small. Usually, the value of flux density on the center of the element is taken as average flux value for the whole element. Note that within the finite element algorithm, the flux Q represents the concentrated value of the flux assigned to the node.

Flux on a finite element located on a part of the boundary can be written as follows

$$\{Q\}^e = \int_{\Delta\Gamma} q_e \left\{ \begin{matrix} N_1 \\ N_2 \end{matrix} \right\} d\Gamma \tag{4.57}$$

Furthermore, the concentrated flux on i-th node consists of contributions from neighboring (adjacent) elements, as indicated in Fig. 4.12, is given by expression:

$$Q_i = \int_{\Delta\Gamma_1} q_1 N_2^{\Gamma_1} d\Gamma + \int_{\Delta\Gamma_2} q_2 N_1^{\Gamma_2} d\Gamma \tag{4.58}$$

Assuming the constant densities q_1 and q_2 yields

$$Q_i = q_1 \frac{\Delta\Gamma_1}{2} + q_2 \frac{\Delta\Gamma_2}{2} \tag{4.59}$$

For the case when $\Delta\Gamma_1 = \Delta\Gamma_2 = \Delta\Gamma$, Eq. (4.59) simplifies into

$$Q_i = q_1 \frac{\Delta\Gamma}{2} + q_2 \frac{\Delta\Gamma}{2} = \frac{\Delta\Gamma}{2}(q_1 + q_2) \tag{4.60}$$

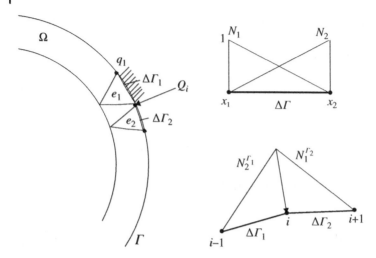

Figure 4.12 Elements on the boundary along which the flux density q is distributed, while the flux Q is concentrated in nodes.

If the boundary consists of three finite elements with constant flux density, i.e. $q_1 = q_2 = q_3 = q_0$, the contributions in nodes simply become

$$Q_1 = q_0 \frac{\Delta \Gamma}{2}; \quad Q_2 = q_0 \Delta \Gamma$$

$$Q_3 = q_0 \Delta \Gamma; \quad Q_4 = q_0 \frac{\Delta \Gamma}{2} \tag{4.61}$$

Therefore, only the first and last contribution represents half the value of other nodes.

Furthermore, contribution of the source on a 2D finite element requires the evaluation of the integral

$$\{p\}^e = \int_{\Omega_e} p_e \{N\} d\Omega = \int_{\Omega_e} p_e \left\{ \begin{array}{c} N_1 \\ N_2 \\ N_3 \end{array} \right\} d\Omega \tag{4.62}$$

where p denotes the source density inside the domain Ω.

Performing sufficiently fine domain discretization, the constant value of the source density over a triangle can be assumed. According to the Bubnov–Galerkin procedure, $W_j = N_j$ the right-hand side is then:

$$\{p\}^e = p_e \int_{\Omega} \{N\} d\Omega \tag{4.63}$$

Integral of the shape functions over a triangular element:

$$\int_{\Omega_e} N_i d\Omega \Rightarrow \int\int_{\Omega} N_i(x, y) dxdy \qquad (4.64)$$

corresponds to the volume of the pyramid with the height equal to 1 and the area of the base (triangle) is A, as depicted in Fig. 4.13.

The right-hand side is now given by local vector:

$$\{p\}^e = \frac{1}{3} p_e \left\{ \begin{array}{c} 1 \\ 1 \\ 1 \end{array} \right\} A \qquad (4.65)$$

The assembling of the global system is carried out by using standard FEM algorithm.

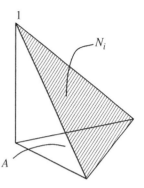

Figure 4.13 Integral of shape function over a finite element.

4.2.3 Three-Dimensional FEM

Four-nodes tetrahedral element is simplest 3D element, as shown in Fig. 4.14 [5].

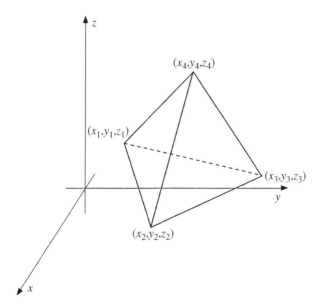

Figure 4.14 Four-node tetrahedral element.

Shape functions can be derived through the same procedure as in the case of 2D triangle elements. The solution on the tetrahedral element can be written as

$$\tilde{u}^e = a'_0 + a'_1 x + a'_2 y + a'_3 z; \qquad (x, y, z) \in e \tag{4.66}$$

Elements a'_0, a'_1, a'_2, and a'_3 are determined from the criterion of function collocation on the tops of tetrahedron resulting in the following linear equation system

$$
\begin{aligned}
\tilde{u}_1 &= a'_0 + a'_1 x_1 + a'_2 y_1 + a'_3 z_1 \\
\tilde{u}_2 &= a'_0 + a'_1 x_2 + a'_2 y_2 + a'_3 z_2 \\
\tilde{u}_3 &= a'_0 + a'_1 x_3 + a'_2 y_3 + a'_3 z_3 \\
\tilde{u}_4 &= a'_0 + a'_1 x_4 + a'_2 y_4 + a'_3 z_4
\end{aligned}
\tag{4.67}
$$

Usage of Cramer's rule yields

$$\tilde{u}^e = \frac{1}{D}[D_0 + D_1 x + D_2 y + D_3 z] \tag{4.68}$$

where

$$
D = \begin{vmatrix}
1 & x_1 & y_1 & z_1 \\
1 & x_2 & y_2 & z_2 \\
1 & x_3 & y_3 & z_3 \\
1 & x_4 & y_4 & z_4
\end{vmatrix}
\tag{4.69a}
$$

$$
D_0 = \begin{vmatrix}
\tilde{u}_1 & x_1 & y_1 & z_1 \\
\tilde{u}_2 & x_2 & y_2 & z_2 \\
\tilde{u}_3 & x_3 & y_3 & z_3 \\
\tilde{u}_4 & x_4 & y_4 & z_4
\end{vmatrix}
= \tilde{u}_1 \underbrace{\begin{vmatrix} x_2 & x_3 & x_4 \\ y_2 & y_3 & y_4 \\ z_2 & z_3 & z_4 \end{vmatrix}}_{V_1} - \tilde{u}_2 \underbrace{\begin{vmatrix} x_1 & x_3 & x_4 \\ y_1 & y_3 & y_4 \\ z_1 & z_3 & z_4 \end{vmatrix}}_{-V_2} + \tilde{u}_3 \underbrace{\begin{vmatrix} x_1 & x_2 & x_4 \\ y_1 & y_2 & y_4 \\ z_1 & z_2 & z_4 \end{vmatrix}}_{V_3} - \tilde{u}_4 \underbrace{\begin{vmatrix} x_1 & x_2 & x_3 \\ y_1 & y_2 & y_3 \\ z_1 & z_2 & z_3 \end{vmatrix}}_{-V_4}
$$
$$\tag{4.69b}$$

$$
D_1 = \begin{vmatrix}
1 & \tilde{u}_1 & y_1 & z_1 \\
1 & \tilde{u}_2 & y_2 & z_2 \\
1 & \tilde{u}_3 & y_3 & z_3 \\
1 & \tilde{u}_4 & y_4 & z_4
\end{vmatrix}
= -\tilde{u}_1 \underbrace{\begin{vmatrix} 1 & y_2 & z_2 \\ 1 & y_3 & z_3 \\ 1 & y_4 & z_4 \end{vmatrix}}_{-a_1} + \tilde{u}_2 \underbrace{\begin{vmatrix} 1 & y_1 & z_1 \\ 1 & y_3 & z_3 \\ 1 & y_4 & z_4 \end{vmatrix}}_{a_2} - \tilde{u}_3 \underbrace{\begin{vmatrix} 1 & y_1 & z_1 \\ 1 & y_2 & z_2 \\ 1 & y_4 & z_4 \end{vmatrix}}_{-a_3} + \tilde{u}_4 \underbrace{\begin{vmatrix} 1 & y_1 & z_1 \\ 1 & y_2 & z_2 \\ 1 & y_3 & z_3 \end{vmatrix}}_{a_4}
$$
$$\tag{4.69c}$$

$$D_2 = \begin{vmatrix} 1 & x_1 & \tilde{u}_1 & z_1 \\ 1 & x_2 & \tilde{u}_2 & z_2 \\ 1 & x_3 & \tilde{u}_3 & z_3 \\ 1 & x_4 & \tilde{u}_4 & z_4 \end{vmatrix} = \tilde{u}_1 \underbrace{\begin{vmatrix} 1 & x_2 & z_2 \\ 1 & x_3 & z_3 \\ 1 & x_4 & z_4 \end{vmatrix}}_{b_1} - \tilde{u}_2 \underbrace{\begin{vmatrix} 1 & x_1 & z_1 \\ 1 & x_3 & z_3 \\ 1 & x_4 & z_4 \end{vmatrix}}_{-b_2} + \tilde{u}_3 \underbrace{\begin{vmatrix} 1 & x_1 & z_1 \\ 1 & x_2 & z_2 \\ 1 & x_4 & z_4 \end{vmatrix}}_{b_3} - \tilde{u}_4 \underbrace{\begin{vmatrix} 1 & x_1 & z_1 \\ 1 & x_2 & z_2 \\ 1 & x_3 & z_3 \end{vmatrix}}_{-b_4}$$

$$(4.69d)$$

$$D_3 = \begin{vmatrix} 1 & x_1 & y_1 & \tilde{u}_1 \\ 1 & x_2 & y_2 & \tilde{u}_2 \\ 1 & x_3 & y_3 & \tilde{u}_3 \\ 1 & x_4 & y_4 & \tilde{u}_4 \end{vmatrix} = -\tilde{u}_1 \underbrace{\begin{vmatrix} 1 & x_2 & y_2 \\ 1 & x_3 & y_3 \\ 1 & x_4 & y_4 \end{vmatrix}}_{-c_1} + \tilde{u}_2 \underbrace{\begin{vmatrix} 1 & x_1 & y_1 \\ 1 & x_3 & y_3 \\ 1 & x_4 & y_4 \end{vmatrix}}_{c_2} - \tilde{u}_3 \underbrace{\begin{vmatrix} 1 & x_1 & y_1 \\ 1 & x_2 & y_2 \\ 1 & x_4 & y_4 \end{vmatrix}}_{-c_3} + \tilde{u}_4 \underbrace{\begin{vmatrix} 1 & x_1 & y_1 \\ 1 & x_2 & y_2 \\ 1 & x_3 & y_3 \end{vmatrix}}_{c_4}$$

$$(4.69e)$$

Now, the solution over element can be written as

$$\tilde{u}^e = \frac{1}{D}\left[\sum_{i=1}^{4}(V_i + a_i x + b_i y + c_i z)\tilde{u}_i\right] \tag{4.70}$$

where the shape functions are of the form

$$N_i(x,y,z) = \frac{1}{D}(V_i + a_i x + b_i y + c_i z); \qquad i = 1, 2, 3, 4 \tag{4.71}$$

Finally, the solution to the element is expressed as follows:

$$\tilde{u}^e = \sum_{i=1}^{4} \tilde{u}_i N_i \tag{4.72}$$

The derivations of the shape functions are simply given by:

$$\frac{\partial N_i}{\partial x} = \frac{a_i}{D}; \quad \frac{\partial N_i}{\partial y} = \frac{b_i}{D}; \quad \frac{\partial N_i}{\partial z} = \frac{c_i}{D} \qquad i = 1, 2, 3, 4 \tag{4.73}$$

Note that consistent order of numbering should be used, usually in anticlockwise direction.

Thus, for the simple case of Laplace equation

$$\nabla^2 \varphi = 0 \tag{4.74}$$

the approximate solution in the matrix notation is given by

$$\varphi^e = \{N\}^T \{\alpha\} = [N_1 \quad N_2 \quad N_3 \quad N_4] \begin{bmatrix} \alpha_1 \\ \alpha_2 \\ \alpha_3 \\ \alpha_4 \end{bmatrix} \tag{4.75}$$

where $\{\alpha\}$ stands for the unknown solution coefficients.

The gradient of scalar potential in 3D is given by

$$\nabla \varphi = \frac{\partial \varphi}{\partial x} \vec{e}_x + \frac{\partial \varphi}{\partial y} \vec{e}_y + \frac{\partial \varphi}{\partial z} \vec{e}_z \tag{4.76}$$

Inserting (4.75) into (4.76), one has

$$\nabla \varphi = \begin{bmatrix} \dfrac{\partial \varphi}{\partial x} \\ \dfrac{\partial \varphi}{\partial y} \\ \dfrac{\partial \varphi}{\partial z} \end{bmatrix} = \begin{bmatrix} \dfrac{\partial N_1}{\partial x} & \dfrac{\partial N_2}{\partial x} & \dfrac{\partial N_3}{\partial x} & \dfrac{\partial N_4}{\partial x} \\ \dfrac{\partial N_1}{\partial y} & \dfrac{\partial N_2}{\partial y} & \dfrac{\partial N_3}{\partial y} & \dfrac{\partial N_4}{\partial y} \\ \dfrac{\partial N_1}{\partial z} & \dfrac{\partial N_2}{\partial z} & \dfrac{\partial N_3}{\partial z} & \dfrac{\partial N_4}{\partial z} \end{bmatrix} \begin{bmatrix} \alpha_1 \\ \alpha_2 \\ \alpha_3 \\ \alpha_4 \end{bmatrix} \tag{4.77}$$

Using the Bubnov–Galerkin procedure $(W_j = N_j)$ leads to following finite element matrix

$$[a]^e = \int_{\Omega_e} \begin{bmatrix} \dfrac{\partial N_1}{\partial x} & \dfrac{\partial N_1}{\partial y} & \dfrac{\partial N_1}{\partial z} \\ \dfrac{\partial N_2}{\partial x} & \dfrac{\partial N_2}{\partial y} & \dfrac{\partial N_2}{\partial z} \\ \dfrac{\partial N_3}{\partial x} & \dfrac{\partial N_3}{\partial y} & \dfrac{\partial N_3}{\partial z} \\ \dfrac{\partial N_4}{\partial x} & \dfrac{\partial N_4}{\partial y} & \dfrac{\partial N_4}{\partial z} \end{bmatrix} \begin{bmatrix} \dfrac{\partial N_1}{\partial x} & \dfrac{\partial N_2}{\partial x} & \dfrac{\partial N_3}{\partial x} & \dfrac{\partial N_4}{\partial x} \\ \dfrac{\partial N_1}{\partial y} & \dfrac{\partial N_2}{\partial y} & \dfrac{\partial N_3}{\partial y} & \dfrac{\partial N_4}{\partial y} \\ \dfrac{\partial N_1}{\partial z} & \dfrac{\partial N_2}{\partial z} & \dfrac{\partial N_3}{\partial z} & \dfrac{\partial N_4}{\partial z} \end{bmatrix} d\Omega \tag{4.78}$$

Performing certain mathematical manipulations yields

$$[a]^e = \frac{1}{D} \begin{bmatrix} a_1^2 + b_1^2 + c_1^2 & a_1 a_2 + b_1 b_2 + c_1 c_2 & a_1 a_3 + b_1 b_3 + c_1 c_3 & a_1 a_4 + b_1 b_4 + c_1 c_4 \\ a_1 a_2 + b_1 b_2 + c_1 c_2 & a_2^2 + b_2^2 + c_2^2 & a_2 a_3 + b_2 b_3 + c_2 c_3 & a_2 a_4 + b_2 b_4 + c_2 c_4 \\ a_1 a_3 + b_1 b_3 + c_1 c_3 & a_2 a_3 + b_2 b_3 + c_2 c_3 & a_3^2 + b_3^2 + c_3^2 & a_3 a_4 + b_3 b_4 + c_3 c_4 \\ a_1 a_4 + b_1 b_4 + c_1 c_4 & a_2 a_4 + b_2 b_4 + c_2 c_4 & a_3 a_4 + b_3 b_4 + c_3 c_4 & a_4^2 + b_4^2 + c_4^2 \end{bmatrix} \tag{4.79}$$

The global matrix is obtained by assembling the local contributions.

4.3 The Boundary Element Method

The basic idea of BEM is to discretize the integral equation using boundary elements [2]. BEM can be regarded as a combination of classical boundary integral equation method and the discretization concepts that originated.

The first step in solving a problem via BEM is to derive the integral formulation of the differential equation governing the problem. The method is illustrated in the case of Laplace and Poisson equations, respectively.

The governing differential equation for static field problem, either electrostatic or magnetostatic, for source-free domains, is defined by the Laplace equation:

$$\nabla^2 u = 0 \tag{4.80}$$

or the Poisson equation, if sources p exist within the domain

$$\nabla^2 u = -p \tag{4.81}$$

A calculation domain Ω with the related boundary Γ is shown in Fig. 4.15a, where n is external normal vector to the boundary, and R denotes the distance from the source to the observation point.

It is worth mentioning that the observation point P can be also located on the boundary itself, as indicated in Fig. 4.15b.

The boundary conditions associated with these problems can be divided into essential condition (Dirichlet): whereas $u = u|_\Gamma$, defined on Γ_1, and natural condition (Neumann): $\dfrac{\partial u}{\partial n}|_\Gamma = q$, defined on Γ_2, as shown in Fig. 4.16.

The total boundary is then given by $\Gamma_1 \cup \Gamma_2$.

For simplicity, the case of the Laplace Eq. (4.80) is considered first and subsequently this procedure can be extended to the solution of the Poisson equation (4.81).

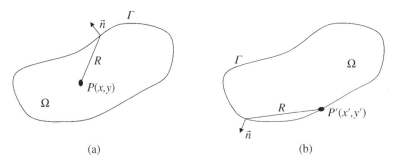

(a) (b)

Figure 4.15 The geometry of the problem: (a) interior point and (b) boundary point.

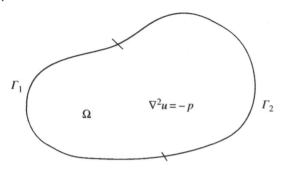

Figure 4.16 Calculation of domain with boundary conditions.

Applying the weighted residual approach, Eq. (4.80) can be integrated over the calculation domain Ω

$$\int_{\Omega} \nabla^2 u \cdot W d\Omega = 0 \tag{4.82}$$

where W is the weighting function.

Performing some mathematical manipulations and applying the generalized Gauss's theorem yields

$$\int_{\Omega} W \cdot \nabla^2 u d\Omega = \int_{\Gamma} W \frac{\partial u}{\partial n} d\Gamma - \int_{\Gamma} u \frac{\partial W}{\partial n} d\Gamma + \int_{\Omega} u \cdot \nabla^2 W d\Omega \tag{4.83}$$

and can be also rewritten as follows

$$\int_{\Gamma} W \frac{\partial u}{\partial n} d\Gamma - \int_{\Gamma} u \frac{\partial W}{\partial n} d\Gamma + \int_{\Omega} \nabla^2 W \cdot u d\Omega = 0 \tag{4.84}$$

The weighting function W can be chosen to be the solution of the differential equation, i.e.

$$\nabla^2 W - \delta\left(\vec{r} - \vec{r}'\right) = 0 \tag{4.85}$$

where δ is the Dirac delta function, r denotes the observation points and r' denotes the source points.

The solution of (4.85) represents the fundamental solution or Green function. Thus, the domain integral in (4.82) simplifies into

$$\int_{\Omega} u \nabla^2 W d\Omega = -\int_{\Omega} u \delta\left(\vec{r} - \vec{r}'\right) d\Omega = -u_i \tag{4.86}$$

and the following integral relation is obtained

$$u_i = \int_\Gamma \Psi \frac{\partial u}{\partial n} d\Gamma - \int_\Gamma u \frac{\partial \Psi}{\partial n} d\Gamma \qquad (4.87)$$

The integral expression (4.87) is the Green representation of the function u, where W can be replaced by function ψ.

The function ψ is the fundamental solution of (4.85). For two-dimensional problems the fundamental solution is

$$\Psi = -\frac{1}{2\pi} \ln R \qquad (4.88)$$

while for three dimensional problems the fundamental solution is

$$\Psi = \frac{1}{4\pi R} \qquad (4.89)$$

where $R = \left| \vec{r} - \vec{r}' \right|$ denotes the distance from the source point (boundary point) to the observation point.

The corresponding Green integral representation of the Poisson equation type (4.81) can be obtained starting from the weighted residual integral:

$$\int_\Omega [\nabla^2 u + p] \cdot W d\Omega = 0 \qquad (4.90)$$

and performing similar mathematical manipulations, one obtains

$$u_i = \int_\Gamma \Psi \frac{\partial u}{\partial n} d\Gamma - \int_\Gamma u \frac{\partial \Psi}{\partial n} d\Gamma - \int_\Omega p\psi d\Omega \qquad (4.91)$$

When the observation point i (P) is located on the boundary Γ, the boundary integral becomes singular as R approaches zero. Performing certain procedures to extract the singularity, relation (4.91) becomes

$$c_i u_i = \int_\Gamma \Psi \frac{\partial u}{\partial n} d\Gamma - \int_\Gamma u \frac{\partial \Psi}{\partial n} d\Gamma + \int_\Omega p\psi d\Omega \qquad (4.92)$$

where

$$c_i = \begin{cases} 1 & i \in \Omega \\ 1 - \dfrac{\theta_2 - \theta_1}{2\pi} & i \in \Gamma \\ 0 & i \notin \Omega \end{cases} \qquad (4.93)$$

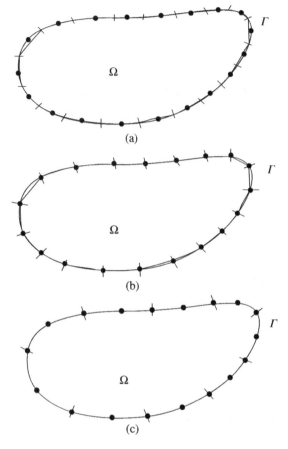

Figure 4.17 (a) Constant element approximation, (b) linear element approximation, (c) quadratic element approximation.

For a well-posed static field problem, either u or $\dfrac{\partial u}{\partial n}$ on the boundary Γ must be known, which is described by the forced (Dirichlet), natural (Neumann), or mixed (Cauchy) boundary condition.

Knowing all values of potential u and its normal derivative $\dfrac{\partial u}{\partial n}$ on the boundary, the potential at an arbitrary point of the domain can be calculated.

The boundary can be discretized into a series of constant, linear, or quadratic elements (Fig. 4.17).

4.3.1 Constant Boundary Elements

The simplest solution can be obtained by using constant boundary elements. The geometry of the constant boundary element for two-dimensional problems is shown in Fig. 4.18.

Figure 4.18 Constant boundary element approximation.

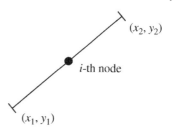

i-th node

(x_2, y_2)

(x_1, y_1)

j-th element

Figure 4.19 Global and local coordinates.

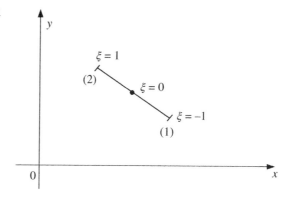

The next step in the boundary element procedure is the transformation of the global coordinates into the local ones, as indicated in Fig. 4.19.

This transformation of coordinates is given by the following set of $x = x(\xi)$, and $y = y(\xi)$, i.e.

$$x(\xi) = \frac{x_1 - x_2}{2}\xi + \frac{x_1 + x_2}{2} \tag{4.94a}$$

$$y(\xi) = \frac{y_1 - y_2}{2}\xi + \frac{y_1 + y_2}{2} \tag{4.94b}$$

where (x_1, y_1) and (x_2, y_2) are the global coordinates of the element.

Furthermore, it follows that

$$d\Gamma = \sqrt{dx^2 + dy^2} = \sqrt{\left(\frac{dx}{d\xi}\right)^2 + \left(\frac{dy}{d\xi}\right)^2}\, d\xi = \frac{\Delta\Gamma}{2}\, d\xi \tag{4.95}$$

where $\Delta\Gamma$ is the segment length defined by

$$\Delta\Gamma = \sqrt{(x_2 - x_1)^2 + (y_2 - y_1)^2} \tag{4.96}$$

Using the constant boundary element approximation, the integral equation formulation becomes

$$c_i u_i = \sum_{j=1}^{M} \left[Q_j \int_{\Gamma_j} \Psi d\Gamma - U_j \int_{\Gamma_j} \frac{\partial \Psi}{\partial n} d\Gamma \right] + \int_{\Omega_S} p \Psi d\Omega \tag{4.97}$$

where i denotes the i-th boundary node and j stands for j-th and p is the constant value of the source on the segment of the domain containing sources.

The resulting algebraic equation system is

$$P_i + \sum_{j=1}^{M} H_{ij} U_j = \sum_{j=1}^{M} Q_j G_{ij} \tag{4.98}$$

or in the matrix form

$$\{P\} + [H]\{U\} = [G]\{Q\} \tag{4.99}$$

where

$$P_i = \int_{\Omega_S} p \Psi d\Omega \tag{4.100}$$

$$\frac{\partial u}{\partial n}\Big|_j = Q_j \tag{4.101}$$

$$u = U_j \tag{4.102}$$

$$G_{ij} = \int_{\Gamma_j} \psi_{ij} d\Gamma = \begin{cases} \dfrac{1}{2\pi} \displaystyle\int_{\Gamma_j} \ln\left(\dfrac{1}{R_{ij}}\right) d\Gamma & 2D\cdots problems \\[4mm] \displaystyle\int_{\Gamma_j} \dfrac{1}{4\pi R_{ij}} d\Gamma & 3D\cdots problems \end{cases} \tag{4.103}$$

$$H_{ij} = \int_{\Gamma_j} \frac{\partial \psi_{ij}}{\partial n} d\Gamma = \begin{cases} \dfrac{1}{2\pi} \displaystyle\int_{\Gamma_j} \dfrac{\partial}{\partial n} \ln\left(\dfrac{1}{R_{ij}}\right) d\Gamma & 2D\cdots problems \\[4mm] \dfrac{1}{4\pi} \displaystyle\int_{\Gamma_j} \dfrac{\partial}{\partial n} \left(\dfrac{1}{R_{ij}}\right) d\Gamma & 3D\cdots problems \end{cases} \tag{4.104}$$

The matrix system (4.99) can be solved once the set of boundary conditions is prescribed. If the domain of interest contains unknown sources are not known, then a coupling of BEM with some domain discretization method, such as FEM, is required which leads to hybrid methods.

4.3.2 Linear and Quadratic Elements

A higher accuracy and faster convergence can be achieved by applying linear or quadratic elements. Note that the geometry of the elements is also modeled by means of quadratic functions. Such elements are then referred to as isoparametric elements [2, 6]. When using isoparametric elements, the global coordinate x is a function of the local parametric coordinate ξ on the element.

Function $x(\xi)$ can be written as

$$x = \sum_{i=1}^{N} x_i f_i(\xi) \tag{4.105}$$

where approximating functions $f_i(\xi)$ are usually polynomials.

Furthermore, the unknowns along elements are interpolated as follows:

$$u = \sum_{j=1}^{n_e} f_j(\xi) U_j \tag{4.106}$$

$$\frac{\partial u}{\partial n} = \sum_{j=1}^{n_e} f_j(\xi) \frac{\partial u}{\partial n}\Big|_j = \sum_{j=1}^{n_e} f_j(\xi) Q_j \tag{4.107}$$

where U_j denotes the unknown coefficients of the potential distribution and Q_j is the value of the normal derivative at the j-th node.

Hence, for a linear approximation, it follows:

$$u = f_1(\xi) U_1 + f_2(\xi) U_2 \tag{4.108}$$

$$\frac{\partial u}{\partial n} = f_1(\xi) Q_1 + f_2(\xi) Q_2 \tag{4.109}$$

where U_1, U_2, Q_1, Q_2 are the values of the vector potential and its normal derivative on the node $j = 1$ and $j = 2$, respectively.

The linear shape functions are given by

$$f_1(\xi) = \frac{1}{2}(1 - \xi) \tag{4.110}$$

$$f_1(\xi) = \frac{1}{2}(1 + \xi) \tag{4.111}$$

For linear elements (Fig. 4.20), the geometry is a linear function of the coordinates, i.e.

$$x = f_1(\xi) x_1 + f_2(\xi) x_2 \tag{4.112}$$

$$y = f_1(\xi) y_1 + f_2(\xi) y_2 \tag{4.113}$$

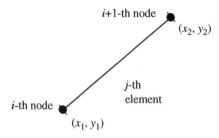

i+1-th node
(x_2, y_2)

j-th
element

i-th node
(x_1, y_1)

Figure 4.20 Linear boundary element approximation.

4.3.3 Quadratic Elements

For a quadratic interpolation, it follows

$$u = f_1(\xi)U_1 + f_2(\xi)U_2 + f_3(\xi)U_3 \tag{4.114}$$

$$\frac{\partial u}{\partial n} = f_1(\xi)Q_1 + f_1(\xi)Q_2 + f_3(\xi)Q_3 \tag{4.115}$$

where the shape functions are defined as

$$f_1(\xi) = \frac{1}{2}\xi(\xi - 1) \tag{4.116}$$

$$f_2(\xi) = \frac{1}{2}(1 + \xi)(1 - \xi) \tag{4.117}$$

$$f_3(\xi) = \frac{1}{2}\xi(\xi + 1) \tag{4.118}$$

and U_j, and Q_j is the value of the vector potential and its normal derivative at the given node j, respectively.

For the case of quadratic elements, Fig. 4.21, the geometry is represented by following functions

$$x = f_1(\xi)x_1 + f_2(\xi)x_2 + f_3(\xi)x_3 \tag{4.119}$$

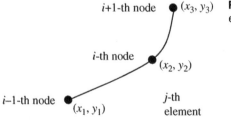

i+1-th node (x_3, y_3)

i-th node
(x_2, y_2)

i−1-th node
(x_1, y_1)

j-th
element

Figure 4.21 Quadratic boundary element approximation.

$$y = f_1(\xi)y_1 + f_2(\xi)y_2 + f_3(\xi)y_3 \tag{4.120}$$

The resulting matrix equation is

$$\{P\} + [H]\{U\} = [G]\{Q\} \tag{4.121}$$

while the corresponding coefficients are now given by

$$G_{ij} = \int\limits_{\Gamma_j} f_j \psi_{ij} d\Gamma = \int\limits_{\Gamma_j} f_j \psi_{ij} \frac{d\Gamma}{d\xi} d\xi \tag{4.122}$$

$$H_{ij} = \int\limits_{\Gamma_j} f_j \frac{\partial \psi_{ij}}{\partial n} d\Gamma = \int\limits_{\Gamma_j} f_j \frac{\partial \psi_{ij}}{\partial n} \frac{d\Gamma}{d\xi} d\xi \tag{4.123}$$

where $\{f\}$ denotes the corresponding linear or quadratic shape functions vector.

The BEM procedures presented so far pertain to the solution of two-dimensional potential problems. If three-dimensional problems are analyzed, then triangular or quadrilateral surface elements have to be applied [2].

4.3.4 Numerical Solution of Integral Equations Over Unknown Sources

An integral equation can be written in an operator form

$$K(u) = E \tag{4.124}$$

where K is a linear integral operator, u is the unknown function to be found for a given excitation E.

The Bubnov–Galerkin Indirect BEM (GB-IBEM) of solution starts by expanding the unknown $u(x)$ into finite sum of linearly independent basis functions $\{f_i\}$ with unknown complex coefficients α_i, i.e.

$$u_n(x') = \sum_{n=1}^{N_g} U_n f_n(x') \tag{4.125}$$

where $f_n(x)$ is the linear elements shape functions, and U_n stands for the unknown coefficients of the solution, and N_g denotes the total number of basis functions.

Substituting (4.125) into (4.124) yields

$$K(u) = \sum_{n=1}^{N_g} U_n K(f_n) \tag{4.126}$$

The residual R is given by

$$R = \sum_{n=1}^{N_g} U_n K(f_n) - E \tag{4.127}$$

According to the definition of the scalar product of functions in Hilbert function space, the error R is weighted to zero with respect to certain weighting functions $\{W_j\}$, i.e.

$$\int_L R W_m^* dx = 0, m = 1, 2, ..., N_g \tag{4.128}$$

where $(^*)$ is related to complex conjugate.

As the operator K is linear, performing some mathematical manipulation and choosing $W_m = f_m$, (the Bubnov–Galerkin procedure), the system of algebraic equations is obtained

$$\sum_{n=1}^{N_g} U_n \int_L K(f_n) f_m dx = \int_L E f_m dx, m = 1, 2, ..., N_g \tag{4.129}$$

As K is an integral operator, determination of the left-hand side term implies evaluation of double integrals.

References

1 L. Suskind and A. Friedman, Special Relativity and Classical Field Theory, New York, UK: Penguin, 2017.

2 D. Poljak, Advanced Modeling in Computational Electromagnetic Compatibility, New Jersey, USA: John Wiley & Sons, Inc., 2007.

3 D. Poljak and K. El Khamlichi Drissi, Computational Methods in Electromagnetic Compatibility: Antenna Theory Approach versus Transmission Line Models, Hoboken, NJ, USA: John Wiley & Sons, 2018.

4 D.E. Neuenschwander, Emmy Noether's Wonderful Theorem, The Johns Hopkins University Press, Baltimore, USA, 2011.

5 D. Poljak and M. Cvetkovic, Human Interaction with Electromagnetic Fields: Computational Models in Dosimetry, St. Louis, USA: Elsevier, Academic Press, 2019.

6 J. G. Van Bladel, Electromagnetic Fields, Second Edition, Wiley, 2007.

Part II

Deterministic Modeling

5

Wire Configurations – Frequency Domain Analysis

Time-harmonic or transient analysis of radiation and scattering from straight thin wires via integral equations is crucial in many areas of computational electromagnetics (CEM) and applications [1, 2], such as antenna arrays, wire-grid configurations, and in electromagnetic compatibility (EMC) areas, finite length transmission lines (TLs), lightning protection, or grounding systems [1, 2]. The analysis of thin wires is based on the numerical solution of integral equations of Pocklington and Hallén type carried out in either frequency or time domain. This chapter deals with the frequency domain analysis.

Of particular interest are the studies of antenna configurations over lossy media, pertaining to homogeneous or multilayered half-space.

First single wire is considered, and then more complex wire configurations are analyzed. The numerical solution procedures are based on the Galerkin–Bubnov scheme of the Indirect Boundary Element Method (GB-IBEM) in the frequency domain. A comprehensive description of GB-IBEM is available in [1] and rather recently in [2].

5.1 Single Wire in the Presence of a Lossy Half-Space

This section deals with a straight thin wire above a half-space and a straight thin wire buried in a lossy ground. The formulation is based on the Pocklington integro-differential equations. The influence of a lossy ground is taken into account via corresponding reflection coefficient (RC). The numerical solution is carried out by means of GB-IBEM.

5.1.1 Horizontal Dipole Above a Homogeneous Lossy Half-Space

Dipole antenna of length $2L$ and radius a located at height h above a lossy ground and driven by an equivalent voltage is considered (Fig. 5.1).

Deterministic and Stochastic Modeling in Computational Electromagnetics: Integral and Differential Equation Approaches, First Edition. Dragan Poljak and Anna Šušnjara.
© 2024 The Institute of Electrical and Electronics Engineers, Inc.
Published 2024 by John Wiley & Sons, Inc.

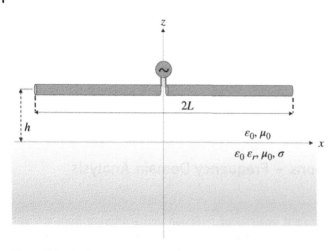

Figure 5.1 Straight thin wire above a lossy half-space.

The thin wire approximation requires wire dimensions to satisfy the following conditions:

$$a \ll \lambda_0 \quad \text{and} \quad a \ll L \tag{5.1}$$

where λ_0 is the wavelength of a plane wave in the free space.

5.1.1.1 Integro-differential Equation Formulation

The current distribution along the thin wire antenna is governed by the frequency domain Pocklington integro-differential equation. The Pocklington equation can be derived starting from Maxwell equation for time-harmonic fields by enforcing the interface conditions for the tangential components of the electric field on the wire surface. Assuming the wire to be perfectly conducting the Pocklington integro-differential equation for the unknown current distribution induced along the wire above a lossy ground is given by [1, 2]

$$E_x^{inc} = j\omega \frac{\mu}{4\pi} \int_{-L}^{L} I(x') g_{tot}(x, x') dx' - \frac{1}{j4\pi\omega\varepsilon} \frac{\partial}{\partial x} \int_{-L}^{L} \frac{\partial I(x')}{\partial x'} g_{tot}(x, x') dx' \tag{5.2}$$

where $\overrightarrow{E}^{inc}$ denotes the incident field, $I(x')$ is the induced current along the line, and $g(x, x')$ stands for the Green's function given by:

$$g_{tot}(x, x') = g_0(x, x') - \Gamma_{ref} g_i(x, x') \tag{5.3}$$

where $g_0(x, x')$ is the free-space Green function:

$$g_0(x, x') = \frac{e^{-jk_0 R_0}}{R_0} \tag{5.4}$$

while $g_i(x, x')$ arises from the image theory and is given by:

$$g_i(x, x') = \frac{e^{-jk_0 R_i}}{R_i} \tag{5.5}$$

where k is the wave number of free space

$$k = \frac{2\pi}{\lambda_0} \tag{5.6}$$

and R_0 and R_i denote the corresponding distance from the source to the observation point, respectively.

Finally, Γ_{ref} is the corresponding RC by which the influence of the air-ground interface is taken into account. The rigorous approach to account for the presence of a lossy half-space requires evaluation of Sommerfeld integrals. An efficient Fresnel RC approach was originally proposed in [3, 4] and has been widely used by many authors. In the case of TM polarization, it follows

$$\Gamma_{ref} = R_{TM} = \frac{n\cos\Theta - \sqrt{n - \sin^2\Theta}}{n\cos\Theta + \sqrt{n - \sin^2\Theta}} \tag{5.7}$$

where n is the refraction index given by

$$n = \varepsilon_r - j\frac{\sigma}{\omega\varepsilon_0}, \quad \theta = \text{arctg}\frac{|x - x'|}{2h} \tag{5.8}$$

In general, RC approximation provides the accuracy of the results within 10% accuracy compared to the results obtained by using rigorous Sommerfeld approach as far as the wire is at least quarter wavelength away from the interface [3, 4]. The Green's function could be further simplified by using RC arising from the modified image theory (MIT), originally proposed for quasi-static applications [5, 6].

$$\Gamma_{ref} = \Gamma_{MIT} = \frac{n - 1}{n + 1} \tag{5.9}$$

and subsequently used in thin wire modeling [2].

Once determining the axial current on the antenna, other important parameters, such as radiated field, radiation pattern, or input impedance can be calculated [1, 2]. An extension to the case of imperfectly conducting wires, based on the concept of surface impedance, is straightforward.

5.1.1.2 Numerical Solution of the Pocklington Equation

Numerical solution of the Pocklington equation (5.2) is carried out using GB-IBEM [1].

It is convenient to write (5.2) in an operator form

$$K(I) = E \tag{5.10}$$

where K is a linear operator, I is the unknown current to be found for a given excitation E.

The GB-IBEM of solution starts by expanding the unknown current $I(x)$ into finite sum of linearly independent basis functions $f_n(x)$ with unknown complex coefficients I_n, i.e.

$$I_n(x') = \sum_{n=1}^{N_g} I_n f_n(x') \tag{5.11}$$

where N_g denotes the total number of basis functions.

Substituting (5.11) into (5.10) yields

$$KI \cong KI_n = \sum_{i=1}^{n} a_i KN_i = Y_n = P_n(Y) \tag{5.12}$$

The residual R is defined as follows

$$R = \sum_{n=1}^{N_g} I_n K(f_n) - E \tag{5.13}$$

According to the definition of the scalar product of functions in Hilbert function space, the error R is weighted to zero with respect to certain weighting functions $\{W_j\}$, i.e.

$$\int_L R W_m^* dx = 0, \quad m = 1, 2, ..., N_g \tag{5.14}$$

where (*) assigns the complex conjugate.

As the operator K is linear, performing some mathematical manipulation and by choosing $W_m = f_m$ (the Galerkin–Bubnov procedure), the system of algebraic equations is obtained

$$\sum_{n=1}^{N_g} I_n \int_L K(f_n) f_m dx = \int_L E f_m dx, \quad m = 1, 2, ..., N_g \tag{5.15}$$

Utilizing the weak formulation by carefully performing the integration by parts, one obtains

$$\sum_{n=1}^{N_g} I_n \left[\int_{-L}^{L} \int_{-L}^{L} \frac{df_m(x)}{dx} \frac{df_n(x')}{dx'} g_{tot}(x, x') dx' dx + k^2 \int_{-L}^{L} \int_{-L}^{L} f_m(x) f_n(x') g_{tot}(x, x') dx' dx \right]$$

$$= -j4\pi\omega\varepsilon \int_{-L}^{L} E_x^{inc}(x) f_m(x) dx, \quad m = 1, 2, ..., N_g$$

$$\tag{5.16}$$

Equation (5.16) represents the weak Galerkin–Bubnov formulation of the integral equation (5.2). The differential operator is replaced by straightforward differentiation over basis and weight functions which must be chosen from the class of order-one differentiable functions.

The weak formulation is convenient for implementation of GB-IBEM and boundary conditions are subsequently incorporated into the global matrix of linear equation system. This is an important advantage compared to numerical techniques such as method of moments (MoM) where bases and weights must be chosen in a way to satisfy prescribed boundary conditions [1].

Applying the GB-IBEM algorithm and discretizing the wire yields the global system of equations assembled from local ones in a manner similar to finite element step-by-step procedures

$$\sum_{i=1}^{M} [Z]_{ji}\{I\}_i = \{V\}_j, \quad j = 1, 2, ..., M \tag{5.17}$$

where M is the total number of finite elements, $[Z]_{ji}$ is the mutual impedance matrix representing the interaction of the i-th source to the j-th observation segment, respectively and $\{V\}_j$ is the right-side voltage vector for the j-th observation segment.

As functions $f(x)$ are required to be once differentiable, a reasonable choice for the shape of functions over the segments is the family of Lagrange's polynomials and one has

$$f_1(x) = \frac{x_2 - x}{\Delta x}, f_2(x) = \frac{x - x_1}{\Delta x} \tag{5.18}$$

where x_1 and x_2 are the coordinates of the segment nodes and $\Delta x = x_2 - x_1$ is the segment length.

Now matrix $[Z]_{ji}$ and vector $\{V\}_j$ are given by

$$[Z]_{ji} = \int\limits_{\Delta l_j} \int\limits_{\Delta l_i} \begin{bmatrix} \dfrac{df_1(x)}{dx}\dfrac{df_1(x')}{dx'} & \dfrac{df_1(x)}{dx}\dfrac{df_2(x')}{dx'} \\ \dfrac{df_2(x)}{dx}\dfrac{df_1(x')}{dx'} & \dfrac{df_2(x)}{dx}\dfrac{df_2(x')}{dx'} \end{bmatrix} g_{tot}(x,x')dx'dx +$$

$$+ k^2 \int\limits_{\Delta l_j} \int\limits_{\Delta l_i} \begin{bmatrix} f_1(x)f_1(x') & f_1(x)f_2(x') \\ f_2(x)f_1(x') & f_2(x)f_2(x') \end{bmatrix} g_{tot}(x,x')dx'dx =$$

$$= \frac{1}{\Delta x^2}\frac{df_1(x')}{dx'} \int\limits_{x_1}^{x_2}\int\limits_{x_1}^{x_2} \begin{bmatrix} 1 & -1 \\ -1 & 1 \end{bmatrix} g_0(x,x')dx'dx$$

$$+ \frac{k^2}{\Delta x^2} \int\limits_{x_1}^{x_2}\int\limits_{x_1}^{x_2} \begin{bmatrix} (x_2-x)(x_2-x') & (x_2-x)(x'-x_1) \\ (x-x_1)(x_2-x') & (x-x_1)(x'-x_1) \end{bmatrix} g_0(x,x')dx'dx$$

$$\tag{5.19}$$

$$\{V\}_j = -j4\pi\omega\varepsilon \int\limits_{\Delta l_j} E_x^{inc}(x) \begin{bmatrix} f_1(x) \\ f_2(x) \end{bmatrix} dx = -\frac{j4\pi\omega\varepsilon}{\Delta x} \int\limits_{x_1}^{x_2} E_x^{inc}(x) \begin{bmatrix} (x_2 - x) \\ (x - x_1) \end{bmatrix} dx$$

(5.20)

where Δl_i, Δl_j assign the widths of i-th and j-th segments.

The evaluation of the right-hand side vector is carried out analytically if the delta-function voltage generator is used (antenna mode), or the plane wave excitation (scatterer mode). In the antenna mode, right-side vector is different from zero only in the feed gap area.

The x-component of the impressed (incident) electric field is given by

$$E_x^{inc}(x) = \frac{V_g}{\Delta l_g}$$

(5.21)

where V_g is the feed voltage and $\Delta l_g = \Delta x$ (for convenience) is the feed-gap width.

Using the linear shape functions yields

$$\{V\}_j = -\frac{j4\pi\omega\varepsilon}{\Delta l_g} \int\limits_{x_1 = -\frac{\Delta l_g}{2}}^{x_2 = \frac{\Delta l_g}{2}} \frac{V_g}{\Delta l_g} \begin{bmatrix} (x_2 - x) \\ (x - x_1) \end{bmatrix} dx = -j2\pi\omega\varepsilon V_g \begin{pmatrix} 1 \\ 1 \end{pmatrix}$$

(5.22)

If the scattering mode for the simple case of normal incidence is considered, the wire is illuminated by the plane wave, i.e.

$$E_x^{inc}(x) = E_0$$

(5.23)

And the voltage vector is given by:

$$\{V\}_j = -\frac{j4\pi\omega\varepsilon}{\Delta l} \int\limits_{x_1 = -\frac{\Delta l}{2}}^{x_2 = \frac{\Delta l}{2}} E_0 \begin{bmatrix} (x_2 - x) \\ (x - x_1) \end{bmatrix} dx = -j2\pi\omega\varepsilon E_0 \Delta l \begin{pmatrix} 1 \\ 1 \end{pmatrix}$$

(5.24)

More mathematical details on the method could be found elsewhere, e.g. in [1].

It is worth addressing some conceptual differences between GB-IBEM, as a variant of BEM, used in this work and standard MoM approach. The theory of the BEM can be derived from the mixed variational principles, while the use of Greens Functions/Fundamental Solutions provides a robustness of the method.

Thus, for example, the use of sinusoidal basis functions can be regarded as MoM but not as BEM (or FEM for integral operators) because these basis functions do not satisfy the requirements for shape functions [7] by which the local basis functions used in FEM and/or BEM are assembled.

An alternative is to formulate the overhead wire problem in terms of the Hallén integral equation.

The principal advantage of the Pocklington integro-differential equation compared to Hallén's integral equation is the simplicity of formulation and related numerical solution. Moreover, the Pocklington equation could be readily applied to many specific EMC problems, e.g. in lightning channel modeling, analysis of grounding systems where the corresponding equation is homogeneous, i.e. does not contain excitation term in the form of voltage source, or plane wave [2]. The serious drawback is the quasi-singularity of the Pocklington equation kernel due to the differential operator. This problem could be overcome by the use of the weak formulation within GB-IBEM solution procedure [1]. Finally, the use of piece-wise constant element is possible only in combination with finite difference approximation [1].

On the other hand, the derivation of the Hallén's integral equation requires additional analytical effort and one also has to tackle the unknown constant. The main benefit is that the problem of quasi-singularity is avoided as there is no differential operator, thus enabling one to use piece-wise constant elements without difficulties arising in the case in the Pocklington equation modeling.

Finally, the use of Hallén's equation in direct time domain modeling has been reported in many applications, which is discussed in Chapter 6.

5.1.1.3 Computational Example

Computational example deals with the classical problem of a dipole antenna radiating above a lossy half-space which has numerous applications not only in engineering practice but also in CEM and EMC applications. Dipole antenna also represents widely used benchmark for testing different numerical methods. The wire radius is $a = 5$ cm, length $2L = 10$ m, height over LHS $h = 5$ m, and operating wavelength is $\lambda_0 = 5$ m. The electrical parameters of the lower LHS are: $\varepsilon_r = 10$, $\sigma = 0.01$ S/m. The results for real and imaginary parts of the current distribution obtained via different approaches (Pocklington equation/Fresnel coefficient, Pocklington equation/MIT coefficient, and Hallén's equation) are depicted in Fig. 5.2 and seem to be in very good agreement [8].

The results obtained via different approaches seem to be in satisfactory agreement.

A rather slight distortion of the results in the smooth current distribution waveform can be noticed as the current moves toward the free wire end. This is due to the integro-differential operator of the Pocklington equation and includes forced zero-current edge condition for the current at the wire end within the numerical solution procedure.

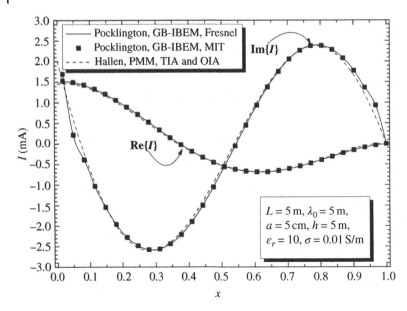

Figure 5.2 Current distribution along the dipole antenna ($2L$ = 10 m, a = 5 cm, h = 5 m, ε_r = 10, σ = 0.01 S/m).

5.1.2 Horizontal Dipole Buried in a Homogeneous Lossy Half-Space

Buried wire configurations may represent telephone or power cables, cylindrical antenna operating at a very low frequency (VLF), etc. Some important applications include submarine communication (long dipoles submerged in water), geophysical probing, and electromagnetic stimulation of biological tissue. Transient excitation of buried wires, being one of the major causes of malfunction of telecommunication and power lines, often pertains the lightning discharge problems [2].

The electromagnetic field coupling to underground wire structures [1, 2] has been investigated to a lesser extent than coupling to aboveground lines [1, 2]. Though many studies of buried wires are usually based on an approximate TL approach [2], it is generally valid for long wires and lower frequencies. If finite-length wires are considered, several problems arise within the TL approach, such as the effects at the wire ends, ground–air interface is usually neglected, the resonances are failed to be predicted, etc. [1, 2].

The rigorous antenna theory (AT) approach accounts for the earth–air interface effects via Sommerfeld integral formulation. However, the analytical evaluation of the Sommerfeld integrals is not possible, while the corresponding numerical solution is rather tedious and rather time-consuming [7, 9], even for simple

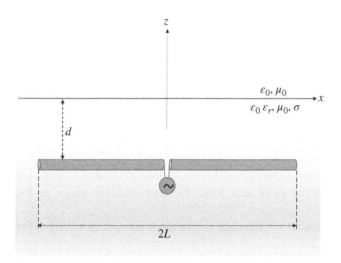

Figure 5.3 Straight thin wire buried in a lossy half-space.

geometries. Consequently, a simplified approach based on the RC approximation is used by some authors [3, 4].

The frequency domain formulation is based on either the Pocklington integro-differential equation or the Hallén integral equation, respectively. A comparison between the Pocklington and the Hallén integral equation approaches has been reported in [10].

Dipole antenna of length $2L$ and radius a located at burial depth d within a lossy medium and excited by an equivalent voltage generator is considered (Fig. 5.3).

The thin wire approximation requires wire dimensions to satisfy conditions $a \ll \lambda_0$ and $a \ll L$ where λ_0 is the wavelength of a plane wave in free space.

5.1.2.1 Pocklington Integro-differential Equation Formulation

The current distribution along the straight thin wire antenna buried in a lossy ground is governed by the frequency domain Pocklington integro-differential equation. The Pocklington equation can be derived starting from the Maxwell equation for time-harmonic fields by enforcing the interface conditions for the tangential electric field components.

Assuming the wire to be perfectly conducting the Pocklington integro-differential equation for the unknown current distribution induced along the wire is given by

$$E_x^{exc} = j\omega \frac{\mu}{4\pi} \int_{-L}^{L} I(x')g(x,x')dx' - \frac{1}{j4\pi\omega\varepsilon_{eff}} \frac{\partial}{\partial x} \int_{-L}^{L} \frac{\partial I(x')}{\partial x'} g(x,x')dx' \qquad (5.25)$$

where $\vec{E}^{exc}$ denotes the incident field, $I(x')$ is the induced current along the dipole, $\varepsilon_{eff} = \varepsilon_r \varepsilon_0 - j\frac{\sigma}{\omega}$ is the complex permittivity of the lossy ground, and $g(x, x')$ stands for the Green's function given by:

$$g(x, x') = g_0(x, x') - R_{TM}g_i(x, x') \tag{5.26}$$

where $g_0(x, x')$ and while $g_i(x, x')$ are:

$$g_0(x, x') = \frac{e^{-jk_2R_1}}{R_1}, \quad g_i(x, x') = \frac{e^{-jk_2R_2}}{R_2} \tag{5.27}$$

where k_2 is the propagation constant of the lower medium and R_1 and R_2 are distances from the source point and from the corresponding image to the observation point.

The influence of a ground–air interface is taken into account by means of the Fresnel plane wave RC:

$$R_{TM} = \frac{\frac{1}{n}\cos\theta - \sqrt{\frac{1}{n} - \sin^2\theta}}{\frac{1}{n}\cos\theta + \sqrt{\frac{1}{n} - \sin^2\theta}}; \quad \theta = \text{arctg}\frac{|x - x'|}{2d}; \quad n = \frac{\varepsilon_{eff}}{\varepsilon_0} \tag{5.28}$$

Once the axial current on the antenna is known, other antenna important parameters, such as input or mutual impedance, respectively, can be evaluated [1, 2]. An extension to the case of imperfectly conducting wires, based on the concept of surface impedance, is straightforward [1, 2].

5.1.2.2 Numerical Solution of the Pocklington Equation
Numerical solution is undertaken via GB-IBEM procedure presented in Section 5.2.1 (Eqs. (5.10)–(5.24)).

The resulting system of algebraic equations arising from the wire segmentation is given by:

$$\sum_{j=1}^{M} [Z]_{ji}\{I\}_i = \{V\}_j, \quad \text{and} \quad j = 1, 2, ..., M \tag{5.29}$$

Vector $\{I\}$ contains the unknown coefficients of the solutions, where $[Z]_{ji}$ is the local matrix representing the interaction of the i-th source boundary element with the j-th observation boundary element:

$$[Z]_{ji} = \frac{1}{j4\pi\omega\varepsilon_{eff}}\left[\int_{-L}^{L}\{D\}_j\int_{-L}^{L}\{D\}^T g(x, x')dx'dx + k_2^2\int_{-L}^{L}\{f\}_j\int_{-L}^{L}\{f'\}^T g(x, x')dx'dx\right] \tag{5.30}$$

Matrices $\{f\}$ and $\{f'\}$ contain the shape functions while $\{D\}$ and $\{D'\}$ contain their derivatives, M is the total number of line segments, and Δl_i, Δl_j are the widths of i-th and j-th segments. The $\{V\}_j$ is the local right-side vector for the j-th observation segment

$$\{V\}_j = \int_{\Delta l_j} E_x^{inc}\{f\}_j dx \tag{5.31}$$

representing the local voltage vector at the observation segment.

The evaluation of the right-hand side vector is carried out analytically if the delta-function voltage generator is used (antenna mode), or the plane wave excitation (scatterer mode). In the antenna mode, right-side vector is different from zero only in the feed gap area.

In particular, the x-component of the impressed (incident) electric field is given by

$$E_x^{inc}(x) = \frac{V_g}{\Delta l_g} \tag{5.32}$$

where V_g is the feed voltage and $\Delta l_g = \Delta x$ (for convenience) is the feed-gap width.

Linear approximation over a boundary element is used as it has been shown that this choice provides accurate and stable results [1].

More mathematical details on the method could be found elsewhere, e.g. in [1].

5.1.2.3 Computational Example

A classical problem of a dipole antenna immersed in a lossy ground is considered. A centrally fed dipole antenna of radius $a = 5\,\text{mm}$, and length $2L = 10\,\text{m}$ is of interest. The burial depth is $d = 0.1\,\text{m}$, while operating frequency is $f = 10\,\text{MHz}$. The electrical parameters of the lower lossy medium are: $\varepsilon_r = 10$, $\sigma = 0.01\,\text{S/m}$. The results for the current magnitude along the antenna conductor obtained via different approaches (Pocklington equation/GB-IBEM/Fresnel coefficient and Hallén equation) are depicted in Fig. 5.4 and qualitatively seem to be in good agreement.

Similarly, for the case of integral equations for buried wires, a trade-off between the Pocklington equation and the Hallén equation can be done. Thus, the main advantage of the Pocklington equation is the simplicity of formulation and numerical solution, while the serious drawback is the kernel quasi-singularity. On the other hand, analytically more demanding Hallén equation does not contain differential operator and therefore does not suffer from quasi-singularity.

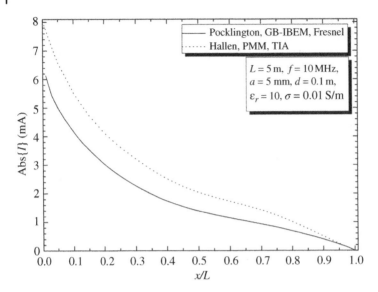

Figure 5.4 Current distribution along one arm of the dipole antenna.

5.2 Horizontal Dipole Above a Multi-layered Lossy Half-Space

There are a number of applications such as ground-penetrating radar (GPR), instrumental landing system (ILS), geophysical prospecting or remote sensing, involving the assessment of the electromagnetic field in the presence of a lossy half-space (simplified scenario) [1, 2], or a multilayered medium (more realistic case) [11–13], radiated by dipole antennas.

A rigorous integro-differential equation formulation of a straight wire above layered media requires evaluation of the spatial domain Green functions containing computationally rather demanding Sommerfeld integrals [14].

Reviews of the various frequency domain techniques to handle this problem are available elsewhere, e.g. in [11, 14]. Among the variety of methods developed to efficiently compute Sommerfeld integrals of particular interest are those approximating the Green function by means of several simple functions. These techniques are mostly related to discrete complex image method (DCIM) and rational function fitting method (RFFM). Some authors tackle a time domain counterpart of thin wire radiation in the presence of a lossy medium, e.g. [15] (dipole antenna above a lossy homogeneous half-space) or [16] (thin wires above a multilayered medium).

The majority of the existing methods to treat the problems involving multilayered media are rather demanding in a sense of formulation and computational cost [13], and sometimes also suffering from hypersingular behavior of integral equation kernel [11, 16, 17].

Contrary to such analysis of dipole antennas above layered media the use of simplified RC approach based on the MIT has been featured in [18]. The analysis presented in this subsection is based on the space-frequency integro-differential equation of the Pocklington type and corresponding field integral relations. The space-frequency Pocklington equation is numerically solved via the GB-IBEM while the corresponding field reflected from interface is obtained by numerically computing related integrals.

5.2.1 Integral Equation Formulation

Geometry of interest pertains to the dipole radiating above a layered medium, as it is shown in Fig. 5.5.

An extension of the MIT, originally proposed in [12], to account for the presence of a layered medium as a part of the Green function within the Pocklington integro-differential equation and related integral formula for the field reflected from interface, has been carried out in [18].

The current induced along the dipole antenna is governed by the integro-differential equation of Pocklington type which is derived by enforcing the interface conditions for the tangential components of the electric field at the wire surface

$$\vec{e}_x \cdot \left(\vec{E}^{exc} + \vec{E}^{sct} \right) = 0 \tag{5.33}$$

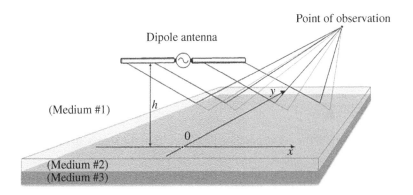

Figure 5.5 Dipole antenna radiating above a layered medium.

where the excitation is composed of the incident field $\vec{E}^{inc}$ and the field reflected from the lossy ground

$$\vec{E}^{exc} = \vec{E}^{inc} + \vec{E}^{ref} \tag{5.34}$$

The scattered electric field can be written in terms of magnetic vector potential $\vec{A}$ and electric scalar potential φ

$$\vec{E}^{sct} = -j\omega\vec{A} - \nabla\varphi \tag{5.35}$$

According to the thin wire approximation [19], expression (5.35) becomes

$$E_x^{sct} = -j\omega A_x - \frac{\partial\varphi}{\partial x} \tag{5.36}$$

with:

$$A_x = \frac{\mu}{4\pi} \int_{-L/2}^{L/2} I(x')g(x,x')dx' \tag{5.37}$$

$$\varphi(x) = -\frac{1}{j4\pi\omega\varepsilon_0} \int_{-L/2}^{L/2} \frac{\partial I(x')}{\partial x'} g(x,x')dx' \tag{5.38}$$

where $I(x')$ is the induced current along the antenna, while $g^{tot}(x,x')$ represents the total Green function

$$g^{tot}(x,x') = g_0(x,x') - R^{tot}g_i(x,x') \tag{5.39}$$

where $g_0(x,x')$ is the free-space Green function

$$g_0(x,x') = \frac{e^{-jk_oR_o}}{R_0} \tag{5.40}$$

while $g_i(x,x')$ arises from the image theory:

$$g_i(x,x') = \frac{e^{-jk_oR_i}}{R_i} \tag{5.41}$$

where R_o and R_i denote the corresponding distances from the source to the observation point.

The total RC to account for the reflection between air and layered lower medium arises from the extended use of MIT [18].

Figure 5.6 shows the tangential components of the incident, reflected and transmitted field, respectively, and the propagation directions.

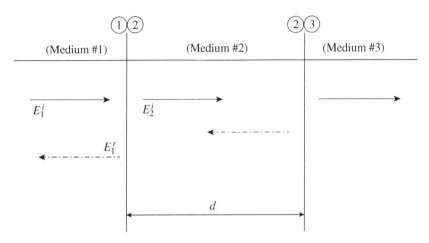

Figure 5.6 Propagating of electric field (normal incidence) through a layered medium.

As the incident, reflected and transmitted fields are simply given by

$$E^i + E^r = E^t \tag{5.42}$$

It can be written as

$$1 + R = T \tag{5.43}$$

where R and T are the reflection coefficient and transmission coefficient, respectively:

$$R = \frac{E^r}{E^i} \tag{5.44}$$

$$T = \frac{E^t}{E^i} \tag{5.45}$$

At interface 1–2, the reflected and incident fields are given by:

$$E_1^r = \frac{Z_2 - Z_1}{Z_2 + Z_1} E_1^i = R_{12} E_1^i \tag{5.46}$$

$$E_2^i = \frac{2Z_2}{Z_2 + Z_1} E_1^i = T_{12} E_1^i \tag{5.47}$$

while at interfaces 2–3, it follows

$$E_2^r = R_{23} T_{12} E_1^i e^{-\gamma_2 d} \tag{5.48}$$

$$E_3^i = T_{23} T_{12} E_1^i e^{-\gamma_2 d} \tag{5.49}$$

where:

$$T_{12} = 1 + R_{12}, \tag{5.50}$$

$$T_{21} = 1 + R_{12} = 1 - R_{12} \Rightarrow R_{12} = -R_{21} \tag{5.51}$$

and γ_{2d} is the complex propagation constant in medium 2.

At the earth-layered earth interface, one obtains:

$$\frac{E_1^{r,tot}}{E_1^i} = R_{12} + T_{12}e^{-\gamma_2 d}R_{23}T_{21} + T_{12}e^{-\gamma_2 d}R_{23}e^{-\gamma_2 d}R_{21}e^{-\gamma_2 d}T_{21} + \ldots$$

$$= R_{12} + \frac{T_{12}T_{21}R_{23}e^{-2\gamma_2 d}}{1 - R_{21}R_{23}e^{-2\gamma_2 d}} = R_{12} + \frac{T_{12}T_{21}R_{23}e^{-2\gamma_2 d}}{1 + R_{12}R_{23}e^{-2\gamma_2 d}} \tag{5.52}$$

Now, expressing the transmission coefficients T_{mn} in terms of reflection coefficients R_{mn} having taken into account (5.50) and (5.51), expression (5.52) can be written as:

$$\frac{E_1^{r,tot}}{E_1^i} = R_{12} + \frac{(1 + R_{12})(1 - R_{12})R_{23}e^{-2\gamma_2 d}}{1 + R_{12}R_{23}e^{-2\gamma_2 d}}$$

$$= \frac{R_{12}(1 + R_{12})R_{23}e^{-2\gamma_2 d} + \left(1 - R_{12}^2\right)R_{23}e^{-2\gamma_2 d}}{1 + R_{12}R_{23}e^{-2\gamma_2 d}} \quad \frac{R_{12} + R_{23}e^{-2\gamma_2 d}}{1 + R_{12}R_{23}e^{-2\gamma_2 d}} \tag{5.53}$$

Therefore, the total RC is given by

$$R^{tot} = \frac{R_{12} + R_{23}e^{-2\gamma_2 d}}{1 + R_{12}R_{23}e^{-2\gamma_2 d}} \tag{5.54}$$

where

$$R_{mn} = \frac{\varepsilon_{eff,m} - \varepsilon_{eff,n}}{\varepsilon_{eff,m} + \varepsilon_{eff,n}} \tag{5.55}$$

and $\varepsilon_{eff,mn}$ is related to the effective permittivity of a given layer.

An alternative approach for the reflection and transmission coefficients for three-layered media is the Fresnel plane wave approach.

Using Snell's law for wave propagation, the continuity conditions for the tangential components of electric and magnetic fields at the two interfaces (01 and 12) can be written as follows

$$E_0^+ = \frac{1}{\tau_{01}}E_1^+ + \frac{\rho_{01}}{\tau_{01}}E_1^- \quad E_0^- = \frac{\rho_{01}}{\tau_{01}}E_1^+ + \frac{1}{\tau_{01}}E_1^- \tag{5.56}$$

$$E_1^+ = \frac{1}{\tau_{12}}e^{\gamma_1 \cos\vartheta_1 d_1}E_2^+ \quad E_1^- = \frac{\rho_{12}}{\tau_{12}}e^{-\gamma_1 \cos\vartheta_1 d_1}E_2^+ \tag{5.57}$$

where the superscripts + and − denote the magnitude of electric field at the layer interfaces for the wave propagating in the negative and positive directions of z axis.

The reflection and transmission coefficients for the two adjacent media indexed with m and n are given as:

$$\rho_{mn} = \frac{Z_n \cos \vartheta_n - Z_m \cos \vartheta_m}{Z_n \cos \vartheta_n + Z_m \cos \vartheta_m} \quad n = 0, 1 \tag{5.58}$$

$$\tau_{mn} = \frac{2Z_n \cos \vartheta_m}{Z_n \cos \vartheta_n + Z_m \cos \vartheta_m} \quad n = 1, 2 \tag{5.59}$$

where ϑ_m is the incident angle of the wave in the layer m on interface mn and Z_m is the impedance of the medium m:

$$Z_k = \sqrt{\frac{j\omega\mu_0}{\sigma_k + j\omega\varepsilon_0\varepsilon_{rk}}} \quad k = 0, 1, 2 \tag{5.60}$$

The complex propagation constant pertaining to each medium is given as:

$$\gamma_k = \sqrt{j\omega\mu_0(\sigma_k + j\omega\varepsilon_0\varepsilon_{rk})} \quad k = 0, 1, 2 \tag{5.61}$$

Note that the incidence angle is calculated iteratively.

The exponential term $e^{\gamma 1\cos(\vartheta 1)d1}$ in (5.57) appears due to the different reference points for the field calculation in each layer: layer0 and layer1 have the reference point at the interface 01, while the field in layer2 has the reference point at the interface 12, while thickness of layer1 is d_1.

It is convenient to write (5.56) and (5.57) in the matrix form:

$$\begin{bmatrix} E_0^+ \\ E_0^- \end{bmatrix} = M_{01} \begin{bmatrix} E_1^+ \\ E_1^- \end{bmatrix} \quad \begin{bmatrix} E_1^+ \\ E_1^- \end{bmatrix} = P_{12}M_{12} \begin{bmatrix} E_2^+ \\ E_2^- \end{bmatrix} \tag{5.62}$$

where M_{mn} is the matching matrix containing the reflection/transmission coefficients for each interface, while P_{12} is the propagation matrix [20, 21]:

$$M_{mn} = \frac{1}{\tau_{mn}} \begin{bmatrix} 1 & \rho_{mn} \\ \rho_{mn} & 1 \end{bmatrix} \quad m = 0, 1, n = 1, 2 \tag{5.63}$$

$$P_{12} = \begin{bmatrix} e^{\gamma_1 \cos \vartheta_1 d_1} & 0 \\ 0 & e^{-\gamma_1 \cos \vartheta_1 d_1} \end{bmatrix} \tag{5.64}$$

By setting the value of E_0^+ to 1 V/m, all the other magnitudes E_m^+ and E_m^- are easily determined. Given the observation point is located in one of the three layers, the related electric field can be calculated using the following expressions:

$$\begin{aligned} \vec{E}_0 &= E_1^+ \left(\cos \vartheta_0 \vec{e}_x + \sin \vartheta_0 \vec{e}_z \right) e^{-\gamma_0(x\sin \vartheta_0 - z\cos \vartheta_0)} \\ &+ E_0^+ \left(\cos \vartheta_0 \vec{e}_x - \sin \vartheta_0 \vec{e}_z \right) e^{-\gamma_0(x\sin \vartheta_0 - z\cos \vartheta_0)} \end{aligned} \tag{5.65}$$

$$\vec{E}_1 = E_1^+ \left(\cos \vartheta_1 \vec{e}_x + \sin \vartheta_1 \vec{e}_z \right) e^{-\gamma_1 (x \sin \vartheta_1 - z \cos \vartheta_1)}$$
$$+ E_1^+ \left(\cos \vartheta_1 \vec{e}_x - \sin \vartheta_1 \vec{e}_z \right) e^{-\gamma_1 (x \sin \vartheta_1 - z \cos \vartheta_1)} \tag{5.66}$$

$$\vec{E}_2 = E_2^+ \left(\cos \vartheta_2 \vec{e}_x + \sin \vartheta_2 \vec{e}_z \right) e^{-\gamma_2 (x \sin \vartheta_2 - z \cos \vartheta_2)} \tag{5.67}$$

where $\bar{E}_0$, $\bar{E}_1$, and $\bar{E}_2$ stand for electric field in air, layer1, and layer2.

The reflection and transmission coefficients R_{TM} and T_{TM} are calculated as the ratio between the field value at the ground surface and the field value obtained at given observation point.

Inserting (5.36)–(5.38) into (5.34) yields the Pocklington integro-differential equation for the unknown current distribution induced along the dipole

$$E_x^{exc} = j\omega \frac{\mu}{4\pi} \int\limits_{-L/2}^{L/2} I(x') g^{tot}(x,x') dx' - \frac{1}{j4\pi\omega\varepsilon_0} \frac{\partial}{\partial x} \int\limits_{-L/2}^{L/2} \frac{\partial I(x')}{\partial x'} g^{tot}(x,x') dx' \tag{5.68}$$

The integro-differential equation (5.68) is solved via the GB-IBEM [1] outlined in Section 5.1.1. Once the induced current is determined, the related radiated field can be obtained.

5.2.2 Radiated Field

The radiated electric field components obtained by combining relations (5.35) to (5.38) are given by

$$E_x = \frac{1}{j4\pi\omega\varepsilon_0} \left[- \int\limits_{-L/2}^{L/2} \frac{\partial I(x')}{\partial x'} \frac{\partial g^{tot}(x,x')}{\partial x'} dx' + k^2 \int\limits_{L/2}^{L/2} g^{tot}(x,x') I(x') dx' \right] \tag{5.69}$$

$$E_y = \frac{1}{j4\pi\omega\varepsilon_0} \int\limits_{-L/2}^{L/2} \frac{\partial I(x')}{\partial x'} \frac{\partial g^{tot}(x',y)}{\partial y} dx' \tag{5.70}$$

$$E_z = \frac{1}{j4\pi\omega\varepsilon_0} \int\limits_{-L/2}^{L/2} \frac{\partial I(x')}{\partial x'} \frac{\partial g^{tot}(x',z)}{\partial z} dx' \tag{5.71}$$

For the case of the field reflected above the interface, the total Green function is of the form (5.39). Furthermore, for the case of the field transmitted into the multi-layered ground, the Green function inside the field integrals is given by [21]

$$g(x,x') = T_{TM} g_0(x,x') \tag{5.72}$$

where T_{TM} is the transmission coefficient for the transverse magnetic (TM) polarization and three media configurations, as discussed in Section 5.2.1.

Numerical solution procedure for the determination of field components (5.69)–(5.71) is based on GB-IBEM formalism as well.

Using the local expansion for the current on segment

$$I(x') = I_{1i} \frac{x_{2i} - x'}{\Delta x} + I_{2i} \frac{x' - x_{1i}}{\Delta x} \tag{5.73}$$

where I_{1i} and I_{2i} are current values at the local nodes of the i-th wire segment Δx, with coordinates x_{1i} and x_{2i}, $\Delta x = x_{2i} - x_{1i}$, and substituting (5.73) into field formulas (5.69) to (5.71) results in the following expressions for the y and z components of the electric field at an arbitrary point in the air

$$E_x = \frac{1}{j4\pi\omega\varepsilon_0} \sum_{j=1}^{M} \sum_{i=1}^{N_j} \left[-\frac{I_{2ij} - I_{1ij}}{\Delta x_j} \int_{x_{1ij}}^{x_{2ij}} \frac{\partial G(x,x')}{\partial x'} dx' + k^2 \int_{x_{1ij}}^{x_{2ij}} G(x,x') I(x') dx' \right] \tag{5.74}$$

$$E_y = \frac{1}{j4\pi\omega\varepsilon_0} \sum_{j=1}^{M} \sum_{i=1}^{N_j} \frac{I_{2ij} - I_{1ij}}{\Delta x_j} \int_{x_{1ij}}^{x_{2ij}} \frac{\partial G(x',y)}{\partial y} dx' \tag{5.75}$$

$$E_z = \frac{1}{j4\pi\omega\varepsilon_0} \sum_{j=1}^{M} \sum_{i=1}^{N_j} \frac{I_{2ij} - I_{1ij}}{\Delta x_j} \int_{x_{1ij}}^{x_{2ij}} \frac{\partial G(x',z)}{\partial z} dx' \tag{5.76}$$

where M is the total number of wires and N_j denotes the total number of boundary elements on the j-th wire.

The integrals (5.73)–(5.75) are numerically calculated via Gaussian quadrature. Due to quasi-singularity of the Green function, the first-order differential operator appearing in the integral equation kernel is likely to cause numerical instability. The problem of quasi-singularity of the Green function is avoided by approximating the derivative of the Green function by means of a central finite difference formula, as explained in [13].

Though the model presented pertains to the finite thickness slab, a proper use of RC provides the treatment of arbitrary number of layers. Also, with many thin layers, one can simulate a continuous variation of the permittivity which is of practical interest, e.g. for GPR applications.

5.2.3 Numerical Results

First set of numerical results is related to current distribution along the dipole antenna, while the next set deals with the corresponding radiation pattern. The computational example is related to the dipole antenna of length $L = 0.4$ m and radius $a = 0.003$ m horizontally located at different heights above a two-layer media (vegetation + lossy ground) with vegetation permittivity $\varepsilon_{r1} = 12$ and conductivity $\sigma_1 = 2$ mS/m. The lossy ground permittivity and conductivity are $\varepsilon_{r2} = 10$ and $\sigma_2 = 1$ mS/m, respectively. Terminal voltage is $V_T = 1$ V and the operating frequency is 300 MHz. The depth of the vegetation layer is assumed to be $d_1 = 0.05$ m. The results obtained via RC arise from the extension of the MIT approach with the results obtained by using the Stratton RC [22].

Figure 5.7a,b shows the current distribution along the half-wave dipole antenna for the height above ground $h = 0.1$ and 0.2 m, respectively.

Figures 5.8 and 5.9 show the corresponding percentage deviation in calculated values between different approaches (MIT and Fresnel RCs) on different heights above ground and operating frequencies.

Figures 5.10 and 5.11 show the corresponding radiation pattern of half-wave dipole for different heights above the layered medium.

The numerical results for the current distribution and the radiated field presented in Figs. 5.7–5.11 obtained by using the Green function containing the RC deriving from MIT are comparable with the results obtained via the Fresnel coefficient in the wide frequency range (0–500 MHz) with the highest percentage deviation at low heights.

Also, the numerical results are not appreciably dependent on depth of the vegetation layer. Deviations are greater at smaller vegetation depths and upon $d_1 = 0.1$ m practically do not vary.

Note that at lower operating frequencies, closer agreement of the numerical results occurs at higher antenna heights while at higher operating frequencies, conversely.

Furthermore, the results for current distribution along the horizontal straight wire above a two-layer medium are analyzed. The simulations are carried out for the antenna height; $h = 0.5$ m, $h = 1.5$ m, and $h = 5$ m, respectively. The thicknesses of the upper layer are assumed to be $d_1 = 0.2$ m and $d_1 = 1.2$ m, respectively. Relative dielectric constant of both layers is set to value: $\varepsilon_r = 10$. The value of voltage source is $V_T = 1$ V, while operating frequency is varied from 0.1 to 10 MHz. The obtained results are compared to the quasistatic model/TL results available in [22].

(a) $h = 0.1$ m

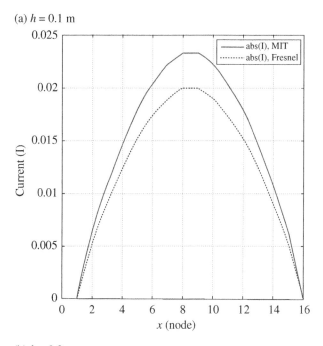

(b) $h = 0.2$ m

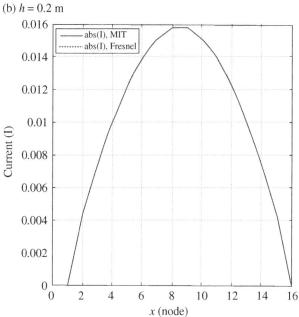

Figure 5.7 The amplitude of current distribution along the dipole. (a) $h = 0.1$ m, (b) $h = 0.2$ m.

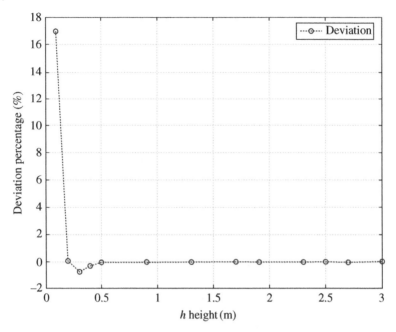

Figure 5.8 The percentage deviation on different heights.

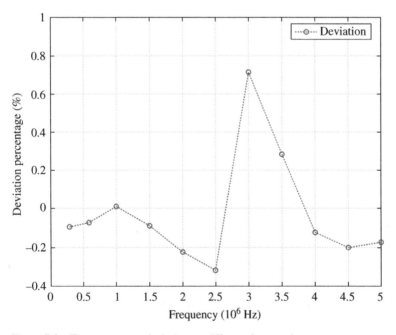

Figure 5.9 The percentage deviation on different frequencies.

(a)

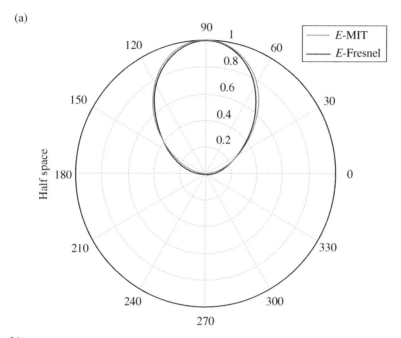

(b)

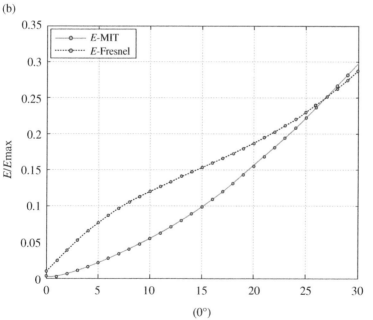

Figure 5.10 (a) Radiation pattern of half-wave dipole above the multilayer (h = 0.2 m),
(b) A view of a part of radiation pattern of half-wave dipole above the multilayer
(h = 0.2 m).

(a)

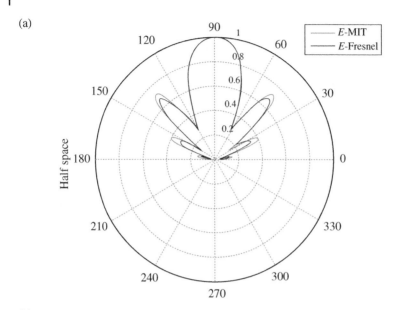

(b)

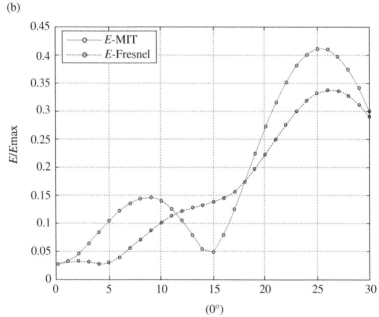

Figure 5.11 (a) Radiation pattern of half-wave dipole above the multilayer ($h = 1.7$ m), (b) A view of a part of radiation pattern of half-wave dipole above the multilayer ($h = 1.7$ m).

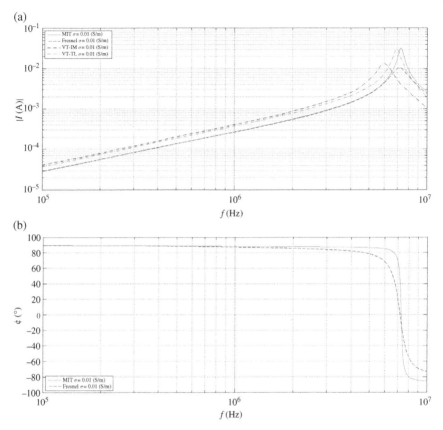

Figure 5.12 The current at the dipole center for the upper layer conductivity $\sigma = 10\,\text{mS/m}$, at height $h = 0.5\,\text{m}$ above the interface. (a) Amplitude spectrum, (b) Phase spectrum.

Figures 5.12 and 5.13 show amplitude and phase spectra at the center of the wire of length $L = 20\,\text{m}$ and radius $a = 0.01\,\text{m}$ located at height $h = 0.5\,\text{m}$. The wire is discretized into 31 linear elements. The upper layer thickness is $d_1 = 0.2\,\text{m}$, and the corresponding conductivity of the lower semi-infinite ground layer are $\sigma = 10\,\text{mS/m}$ and $\sigma = 1\,\text{mS/m}$, respectively.

Figures 5.14 and 5.15 show amplitude and phase spectra at the center of the wire of length $L = 20\,\text{m}$ and radius $a = 0.01\,\text{m}$ located at height $h = 2.5\,\text{m}$. The upper layer thickness is $d_1 = 0.2\,\text{m}$, and the corresponding conductivity of the lower semi-infinite ground layer is $\sigma = 10\,\text{mS/m}$ and $\sigma = 1\,\text{mS/m}$, respectively.

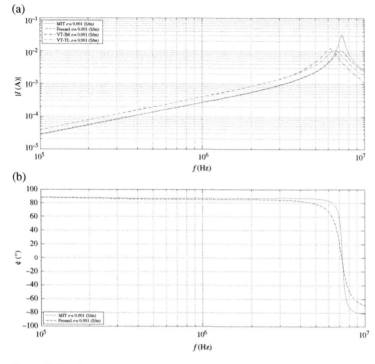

Figure 5.13 The current along the dipole for the upper layer conductivity $\sigma = 1$ mS/m, at height $h = 0.5$ m above the interface. (a) Amplitude spectrum, (b) Phase spectrum.

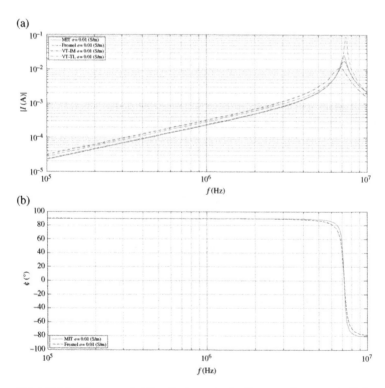

Figure 5.14 The current at the dipole center for the upper layer conductivity $\sigma = 10$ mS/m, at height $h = 2.5$ m above the interface. (a) Amplitude spectrum, (b) Phase spectrum.

(a)

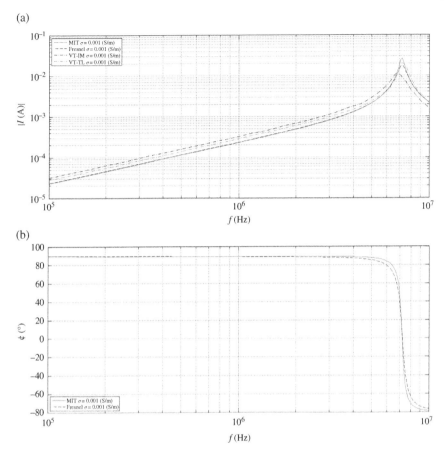

(b)

Figure 5.15 The current at the dipole center for the upper layer conductivity σ = 1 mS/m, at height for the upper layer conductivity h = 2.5 m. (a) Amplitude spectrum, (b) Phase spectrum.

Figures 5.16 and 5.17 show amplitude and phase spectra at the center of the wire of length $L = 20$ m and radius $a = 0.01$ m located at height $h = 5$ m. The upper layer thickness is $d_1 = 0.2$ m, and the corresponding conductivity of the lower semi-infinite ground layer are $\sigma = 10$ mS/m and $\sigma = 1$ mS/m, respectively.

Figures 5.18 and 5.19 show amplitude and phase spectra at the center of the wire of length $L = 200$ m and radius $a = 0.01$ m located at height $h = 2.5$ m. The upper layer thickness is $d_1 = 0.2$ m, and the corresponding conductivity of the lower semi-infinite ground layer is $\sigma = 10$ mS/m and $\sigma = 1$ mS/m, respectively.

(a)

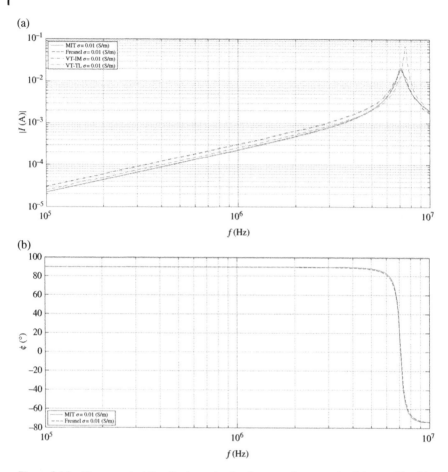

Figure 5.16 The current at the dipole center for the upper layer conductivity σ = 10 mS/m, at height for the upper layer conductivity h = 5 m. (a) Amplitude spectrum, (b) Phase spectrum.

The obtained results agree relatively satisfactorily in whole spectrum with the results available in [22]. For higher frequencies and shorter wires, some discrepancies are noticeable above a few MHz, while for longer wires, discrepancies appear above a few kHz.

The higher discrepancies are visible at lower heights of the wire. Closer agreement between the results can be observed if the conductivity of the lower level decreases and if the upper-level thickness increases. Some appreciable discrepancies at high frequencies are noticeable for longer wires at h = 2.5 m.

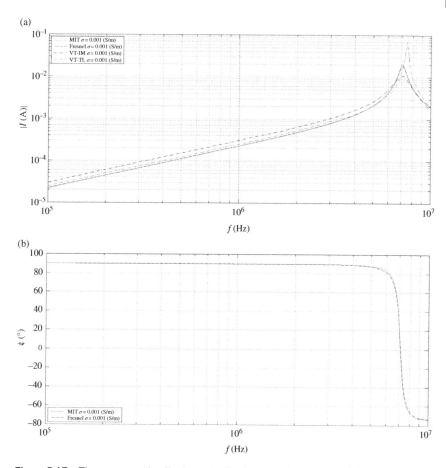

Figure 5.17 The current at the dipole center for the upper layer conductivity $\sigma = 1$ mS/m, at height for the upper layer conductivity, $h = 5$ m. (a) Amplitude spectrum, (b) Phase spectrum.

Next computational example deals with the field in the upper half-space radiated by GPR antenna. A center-fed horizontal dipole is placed at height h above a two-layered lossy half space as shown in Fig. 5.20. The wire is assumed to be perfectly conducting.

The layers and the layer interfaces are indexed in the following way: 0, 1, and 2 for the three layers – air, upper ground layer, and lower ground layer and 01 and 12 for the interfaces air-layer1 and layer1-layer2, respectively. The incident field radiated by the antenna is approximated by a plane wave with the oblique incidence on the two-layered subsurface.

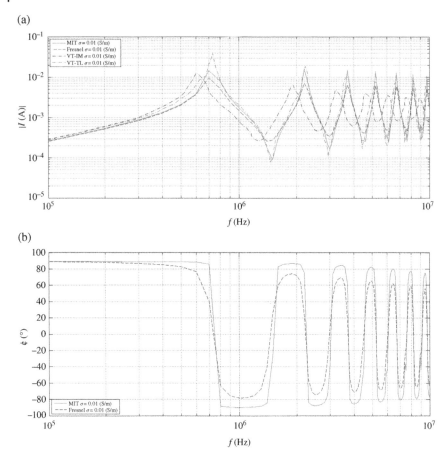

Figure 5.18 The current at the dipole center for the upper layer conductivity σ = 10 mS/m, at height for the upper layer conductivity h = 2.5 m. (a) Amplitude spectrum. (b) Phase spectrum.

The wire length is $L = 1$ m and radius $a = 6.74$ mm is located at height $h = 0.5$ m above the ground, as depicted in Fig. 5.5. The antenna is fed by a generator of 1V. The values for relative permittivity of the ground layers are $\varepsilon_{r1} = 10$, $\varepsilon_{r2} = 15$, respectively, and the conductivity values are $\sigma_1 = 5$ mS/m and $\sigma_2 = 25$ mS/m, respectively. The thickness of layer1 is $d_1 = 0.5$ m. For the purpose of the illustration, the absolute values of normal (E_z) and tangential (E_x) components of the electric field radiated in the air are depicted in Fig. 5.21. The results are obtained via the Fresnel RC approximation (RCA) at frequency $f = 150$ MHz.

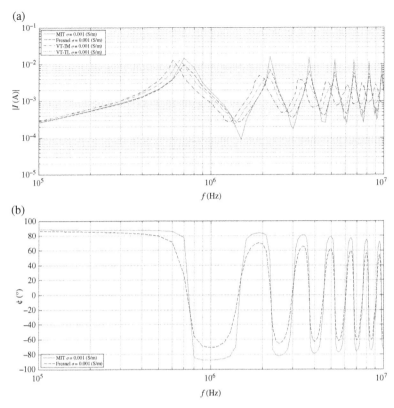

Figure 5.19 The current at the dipole center for the upper layer conductivity $\sigma = 1$ mS/m, at height for the upper layer conductivity $h = 2.5$ m. (a) Amplitude spectrum. (b) Phase spectrum.

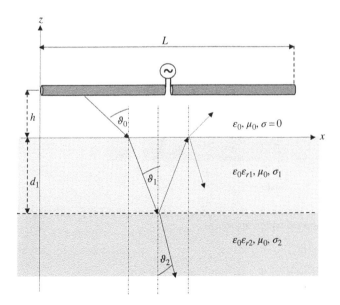

Figure 5.20 Dipole antenna located over a two-layered lossy half-space.

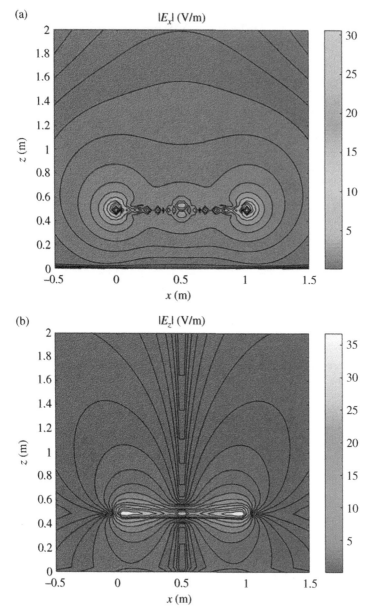

Figure 5.21 Tangential (a) and normal (b) components of electric field at $f = 150\,\text{MHz}$ obtained by plane wave approximation approach.

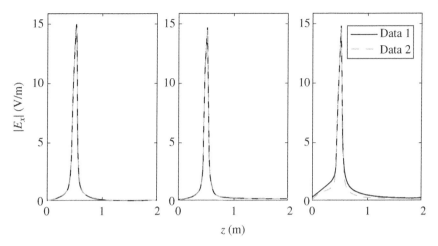

Figure 5.22 The absolute value of the tangential component of electric field (E_x) obtained using RCA and MIT approaches at the broadside antenna direction, $x = 0.5$ m at fixed frequency values: 10, 100, and 300 MHz (from left to right).

Next computations are performed for $z = [0–2]$ m, with the step $\Delta z = 4.23$ cm for $f = [10, 100,$ and $300]$ MHz and for two different positions along the x axis: $x = 0.5$ and 0.75 m.

The results are shown in Figs. 5.22 and 5.23.

The results for the tangential field component presented in Fig. 5.22 for three different frequencies at the fixed position of $x = 0.5$ m show excellent agreement between the two approaches. The value of the normal field component is zero. Hence, the results are not presented in the figure. On the other hand, when the observation point is moved from the broadside direction, the difference between the two approaches is more pronounced as depicted in Fig. 5.23 for the position of $x = 0.75$ m. The values of the normal field component obtained by both approaches show a satisfactory agreement. However, at higher frequencies, the tangential components of electric field obtained by RCA and MIT approaches differ appreciably. Nevertheless, as the magnitude of the normal field component is higher than that of the tangential one, the total field values are in satisfactory agreement.

In the second test case, the heights of the observation points are set to $z = 0.25$ and 1 m, for the frequency range $f = [10–300]$ MHz with the step $\Delta f = 10$ MHz. The results are shown in Figs. 5.24 and 5.25.

Generally, the MIT-based approach underestimates the field values with respect to the values obtained by RCA approach, and the differences increase with frequency. In some of the cases, the MIT approach overestimates the RCA results.

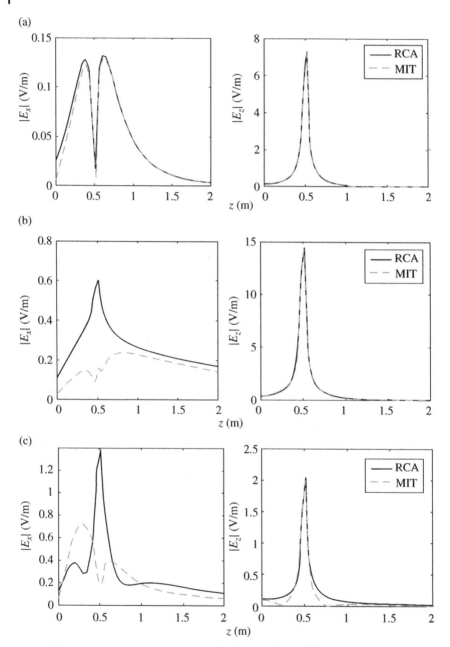

Figure 5.23 The absolute value of the tangential (E_x) and normal (E_z) components of electric field obtained using RCA and MIT approaches at the position of $x = 0.75$ m at fixed frequency values: 10, 100, and 300 MHz (from (a) to (c)).

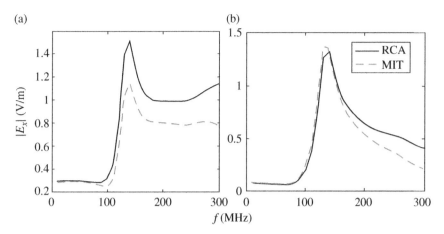

Figure 5.24 The absolute value of the tangential component of electric field (E_x) obtained using RCA and MIT approaches at the broadside antenna direction, $x = 0.5$ m at height $z = 0.25$ m (a) and $z = 1$ m (b).

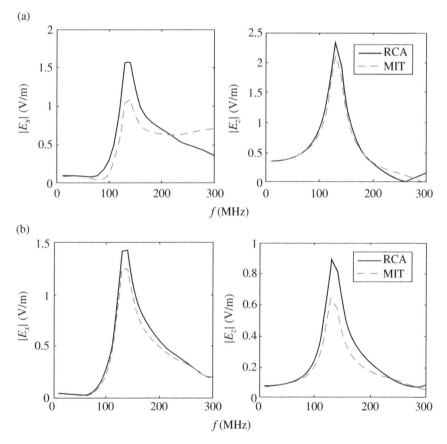

Figure 5.25 The absolute value of the tangential (E_x) and normal (E_z) components of electric field obtained using RCA and MIT approaches at the position of $x = 0.75$ m at height $z = 0.25$ m (a) and $z = 1$ m (b).

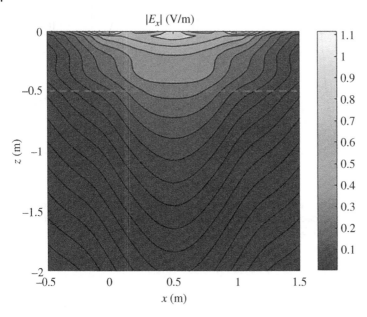

Figure 5.26 Tangential component of electric field at $f = 150$ MHz obtained by plane wave approximation approach.

Computational example to follow deals with the transmitted field in the multi-layered ground due to GPR antenna of length $L = 1$ m and radius $a = 6.74$ mm is located at height $h = 0.1$ m above the ground (Fig. 5.20). The antenna is fed by a generator of 1 V. The values for relative permittivity of the ground layers are $\varepsilon_{r1} = 10$, $\varepsilon_{r2} = 15$, respectively, and the conductivity values are $\sigma_1 = 5$ mS/m and $\sigma_2 = 25$ mS/m, respectively. The thickness of layer1 is $d_1 = 0.5$ m.

The absolute values of normal and tangential components of the electric field transmitted into the ground are depicted in Figs. 5.26 and 5.27, respectively. The results are obtained with Fresnel's approximation of transmission coefficient at $f = 150$ MHz. The dot red line indicates the border between the two ground layers.

As the approximations introduced by MIT approach are expected to cause a large error in the field estimation, the reasonable application is valid only along the broadside direction of the antenna, $x = L/2$. Therefore, to compare the two approaches, the x coordinates of the observation points are set to 0.5 m. This is justified in applications such as GPR modeling where just the field transmitted in the broadside direction carries important information about the buried targets. As the value of normal field component (E_z) at the center of the antenna is rather negligible, the results are shown for the tangential component (E_x) only.

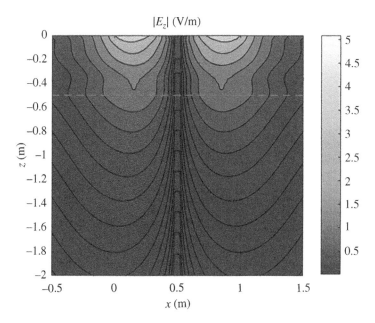

Figure 5.27 Normal component of electric field at f = 150 MHz obtained by plane wave approximation approach.

First, the z coordinates of the observation points are set to $-z = [0-2]$ m with the step $\Delta z = 4.08$ cm. The calculation is carried out for five frequencies, i.e.: $f = [10, 75, 150, 200,$ and $300]$ MHz. The results are shown in Fig. 5.28 as the ratio between the electric field values obtained by MIT approach and those obtained using the plane wave approximation. In the second case, the depths of the observation points are set to $-z = [0.1, 0.3, 1, 1.5,$ and $1.85]$ m, for the frequency range $f = [10–300]$ MHz with the step $\Delta f = 9.31$ MHz. The results are depicted in Fig. 5.29.

Obviously, the MIT-based method overestimates the field values with respect to the values obtained by the Fresnel plane wave approximation. The difference between the two approaches decreases around the resonance frequency of 150 MHz and decays as the depth of the observation point increases.

5.3 Wire Array Above a Multilayer

Majority of engineering applications involve different wire array configurations in the presence of a multilayer medium which requires an extension of the model presented in [18] to the case of wire arrays above a multilayer feature.

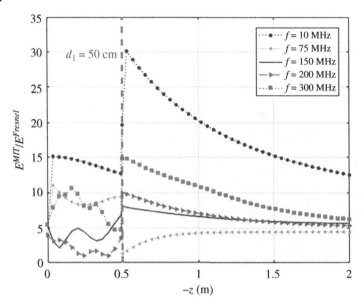

Figure 5.28 The ratio of electric field values obtained by using MIT approach (E^{MIT}) and plane wave approximation approach ($E^{Fresnel}$) along the broadside antenna direction at fixed frequency values: 10, 75, 150, 200, and 300 MHz.

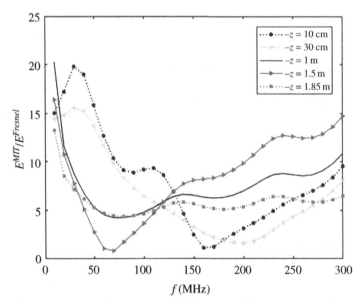

Figure 5.29 The ratio of electric field values obtained by using MIT approach (E^{MIT}) and plane wave approximation approach ($E^{Fresnel}$) versus frequency for five observation point depths: 10, 30, 1, 1.5, and 1.85 m.

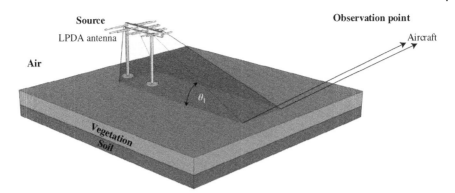

Figure 5.30 Elementary log periodic dipole antenna radiating above a multilayer.

Of particular interest is the use of various dipole arrays in practical engineering scenarios. Application of dipole antenna arrays in ILS has been reported in [23, 24] featuring logarithmic periodic dipole arrays (LPDAs) for localizer antenna system and panel antennas for the glide path system.

The formulation is based on the set of coupled Pocklington equations numerically solved via GB-IBEM. The field above a multilayer is obtained by integrating currents along the wires provided the set of integral equations is solved and the currents are known.

Geometry of single LPDA radiating above a layered medium is shown in Fig. 5.30, while the geometry of a localizer antenna system is presented in Fig. 5.31.

Figure 5.31 Geometry of a localizer.

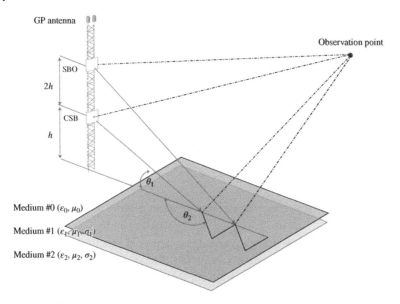

Figure 5.32 A glide path antenna system above two-layered ground.

Additional geometry of interest pertains to the glide path antenna as shown in Figs. 5.32 and 5.33, respectively.

The vegetation is considered a flat layer with finite thickness. Such a geometry is of interest in both GPR and ILS applications. A behavior of LPDAs in the presence of a dissipative half-space can be analyzed by means of the rigorous Sommerfeld integral formulation, or the approximate reflection approach such as proposed in [1–4], or [25]. This work extends the use of the MIT, originally proposed in [5] and used in [18], to account for the presence of a layered medium as a part of the Green function within the Pocklington equation and also within related integral formula for the field reflected from interface, has been presented in [23, 24].

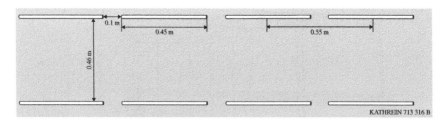

Figure 5.33 Panel antenna for the glide path system.

5.3.1 Formulation

According to the wire AT and *RC* (reflection coefficient) approximation, the set of coupled Pocklington integral equations is given by [1, 2]

$$E_x^{exc} = -\frac{1}{j4\pi\omega\varepsilon_0} \sum_{n=1}^{M} \int_{-L_n/2}^{L_n/2} \left[\frac{\partial^2}{\partial x^2} + k_1^2\right] g_{mn}^{tot}(x,x') I_n(x') dx' \quad m = 1, 2, ...M$$

(5.77)

where $I_n(x')$ is the unknown current distribution induced on the n-th wire axis, while $g^{tot}(x,x')$ denotes the total Green function

$$g_{mn}^{tot}(x,x') = g_{0mn}(x,x') - \Gamma_{ref}^{tot} g_{imn}(x,x')$$

(5.78)

where $g_{0mn}(x,x')$ denotes the free-space Green function

$$g_{0mn}(x,x') = \frac{e^{-jk_1 R_{1mn}}}{R_{1mn}}$$

(5.79)

and $g_{imn}(x,x')$, arising from the image theory, is given by

$$g_{imn}(x,x') = \frac{e^{-jk_1 R_{2mn}}}{R_{2mn}}$$

(5.80)

where k_1 is the propagation constant of free space and R_{1mn} and R_{2mn} are distances from the source point and from the corresponding image to the observation point defined by

$$\begin{aligned} R_{1mn} &= \sqrt{(x-x')^2 + a_m^2}, \\ R_{2mn} &= \sqrt{(x-x')^2 + 4h^2} \end{aligned} \quad m = n$$

(5.81)

$$\begin{aligned} R_{1mn} &= \sqrt{(x-x')^2 + D_{mn}^2}, \\ R_{2mn} &= \sqrt{R_{1mn}^2 + 4h^2} \end{aligned} \quad m \neq n$$

(5.82)

while D_{mn} is the separation between m-th and n-th antenna.

The influence of an imperfectly conducting lower medium is taken into account by means of the certain RC Γ_{ref}^{tot}.

This RC arises from the MIT or Fresnel plane wave approximation, respectively.

The total RC, by which the reflection between air and a multilayer is taken into account, can be obtained by extending the MIT approach originally proposed in [25] and is given by

$$\Gamma_{ref}^{tot} = \frac{R_{12} + R_{23}e^{-2\gamma_2 d}}{1 + R_{12}R_{23}e^{-2\gamma_2 d}}$$

(5.83)

where the reflection between m-th and n-th layers, respectively, is given by the RC:

$$R_{mn} = \frac{\varepsilon_{eff,m} - \varepsilon_{eff,n}}{\varepsilon_{eff,m} + \varepsilon_{eff,n}} \tag{5.84}$$

and $\varepsilon_{effm,n}$ for a particular layer is

$$\varepsilon_{effm,n} = \varepsilon_{rm,n} - j\frac{\sigma_{m,n}}{\omega} \tag{5.85}$$

The details regarding Fresnel plane wave RC could be found elsewhere, e.g. in [25].

Furthermore, the radiated electric field components are given by [1]:

$$E_x\left(\overrightarrow{r}\right) = -\frac{1}{j4\pi\omega\varepsilon_0} \sum_{n=1}^{M} \int_{-L_n/2}^{L_n/2} \left[\frac{\partial I_n(x')}{\partial x'} \frac{\partial g_{mn}^{tot}\left(\overrightarrow{r},x'\right)}{\partial x} + k^2 g_{mn}^{tot}(x,x')I_n(x') \right] dx' \tag{5.86}$$

$$E_y\left(\overrightarrow{r}\right) = -\frac{1}{j4\pi\omega\varepsilon_0} \sum_{n=1}^{M} \int_{-L_n/2}^{L_n/2} \frac{\partial I_n(x')}{\partial x'} \frac{\partial g_{mn}^{tot}\left(\overrightarrow{r},x'\right)}{\partial y} dx' \tag{5.87}$$

$$E_y\left(\overrightarrow{r}\right) = -\frac{1}{j4\pi\omega\varepsilon_0} \sum_{n=1}^{M} \int_{-L_n/2}^{L_n/2} \frac{\partial I_n(x')}{\partial x'} \frac{\partial g_{mn}^{tot}\left(\overrightarrow{r},x'\right)}{\partial y} dx' \tag{5.88}$$

The set of the Pocklington integro-differential equation (5.77) is solved via the GB-IBEM [1]. Once the currents along the wires are obtained the electric field components can be determined by numerical evaluation of field integrals (5.86)–(5.88).

An extension of GB-IBEM to treat wire array above a multilayer is outlined in the subsection below. The evaluation of field integrals via BEM formalism is presented below. It is worth noting that the model presented in this work deals with a finite thickness slab, but the approach could be readily extended to an arbitrary number of layers.

5.3.2 Numerical Procedures

GB-IBEM procedure starts by applying the standard representation of the unknown current along an i-th wire:

$$I_n(x') = \sum_{n=1}^{N_g} I_{ni} f_{ni}(x') \tag{5.89}$$

where $f_{ni}(x)$ denotes the linear elements shape functions, I_{ni} stands for the unknown coefficients of the solution, and N_g denotes the total number of base functions.

The weak Galerkin–Bubnov formulation of the system of integro-differential equations (5.77) for N_w wires is given by:

$$
\sum_{n=1}^{N_g} \sum_{i=1}^{N_w} \left[-\int_{-L_m}^{L_m} \int_{-L_n}^{L_n} \frac{df_{jm}(x)}{dx} \frac{df_{in}(x')}{dx'} g_{mn}^{tot}(x,x') dx' dx + k^2 \int_{-L_m}^{L_m} \int_{-L_n}^{L_n} f_{jm}(x) f_{in}(x') g_{mn}^{tot}(x,x') dx' dx \right]
$$

$$
I_{ni} = j4\pi\omega\varepsilon \int_{-L_m}^{L_m} E_{xj}^{inc}(x) f_{jm}(x) dx \quad m=1,2,...,N_g; \quad j=1,2,...N_w
$$

$$(5.90)$$

Performing the boundary element discretization of the wire array, one obtains the matrix equation

$$
\sum_{k=1}^{M} [Z]_{lk} \{I\}_k = \{V\}_l \quad l = 1, 2, ..., M \tag{5.91}
$$

where M is the total number of elements along the actual multiple wire configuration, and $[Z]_{lk}$ is the interaction matrix representing the mutual impedance between each segment on the i-th (source) wire to each segment on the j-th (observation) wire:

$$
[Z]_{ji} = \int_{\Delta l_j} \int_{\Delta l_i} \begin{bmatrix} \dfrac{df_1(x)}{dx} \dfrac{df_1(x')}{dx'} & \dfrac{df_1(x)}{dx} \dfrac{df_2(x')}{dx'} \\ \dfrac{df_2(x)}{dx} \dfrac{df_1(x')}{dx'} & \dfrac{df_2(x)}{dx} \dfrac{df_2(x')}{dx'} \end{bmatrix} g_{mn}^{tot}(x,x') dx' dx +
$$

$$
+ k^2 \int_{\Delta l_j} \int_{\Delta l_i} \begin{bmatrix} f_1(x)f_1(x') & f_1(x)f_2(x') \\ f_2(x)f_1(x') & f_2(x)f_2(x') \end{bmatrix} g_{mn}^{tot}(x,x') dx' dx
$$

$$
= \frac{1}{\Delta x^2} \frac{df_1(x')}{dx'} \int_{x_1}^{x_2} \int_{x_1}^{x_2} \begin{bmatrix} 1 & -1 \\ -1 & 1 \end{bmatrix} g_{mn}^{tot}(x,x') dx' dx +
$$

$$
+ \frac{k^2}{\Delta x^2} \int_{x_1}^{x_2} \int_{x_1}^{x_2} \begin{bmatrix} (x_2-x)(x_2-x') & (x_2-x)(x'-x_1) \\ (x-x_1)(x_2-x') & (x-x_1)(x'-x_1) \end{bmatrix} g_{mn}^{tot}(x,x') dx' dx
$$

$$(5.92)$$

The local voltage vector $\{V\}_j$ is given by:

$$
\{V\}_j = -j4\pi\omega\varepsilon \int_{\Delta l_j} E_x^{inc}(x) \begin{bmatrix} f_1(x) \\ f_2(x) \end{bmatrix} dx = -\frac{j4\pi\omega\varepsilon}{\Delta x} \int_{x_1}^{x_2} E_x^{inc}(x) \begin{bmatrix} (x_2-x) \\ (x-x_1) \end{bmatrix} dx
$$

$$(5.93)$$

where Δl_i and Δl_j assign the widths of i-th and j-th segments.

The evaluation of the right-hand side vector is carried out analytically as in the radiation mode, the right-side vector is different from zero only in the feed gap area. The x-component of the impressed (incident) electric field is:

$$E_x^{inc}(x) = \frac{V_g}{\Delta l_g} \tag{5.94}$$

where V_g is the feed voltage and $\Delta l_g = \Delta x$ (for convenience) is the feed-gap width. Using the linear shape functions, it follows:

$$\{V\}_j = -\frac{j4\pi\omega\varepsilon}{\Delta l_g} \int\limits_{x_1 = -\frac{\Delta l_g}{2}}^{x_2 = \frac{\Delta l_g}{2}} \frac{V_g}{\Delta l_g} \begin{bmatrix} (x_2 - x) \\ (x - x_1) \end{bmatrix} dx = -j2\pi\omega\varepsilon V_g \begin{pmatrix} 1 \\ 1 \end{pmatrix} \tag{5.95}$$

More mathematical details on the method could be found elsewhere, e.g. in [1]. Provided the currents along the wires are determined, the evaluation of the field integrals is also carried out using GB-IBEM formalism.

Using the local expansion for the current on segment

$$I(x') = I_{1i}\frac{x_{2i} - x'}{\Delta x} + I_{2i}\frac{x' - x_{1i}}{\Delta x} \tag{5.96}$$

where I_{1i} and I_{2i} are current values at the local nodes of the i-th wire segment Δx, with coordinates x_{1i} and x_{2i}, $\Delta x = x_{2i} - x_{1i}$. Substituting local approximation for current distribution (5.89) into field formulas (5.86) to (5.88) results in the following expressions for the y and z components of the electric field at an arbitrary point in the air:

$$E_x = \frac{1}{j4\pi\omega\varepsilon_0} \sum_{j=1}^{M}\sum_{i=1}^{N_j} \left[-\frac{I_{2ij} - I_{1ij}}{\Delta x_j} \int\limits_{x_{1ij}}^{x_{2ij}} \frac{\partial g_{mn}^{tot}(\vec{r}, x')}{\partial x'} dx' + k^2 \int\limits_{x_{1ij}}^{x_{2ij}} g_{mn}^{tot}(\vec{r}, x') I(x') dx' \right] \tag{5.97}$$

$$E_y = \frac{1}{j4\pi\omega\varepsilon_0} \sum_{j=1}^{M}\sum_{i=1}^{N_j} \frac{I_{2ij} - I_{1ij}}{\Delta x_j} \int\limits_{x_{1ij}}^{x_{2ij}} \frac{\partial g_{mn}^{tot}(\vec{r}, x')}{\partial y} dx' \tag{5.98}$$

$$E_z = \frac{1}{j4\pi\omega\varepsilon_0} \sum_{j=1}^{M}\sum_{i=1}^{N_j} \frac{I_{2ij} - I_{1ij}}{\Delta x_j} \int\limits_{x_{1ij}}^{x_{2ij}} \frac{\partial g_{mn}^{tot}(\vec{r}, x')}{\partial z} dx' \tag{5.99}$$

where M is the total number of wires and N_j denotes the total number of boundary elements on the j-th wire.

The integrals (5.97)–(5.99) are numerically calculated via Gaussian quadrature. Due to quasi-singularity of the Green function, the first-order differential operator appearing in the integral equation kernel is likely to cause numerical instability.

The problem of quasi-singularity of the Green function is avoided by approximating the derivative of the Green function derivative by means of a central finite difference formula, as explained in [1].

5.3.3 Computational Examples

The computational example is related to the LPDA antenna where $L_1 = 1.27$ m is the length of the longest of seventh dipole antennas with radius $a = 0.004$ m horizontally located at certain height $h = 1.82$ m above a two-layer media (vegetation + lossy ground) with vegetation permittivity ε_{r1} and conductivity σ_1. The lossy ground permittivity and conductivity are ε_{r2} and σ_2, respectively. LLZ antenna system is composed of 14 LPDA elements energized symmetrically with absolute magnitude but differing in phase. The depth of the vegetation layer is assigned as d_1. The localizer (Fig. 5.34) consists of 14 LPDAs.

The numerical results for localizer antenna system are obtained for the variation in (a) operating frequency, (b) relative permittivity, (c) depth of vegetation layer, (d) conductivity, and (e) antenna height above ground.

All numerical results are obtained via both Fresnel RC approximation and the MIT approach [18].

First set of results deals with the variation in operating frequency. The first data set is given in Table 5.1.

Figure 5.34 shows the LPDA vertical radiation pattern for the case of maximum angle ϕ above ground for the operating frequency $f = 108.50$ MHz.

Figure 5.35 shows the LPDA vertical radiation pattern for the case of maximum angle ϕ above ground for the operating frequency $f = 111.50$ MHz.

It can be noticed that as the operating frequency increases, the maximum angle changes, i.e. there is a down tilt of radiation pattern. Furthermore, side lobes are suppressed.

Generally, the results obtained by means of different RC approaches are in good agreement in the entire domain of interest. Main discrepancies can be noticed at the zeros of radiation pattern, especially at higher frequencies.

In the next set of numerical results, the relative permittivity is varied. The corresponding data set is given in Table 5.2.

Figure 5.36 shows the LPDA vertical radiation pattern for the case of maximum angle ϕ above ground for $\varepsilon_{r1} = 4$.

The figures (Figs. 5.37–5.40) to follow deal with $\varepsilon_{r1} = 12$, 18, 24, and 30, respectively.

The radiation patterns obtained using the Fresnel coefficient approach are not found to be significantly sensitive to the permittivity variations. On the other hand, results obtained by the MIT approach show noticeable deviations in the shape of the main lobe (when angle is close to 180°) for the low values of the vegetation

(a)

Calculation settings: $d_1 = 0.01$ m, $h = 1.82$ m, Freq = 108.5 e6Hz

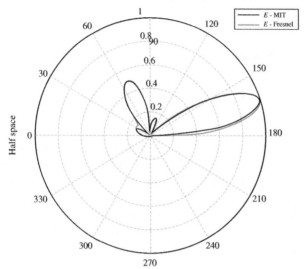

LPDA vertical radiation versus maximum PHI above ground

(b)

Partial radiation pattern view of LPDA above ground
Calculation settings: $d_1 = 0.01$ m, $h = 1.82$ m Epsilon 1 = 12, Freq = 108.50 e6Hz

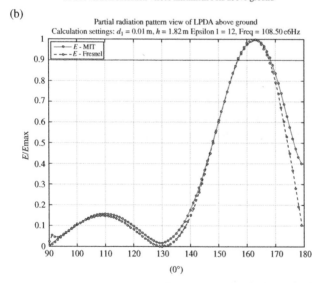

Figure 5.34 (a) LPDA vertical pattern above the multilayer ($h = 1.82$ m). (b) A view of a part of radiation pattern of half-wave dipole above the multilayer ($h = 1.82$ m).

Table 5.1 Input data set number 1.

ε_{r1}	σ_1 (mS/m)	ε_{r2}	σ_2 (mS/m)	d_1 (cm)
12	2	10	1	1

(a)

Calculation settings: d_1 = 0.01 m, h = 1.82 m, σ = 0.002S, ε = 12, Freq = 111.50 e6Hz

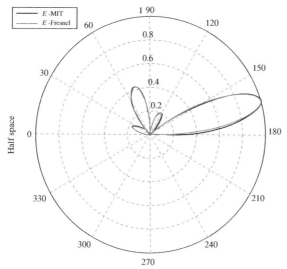

LPDA vertical radiation versus maximum PHI above ground

(b)

Partial radiation pattern view of LPDA above ground
Calculation settings: d_1 = 0.01 m, h = 1.82 m, Epsilon1 =12, Freq = 111.50 e6Hz

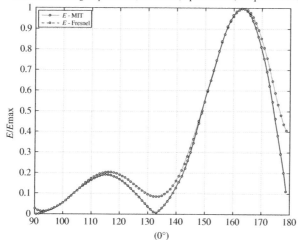

Figure 5.35 (a) LPDA vertical pattern above the multilayer (h = 1.82 m). (b) A view of a part of radiation pattern of half-wave dipole above the multilayer (h = 1.82 m).

Table 5.2 Input data set number 2.

f (MHz)	σ_1 (mS/m)	ε_{r2}	σ_2 (mS/m)	d_1 (m)
108.50	2	10	1	0.2

(a)
Calculation settings: $d_1 = 0.2\,\text{m}$, $h = 1.82\,\text{m}$, Freq $= 108.5\,\text{e6Hz}$

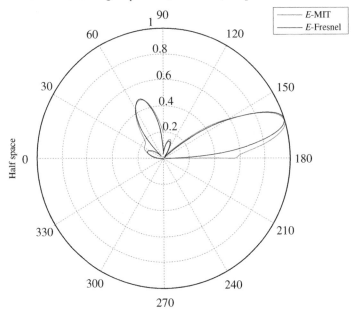

LPDA vertical radiation versus maximum PHI above ground

(b)

Partial radiation pattern view of LPDA above ground
Calculation settings: $d_1 = 0.2\,\text{m}$, $h = 1.82\,\text{m}$ Epsilon 1 $= 4$ Freq $= 108.50\,\text{e6Hz}$

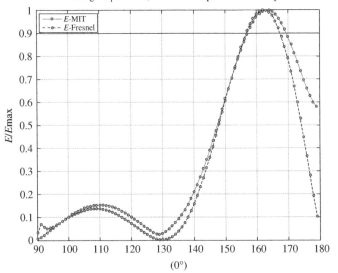

Figure 5.36 (a) LPDA vertical pattern above the multilayer ($h = 1.82$ m). (b) A view of a part of radiation pattern of LPDA above the multilayer ($h = 1.82$ m).

(a)

Calculation settings: $d_1 = 0.2$ m, $h = 1.82$ m, Freq = 108.5 e6Hz

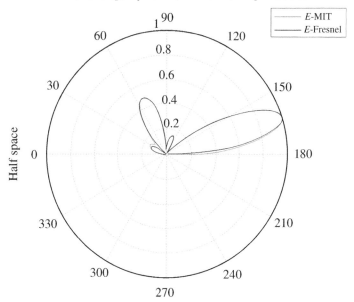

Half space

LPDA vertical radiation versus maximum PHI above ground

(b)

Partial radiation pattern view of LPDA above ground
Calculation settings: $d_1 = 0.2$ m, $h = 1.82$ m, Epsilon1 = 12, Freq = 108.50 e6Hz

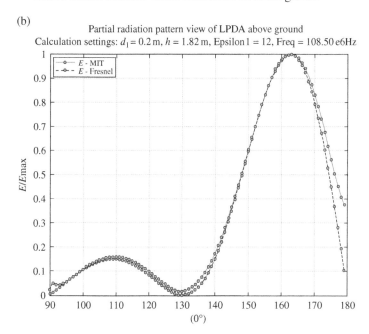

Figure 5.37 (a) LPDA vertical pattern above the multilayer ($h = 1.82$ m). (b) A view of a part of radiation pattern of LPDA above the multilayer ($h = 1.82$ m).

(a)

Calculation settings: $d_1 = 0.2$ m, $h = 1.82$ m, Freq $= 108.5$ e6Hz

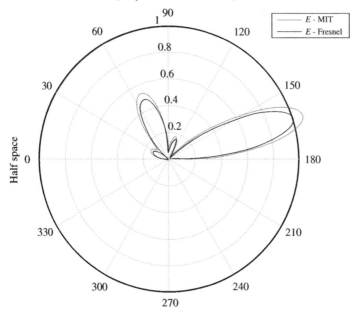

LPDA vertical radiation versus maximum PHI above ground

(b)

Partial radiation pattern view of LPDA above ground
Calculation settings: $d_1 = 0.2$ m, $h = 1.82$ m, Epsilon 1 = 18, Freq = 108.50 e6Hz

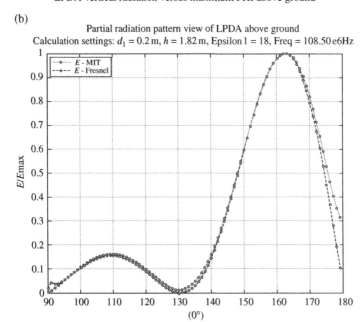

Figure 5.38 (a) LPDA vertical pattern above the multilayer ($h = 1.82$ m). (b) A view of a part of radiation pattern of LPDA above the multilayer ($h = 1.82$ m).

(a)

Calculation settings: $d_1 = 0.2$ m, $h = 1.82$ m, Freq $= 108.5$ e6Hz

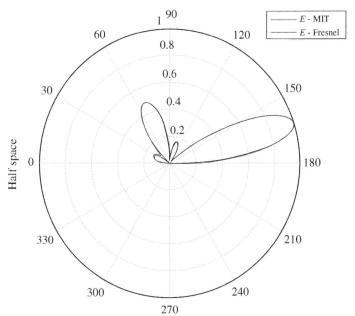

LPDA vertical radiation versus maximum PHI above ground

(b)

Partial radiation pattern view of LPDA above ground
Calculation settings: $d_1 = 0.2$ m, $h = 1.82$ m, Epsilon1 $= 24$, Freq $= 108.50$ e6Hz

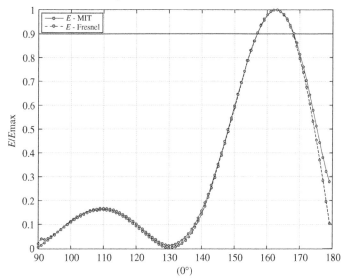

Figure 5.39 (a) LPDA vertical pattern above the multilayer ($h = 1.82$ m). (b) A view of a part of radiation pattern of LPDA above the multilayer ($h = 1.82$ m).

(a)

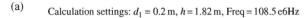

Calculation settings: $d_1 = 0.2\,\text{m}$, $h = 1.82\,\text{m}$, Freq $= 108.5\,\text{e}6\text{Hz}$

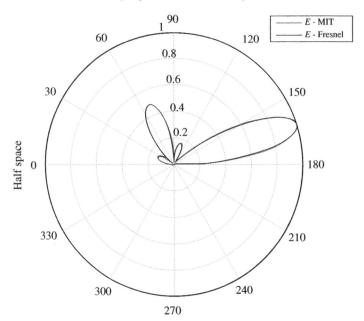

LPDA vertical radiation versus maximum PHI above ground

(b)

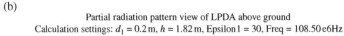

Partial radiation pattern view of LPDA above ground
Calculation settings: $d_1 = 0.2\,\text{m}$, $h = 1.82\,\text{m}$, Epsilon1 $= 30$, Freq $= 108.50\,\text{e}6\text{Hz}$

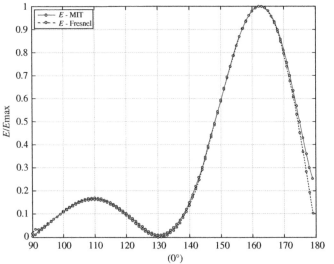

Figure 5.40 (a) LPDA vertical pattern above the multilayer ($h = 1.82\,\text{m}$). (b) A view of a part of radiation pattern of LPDA above the multilayer ($h = 1.82\,\text{m}$).

Table 5.3 Input data set number 3.

f (MHz)	σ_1 (mS/m)	ε_{r2}	σ_2 (mS/m)	ε_{r1}
108.50	2	10	1	12

layer's permittivity. As the permittivity of the vegetation layer reaches and raises above the permittivity of the ground, results obtained by both approaches show excellent agreement.

Next set of numerical results deals with the variation in vegetation layer thickness.

The data set is given in Table 5.3.

Figure 5.41 shows the LPDA vertical radiation pattern for the case of maximum angle ϕ above ground for the layer thickness $d = 0.01$ m.

Figures 5.42 and 5.43 to follow show the results for $d = 0.1$ m and $d = 0.2$ m.

As it could be seen, radiation patterns show almost no change with the growth of vegetation layer, which is a little bit surprising but it is satisfactory from the aspect of the localizer system operation. Agreement between the results obtained by two approaches is found to be satisfactory.

Next set of figures deals with the variation in layer conductivity.

The data set is given in Table 5.4.

Fig. 5.44 shows the LPDA vertical radiation pattern for the case of maximum angle ϕ above ground for the layer conductivity $\sigma_1 = 1$ mS/m.

Next examples deal with $\sigma_1 = 10$ mS/m and $\sigma_1 = 100$ mS/m (Figs. 5.45 and 5.46).

Some discrepancies in the results can be noticed $\sigma_1 = 0.1$ S/m with a slight drop in maximum angle.

Furthermore, the antenna height above ground is varied and the results are presented in the figures.

The data set is given in Table 5.5.

Figure 5.47 to follow show the LPDA vertical radiation pattern on $h = 0.5$ m above ground including numerical results obtained for two different operating frequencies: $f = 108.10$ MHz and $f = 111.50$ MHz.

The vertical pattern is distorted with regard to operational requirements of LLZ antenna system where $h = 1.82$ m.

Next example deals with $\varepsilon_{r1} = 12$ on $f = 111.50$ MHz (Figure 5.48).

Variations in layer thickness and frequency with similar parameters of the ground and vegetation do not cause significant changes compared to the previous case. At smaller heights, LLZ antenna system is practically useless as it does not fulfill the operational requirements.

(a)

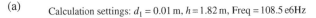

Calculation settings: $d_1 = 0.01$ m, $h = 1.82$ m, Freq = 108.5 e6Hz

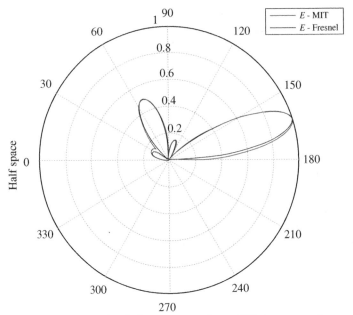

LPDA vertical radiation versus maximum PHI above ground

(b)

Partial radiation pattern view of LPDA above ground

Calculation settings: $d_1 = 0.01$ m, $h = 1.82$ m, Epsilon1 = 12, Freq = 108.50 e6Hz

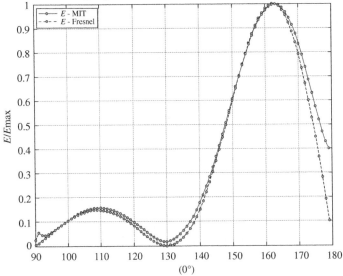

Figure 5.41 (a) LPDA vertical pattern above the multilayer ($d = 0.001$ m, $h = 1.82$ m). (b) A view of a part of radiation pattern of LPDA above the multilayer ($d = 0.001$ m, $h = 1.82$ m).

(a)

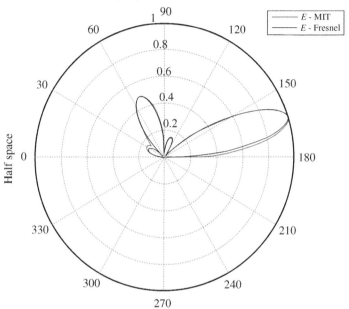

Calculation settings: $d_1 = 0.1$ m, $h = 1.82$ m, Freq = 108.5 e6Hz

LPDA vertical radiation versus maximum PHI above ground

(b)

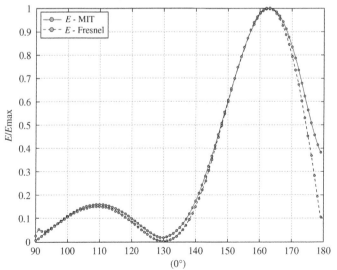

Partial radiation pattern view of LPDA above ground
Calculation settings: $d_1 = 0.1$ m, $h = 1.82$ m, Epsilon 1 = 12, Freq = 108.50 e6Hz

Figure 5.42 (a) LPDA vertical pattern above the multilayer ($d = 0.1$ m, $h = 1.82$ m).
(b) A view of a part of radiation pattern of LPDA above the multilayer ($d = 0.1$ m, $h = 1.82$ m).

(a)

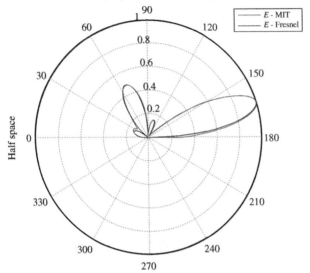

Calculation settings: $d_1 = 0.2$ m, $h = 1.82$ m, Freq = 108.5 e6Hz

LPDA vertical radiation versus maximum PHI above ground

(b)

Partial radiation pattern view of LPDA above ground
Calculation settings: $d_1 = 0.2$ m, $h = 1.82$ m, Epsilon 1 = 12, Freq = 108.50 e6Hz

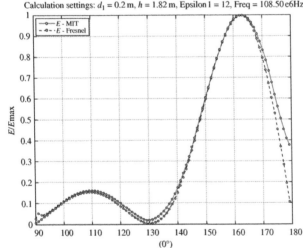

Figure 5.43 (a) LPDA vertical pattern above the multilayer ($d = 0.2$ m, $h = 1.82$ m). (b) A view of a part of radiation pattern of LPDA above the multilayer ($h = 1.82$ m).

Table 5.4 Input data set number 4.

f (MHz)	d_1 (m)	ε_{r2}	σ_2 (mS/m)	ε_{r1}
108.50	0.2	10	1	30

(a)

Calculation settings: $d_1 = 0.2\,\text{m}$, $h = 1.82\,\text{m}$, Freq $= 108.5\,\text{e}6\text{Hz}$

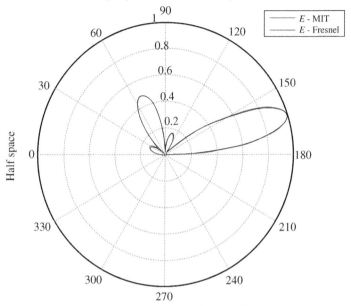

LPDA vertical radiation versus maximum PHI above ground

(b)

Partial radiation pattern view of LPDA above ground

Calculation settings: $d_1 = 0.2\,\text{m}$, $h = 1.82\,\text{m}$, Epsilon 1 = 30, Freq $= 108.50\,\text{e}6\text{Hz}$

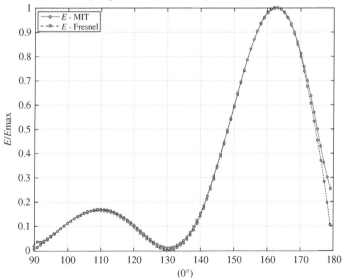

Figure 5.44 (a) LPDA vertical pattern above the multilayer ($\sigma_1 = 1\,\text{mS/m}$, $h = 1.82\,\text{m}$). (b) A view of a part of radiation pattern of LPDA above the multilayer ($\sigma_1 = 1\,\text{mS/m}$, $h = 1.82\,\text{m}$).

(a)   Calculation settings: $d_1 = 0.2$ m, $h = 1.82$ m, Freq $= 108.5$ e6Hz

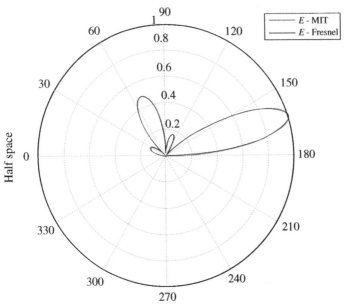

LPDA vertical radiation versus maximum PHI above ground

(b)   Partial radiation pattern view of LPDA above ground
Calculation settings: $d_1 = 0.2$ m, $h = 1.82$ m, Epsilon 1 = 30, Freq $= 108.50$ e6Hz

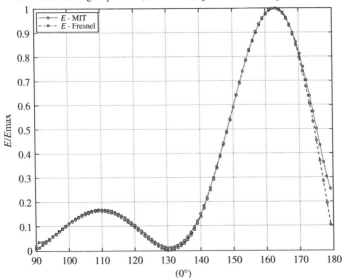

Figure 5.45 (a) LPDA vertical pattern above the multilayer ($\sigma_1 = 10$ mS/m, $h = 1.82$ m).
(b) A view of a part of radiation pattern of LPDA above the multilayer $\sigma_1 = 10$ mS/m,
$h = 1.82$ m).

(a)

Calculation settings: $d_1 = 0.2$ m, $h = 1.82$ m, Freq $= 108.5$ e6Hz

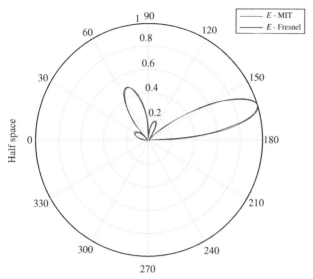

LPDA vertical radiation versus maximum PHI above ground

(b)

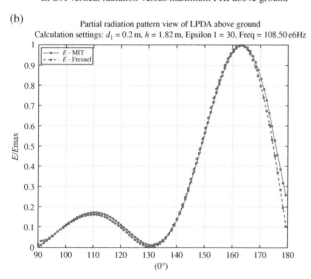

Figure 5.46 (a) LPDA vertical pattern above the multilayer ($\sigma_1 = 100$ mS/m, $h = 1.82$ m). (b) A view of a part of radiation pattern of LPDA above the multilayer ($\sigma_1 = 100$ S/m, $h = 1.82$ m).

Table 5.5 Input data set number 5.

d_1 (m)	ε_{r1}	σ_1 (mS/m)	ε_{r2}	σ_2 (mS/m)
0.05	24	2	10	1

(a)

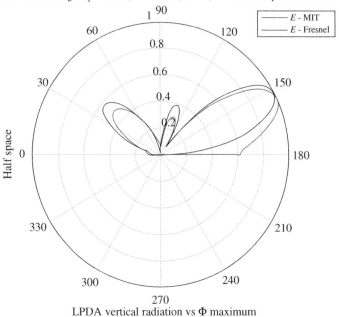

Single LPDA polar radiation view

Calculation settings: $d_1 = 0.05$ m, $\sigma = 0.002$ S, $\varepsilon = 24$, $h = 0.5$ m Freq = 108.10 e6Hz

LPDA vertical radiation vs Φ maximum

(b)

Single LPDA partial radiation view

Calculation settings: $d_1 = 0.05$ m, $h = 0.5$ m, $\sigma = 0.002$S, $\varepsilon = 24$, Freq = 108.10 e6Hz

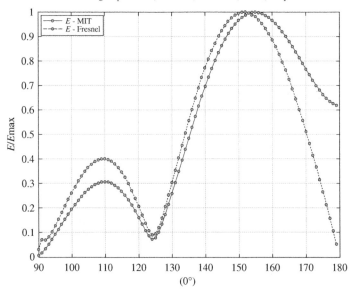

Figure 5.47 (a) LPDA vertical pattern above the multilayer ($h = 1.82$ m). (b) A view of a part of radiation pattern of LPDA above the multilayer ($h = 1.82$ m).

(a)

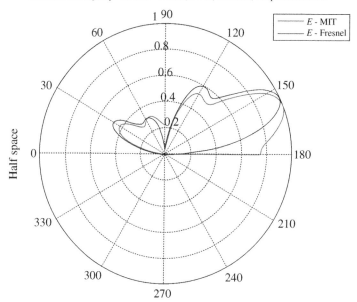

LPDA vertical diagram radiation versus Φ maximum

(b)

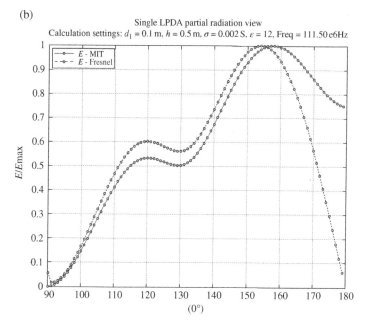

Figure 5.48 (a) LPDA vertical pattern above the multilayer ($h = 0.5$ m, $\varepsilon_{r1} = 12$). (b) A view of a part of radiation pattern of LPDA above the multilayer ($h = 0.5$ m, $\varepsilon_{r1} = 12$).

Figure 5.49 to follow show the LPDA vertical radiation pattern on $h = 1.0$ m above ground with $\varepsilon_{r1} = 24$ on frequency $f = 108.50$ MHz.

Figure 5.50a,b show the LPDA vertical radiation pattern on $h = 1.0$ m above ground with $d_1 = 0.2$ m, and $\varepsilon_{r1} = 12$ on frequency $f = 111.50$ MHz.

Analyzing the numerical results for the radiated field obtained by using the Green function containing the RC derived from MIT are comparable with the results obtained via the Fresnel coefficient.

5.4 Wires of Arbitrary Shape Radiating Over a Layered Medium

Radiation from arbitrarily shaped overhead wires is of great practical interest for numerous CEM applications, e.g. [26], such as transient excitation of antennas, power, or communication cables, respectively.

Moreover, there are a number of specific applications such as GPR ILS, geophysical prospecting, or remote sensing, involving the assessment of the electromagnetic field not only in the presence of a lossy half-space (simplified scenario), e.g. [1–4] but also in presence of a multilayered medium (more realistic case) [11–13].

Generally, the radiation from finite-length overhead wires can be analyzed via a simplified TL model or a more rigorous thin-wire antenna approach in either frequency or time domain [26]. In particular, the transient response of a wire configuration of interest can be computed directly by solving the corresponding time domain equations, or by the indirect approach, i.e. by solving their frequency domain counterparts. Within the framework of the indirect approach, the frequency spectrum in effective bandwidth has to be computed, and then the transient response is evaluated by using the inverse Fourier transform (IFT) algorithm.

Many practical engineering problems pertaining to the electromagnetic field coupling to thin wires can be successfully treated by using the TL models [1, 2, 26] including the analysis of the field exciting the line and the propagation of induced currents and voltages along the line. Generally, TL models give valid results provided the line length is appreciably larger than the separation between the wires, and also larger than the actual height above ground [1].

On the other hand, the TL approximation fails to ensure a complete solution for the excitation of an actual wire configuration by an incident field if the wavelength of the external field is comparable to or less than the transverse electrical dimensions of the wire structure of interest. Namely, TL model often fails to predict resonances and to rigorously take into account the presence of a lossy ground [26]. A serious problem with TL approach is due to the current growing to infinity at

(a)

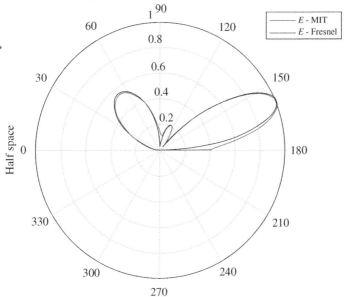

Single LPDA polar radiation view
Calculation settings: $d_1 = 0.05$ m, $\sigma = 0.002$ S, $\varepsilon = 24$, $h = 1.0$ m, Freq $= 108.50$ e6Hz

LPDA vertical diagram radiation versus Φ maximum

(b)

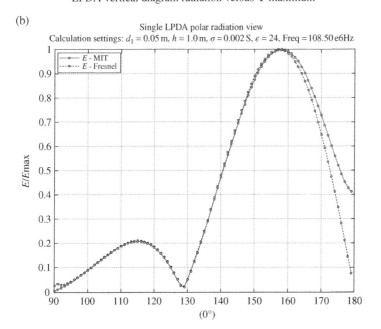

Single LPDA polar radiation view
Calculation settings: $d_1 = 0.05$ m, $h = 1.0$ m, $\sigma = 0.002$ S, $\varepsilon = 24$, Freq $= 108.50$ e6Hz

Figure 5.49 (a) LPDA vertical pattern above the multilayer ($h = 1.0$ m). (b) A view of a part of radiation pattern of LPDA above the multilayer ($h = 1.0$ m).

(a)

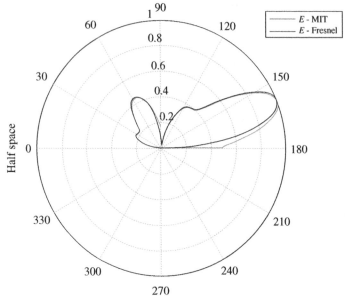

Single LPDA polar radiation view

Calculation settings: $d_1 = 0.2$ m, $\sigma = 0.002$ S, $\varepsilon = 12$, $h = 1.0$ m, Freq $= 111.50$ e6Hz

LPDA vertical diagram radiation versus Φ maximum

(b)

Single LPDA partial radiation view

Calculation settings: $d_1 = 0.2$ m, $h = 1.0$ m, $\sigma = 0.002$ S, $\varepsilon = 12$, Freq $= 111.50$ e6Hz

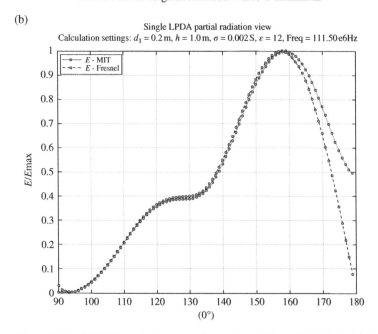

Figure 5.50 (a) LPDA vertical pattern above the multilayer ($h = 1.0$ m). (b) A view of a part of radiation pattern of LPDA above the multilayer ($h = 1.0$ m).

resonant points if losses do not exist as there is no radiation resistance, included within the formulation framework, to limit the current flow [1].

The AT with related governing integral equations is far more rigorous and should be used whenever the above-ground TLs of the finite length are considered. However, what still remains a serious drawback of the AT approach is too long computational time necessary for the calculations pertaining to long wires.

The analysis of radiation and scattering from arbitrarily shaped overhead wires above a multilayer in the frequency domain for different applications has been addressed elsewhere, e.g. [24]. Note that the most rigorous integro-differential equation formulation of straight wires above layered media requires evaluation of the spatial domain Green functions containing computationally rather demanding Sommerfeld integrals [27].

A useful review of the various frequency domain techniques to handle this problem has been presented in [11], or later in [27], respectively. Among the variety of methods developed to efficiently compute Sommerfeld integrals of particular interest are those approximating the Green function by means of several simple functions, thus enabling one to obtain closed-form expressions for Green functions. These techniques are mostly related to DCIM and RFFM. Some authors tackle a time domain counterpart of thin wire radiation in the presence of a lossy medium, e.g. [15] (dipole antenna above a lossy homogeneous half-space) or [13] (thin wires above a multilayered medium). Generally, the majority of the existing methods to treat the problems involving multilayered media are rather demanding in a sense of formulation, applied numerical solution method, and, consequently, computational cost [13], and sometimes also suffering from hypersingular behavior of integral equation kernel [11, 16, 17].

The use of a simplified RC approach was based on MIT to cope with realistic engineering applications. The formulation is based on the space-frequency integro-differential equation of the Pocklington type for the determination of corresponding current distributions and corresponding integral relations to compute radiated fields.

The above-ground wires subjected to an equivalent voltage source inducing current to flow along the wires are the key to understanding the behavior of related radiated fields.

Belowground wires are studied to a lesser extent, e.g. in [28].

The resulting integro-differential expressions are numerically handled via the frequency domain GB-IBEM for arbitrarily shaped wires being discussed elsewhere, e.g. in [1].

This section first deals with a single curved wire in free space, then a single curved wire above a lossy ground and, finally, with a wire array above a multilayer. Thus, the set of Pocklington equations for a configuration of arbitrarily shaped overhead wires can be obtained as an extension of the Pocklington integro-differential equation for a single wire of an arbitrary shape.

5.4.1 Curved Single Wire in Free Space

The Pocklington equation for a single curved wire can be derived by enforcing the continuity conditions for the tangential components of the electric field along the perfectly conducting wire surface [2, 29–31], as shown in Fig. 5.51.

For the PEC wire, the total field composed of the excitation field $\vec{E}^{exc}$ and scattered field $\vec{E}^{sct}$ vanishes, leading to the continuity condition [29–31]

$$\vec{e}_t \cdot \left(\vec{E}^{exc} + \vec{E}^{sct} \right) = 0 \text{ on the wire surface} \tag{5.100}$$

where $\vec{e}_t$ is a unitary vector tangential to the curved wire.

Starting from Maxwell's equations and Lorentz gauge, the scattered field from the wire surface can be expressed in terms of the vector potential $\vec{A}$:

$$\vec{E}^{sct} = -j\omega\vec{A} + \frac{1}{j\omega\mu\varepsilon} \nabla\left(\nabla\vec{A}\right) \tag{5.101}$$

The vector potential is defined by the particular integral over PEC wire configuration

$$\vec{A}(s) = \frac{\mu}{4\pi} \int_{w'} I(s')g_0(s,s',s^*)\vec{s}'\,ds' \tag{5.102}$$

where s′ and s are the source and observation point, respectively; s′ is the vector tangential to the curved wire; and $I(s')$ is the induced current along the wire, while $g_0(s, s')$ is the lossless medium Green function:

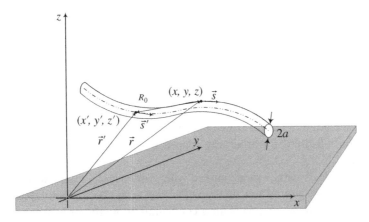

Figure 5.51 Single wire of arbitrary shape in free space.

$$g_0(s, s') = \frac{e^{-jkR}}{R} \tag{5.103}$$

and R is the distance from the source point to the observation point, respectively, while the propagation constant of the lossless homogeneous medium is:

$$k^2 = \omega^2 \mu_0 \varepsilon_0 \tag{5.104}$$

Inserting (5.102) into (5.101), the following integral relation for the scattered electric field is obtained

$$\vec{E}^{sct} = \frac{1}{j4\pi\omega\varepsilon_0} \int\limits_{w'} I(s') \cdot \vec{s}' \cdot [k^2 + \nabla\nabla] g_0(s, s') ds' \tag{5.105}$$

Combining (5.105) and (5.100) yields the Pocklington integral equation for the unknown current distribution along the wire of arbitrary shape insulated in free space:

$$E_{\tan}^{exc}(s) = -\frac{1}{j4\pi\omega\varepsilon_0} \int\limits_{w'} I(s') \cdot \vec{s} \cdot \vec{s}' \cdot [k^2 + \nabla\nabla] g_0(s, s') ds' \tag{5.106}$$

where $E_{\tan}^{exc}$ denotes the tangential component of the electric field illuminating the wire.

5.4.2 Curved Single Wire in the Presence of a Lossy Half-space

The case of curved wire located above an imperfectly conducting half-space can be analyzed by extending the kernel of integro-differential equation (5.106) using the RC approach [4]. The geometry of an arbitrary wire and its image, respectively, is shown in Fig. 5.52.

The excitation function E^{exc} is composed of the incident and reflected field, respectively.

$$E^{exc} = E^{inc} + E^{ref} \tag{5.107}$$

Performing certain mathematical manipulations, the Pocklington integro-differential equation for a curved wire above a lossy ground becomes [24].

$$E_s^{exc}(s) = \frac{j}{4\pi\omega\varepsilon_0} \int\limits_0^L \left\{ \left[k^2 \vec{e}_s \vec{e}_{s'} - \frac{\partial^2}{\partial s \partial s'} \right] g_0(s, s') + R_{TM} \left[k^2 \vec{e}_s \vec{e}_{s*} - \frac{\partial^2}{\partial s \partial s*} \right] g_i(s, s*) + \right.$$

$$\left. + (R_{TE} - R_{TM}) \vec{e}_s \vec{e}_p \cdot \left[k^2 \vec{e}_p \vec{e}_{s*} - \frac{\partial^2}{\partial p \partial s*} \right] g_i(s, s*) \right\} I(s') ds' \tag{5.108}$$

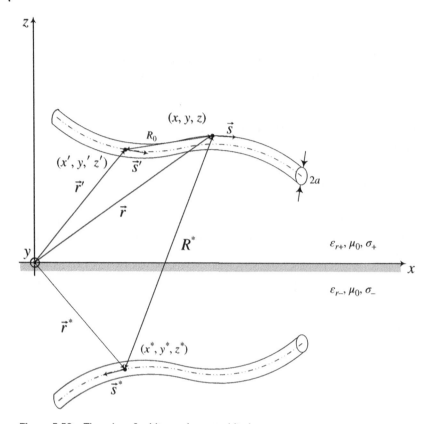

Figure 5.52 The wire of arbitrary shape and its image.

where $\vec{e}_p$ is the unit vector normal to the incident plane, while $g_i(s, s^*)$ arises from the image theory and is given by:

$$g_i(s, s^*) = \frac{e^{-jkR^*}}{R^*} \tag{5.109}$$

and R^* is the distance from the image source point to the observation point, respectively.

5.4.3 Multiple Curved Wires

An extension to the configuration of multiple curved wires results in the system of integral equations [24]

$$E_{sm}^{exc}(s) = \frac{j}{4\pi\omega\varepsilon_0} \sum_{n=1}^{N_w} \int_0^{L_n} \left\{ \left[k^2 \vec{e}_{s_m} \vec{e}_{s'_n} - \frac{\partial^2}{\partial s_m \partial s'_n} \right] g_{0n}\left(s_m, s'_n\right) + \right.$$

$$+ R_{TM} \left[k^2 \vec{e}_{s_m} \vec{e}_{s_n^*} - \frac{\partial^2}{\partial s_m \partial s_n^*} \right] g_{in}\left(s_m, s_n^*\right) + \tag{5.110}$$

$$\left. (R_{TE} - R_{TM}) \vec{e}_{s_m} \vec{e}_p \cdot \left[k^2 \vec{e}_p \vec{e}_{s*} - \frac{\partial^2}{\partial p \partial s^*} \right] g_i\left(s_m, s_n^*\right) \right\} I\left(s'_n\right) ds'$$

where N_w is the total number of wires and $I_n(s'_n)$ is the unknown current distribution induced on the n-th wire. Furthermore, $g_{0mn}(x, x')$ and $g_{imn}(s, s')$ are the Green functions of the form:

$$g_{0mn}\left(s_m, s'_n\right) = \frac{e^{-jkR_{1mn}}}{R_{1mn}}, \quad g_{imn}\left(s_m, s'_n\right) = \frac{e^{-jkR_{2mn}}}{R_{2mn}} \tag{5.111}$$

where and R_{1mn} and R_{2mn} are distances from the source point and from the corresponding image, respectively, to the observation point of interest.

The influence of a homogeneous lossy half-space can be taken into account via the Fresnel plane wave RC for TM and TE polarization, respectively, given by [24]:

$$R'_{TM} = \frac{\underline{n}\cos\theta' - \sqrt{\underline{n} - \sin^2\theta'}}{\underline{n}\cos\theta' + \sqrt{\underline{n} - \sin^2\theta'}} \tag{5.112}$$

$$R_{TE} = \frac{\cos\theta' - \sqrt{\underline{n} - \sin^2\theta'}}{\cos\theta' + \sqrt{\underline{n} - \sin^2\theta'}} \tag{5.113}$$

where θ' is the angle of incidence and $\underline{n}$ is given by:

$$\underline{n} = \frac{\varepsilon_{eff}}{\varepsilon_0}, \quad \varepsilon_{eff} = \varepsilon_r \varepsilon_0 - j\frac{\sigma}{\omega} \tag{5.114}$$

and ε_{eff} is the complex permittivity of the ground.

Though appreciably less demanding than Sommerfeld integral approach, the reflection approximation is still not convenient for handling realistic scenarios with complex wire configurations above a multilayer. Therefore, an alternative approach featuring reflection arising from MIT (originally derived for quasistatic phenomena [5]) has been promoted in [18] where slightly modified MIT coefficient has been proposed.

The total RC to account for the reflection between air and layered lower medium, arising from the extended use of MIT [5], and the approach documented in [18] can be written in the form

$$R^{tot} = \frac{R_{12} + R_{23}e^{-2\gamma_2 d}}{1 + R_{12}R_{23}e^{-2\gamma_2 d}} \tag{5.115}$$

where the reflection between m-th and n-th layers, respectively, is given by

$$R_{mn} = \frac{\varepsilon_{eff,m} - \varepsilon_{eff,n}}{\varepsilon_{eff,m} + \varepsilon_{eff,n}} \tag{5.116}$$

and $\varepsilon_{effm,n}$ denotes the effective permittivity of a particular layer

$$\varepsilon_{effm,n} = \varepsilon_{rm,n} - j\frac{\sigma_{m,n}}{\omega} \tag{5.117}$$

The corresponding simplified variants of integral equation (5.108) and system of coupled integral equations (5.110) are as follows:

$$E_s^{exc}(s) = \frac{j}{4\pi\omega\varepsilon_0} \int_0^L \left\{ \left[k^2 \vec{e}_s \vec{e}_{s'} - \frac{\partial^2}{\partial s \partial s'} \right] g_0(s, s') \right.$$
$$\left. + R^{tot} \left[k^2 \vec{e}_s \vec{e}_{s*} - \frac{\partial^2}{\partial s \partial s*} \right] g_i(s, s^*) \right\} I(s') ds' \tag{5.118}$$

$$E_{sm}^{exc}(s) = \frac{j}{4\pi\omega\varepsilon_0} \sum_{n=1}^{N_w} \int_0^{L_n} \left\{ \left[k^2 \vec{e}_{sm} \vec{e}_{s'_n} - \frac{\partial^2}{\partial s_m \partial s'_n} \right] g_{0n}\left(s_m, s'_n\right) \right.$$
$$\left. + R^{tot} \left[k^2 \vec{e}_{sm} \vec{e}_{s*_n} - \frac{\partial^2}{\partial s_m \partial s*_n} \right] g_{in}\left(s_m, s^*_n\right) \right\} I\left(s'_n\right) ds' \tag{5.119}$$

The influence of the load impedance can be easily included in the integral equation formulation by modifying the continuity condition for the tangential components of the electric field at the wire surface, as follows

$$E_s^{inc} + E_s^{sct} = Z'_L I(s) \tag{5.120}$$

where Z'_L is the corresponding conductor per conductor length impedance.

The modified Pocklington equation for the wire containing the load impedance is now given by

$$E^{inc}(s) = -\frac{1}{j4\pi\omega\varepsilon_0} \int_0^L \left\{ \left[k_1^2 \vec{e}_s \vec{e}_{s'} - \frac{\partial^2}{\partial s \partial s'} \right] g_0(s, s') \right.$$
$$\left. + R^{tot} \left[k_1^2 \vec{e}_s \vec{e}_{s*} - \frac{\partial^2}{\partial s \partial s*} \right] g_i(s, s^*) \right\} I(s') ds' + Z'_L I(s) \tag{5.121}$$

An extension to the curved wire array yields

$$E_{sm}^{exc}(s) = \frac{j}{4\pi\omega\varepsilon_0} \sum_{n=1}^{N_w} \int_0^{L_n} \left\{ \left[k^2 \vec{e}_{s_m} \vec{e}_{s'_n} - \frac{\partial^2}{\partial s_m \partial s'_n} \right] g_{0n}(s_m, s'_n) \right. $$

$$\left. + R^{tot} \left[k^2 \vec{e}_{s_m} \vec{e}_{s_n^*} - \frac{\partial^2}{\partial s_m \partial s_n^*} \right] g_{in}(s_m, s_n^*) \right\} I(s'_n) ds' + Z'_L I(s'_n) $$

$$(5.122)$$

Integral equations (5.121) and (5.122), respectively, are numerically solved using GB-IBEM, e.g. [1, 2].

Due to the very nature of MIT RC, second term in (5.108) and (5.110) vanishes, thus making (5.121) and (5.122) significantly simpler and convenient to use for more demanding geometries.

A trade-off between the rigorous Sommerfeld integral approach and approximate RC approach for homogeneous half-space has been presented elsewhere, e.g. in [4]. Although RC approximation causes certain error, it takes significantly less computational effort than a rigorous Sommerfeld approach, in particular, for the case of a multilayer [1–4].

The total electric field irradiated by some configuration of multiple wires of arbitrary shape is given by [24]

$$\vec{E} = \sum_{n=1}^{N_w} \left[\vec{E}_{0n} + R_{TM} \vec{E}_{in} + (R_{TE} - R_{TM})\left(\vec{E}_{in} \cdot \vec{e}_p\right)\vec{e}_p \right] \tag{5.123}$$

where:

$$\vec{E}_{0n} = \frac{1}{j4\pi\omega\varepsilon_0} \left[k_1^2 \int_0^{L_n} \vec{e}_{s_n'} I(s_n') g_{0n}\left(\vec{r}, \vec{r'}\right) ds_n' + \int_0^L \frac{\partial I(s_n')}{\partial s_n'} \nabla g_{0n}\left(\vec{r}, \vec{r'}\right) ds_n' \right] \tag{5.124}$$

$$\vec{E}_{in} = \frac{1}{j4\pi\omega\varepsilon_0} \left[k_1^2 \int_0^{L_n} \vec{e}_{s_{n*}} I(s_n') g_{in}\left(\vec{r}, \vec{r}^*\right) dw' - \int_0^{L_n} \frac{\partial I(s^*)}{\partial s_n^*} \nabla g_{in}\left(\vec{r}, \vec{r}^*\right) ds' \right] \tag{5.125}$$

Note that indices 0 and i are related to the source and image wire, respectively.

The radiated magnetic field of the curved wire system can be written as follows [24]:

$$\vec{H} = \sum_{n=1}^{N_w} \left[\vec{H}_{Sn} + R_{TE} \vec{H}_{In} + (R_{TM} - R_{TE})\left(\vec{H}_{In} \cdot \vec{e}_p\right)\vec{e}_p \right] \tag{5.126}$$

where:

$$\vec{H}_{Sn} = -\frac{1}{4\pi} \int_0^{L_n} I(s_n')\vec{e}_{s'} \times \nabla g_{0n}\left(\vec{r},\vec{r}'\right) ds' \tag{5.127}$$

$$\vec{H}_{In} = -\frac{1}{4\pi} \int_0^{L_n} I(s_n')\vec{e}_{s*} \times \nabla g_{in}\left(\vec{r},\vec{r}^*\right) ds' \tag{5.128}$$

For the case of curved wires above a multilayer featuring MIT approach, expressions (5.123) and (5.126) simplify into the following set of formulas:

$$\vec{E} = \sum_{n=1}^{N_w} \left[\vec{E}_{0n} + R^{tot} \cdot \vec{E}_{in}\right] \tag{5.129}$$

$$\vec{H} = \sum_{n=1}^{N_w} \left[\vec{H}_{Sn} + R^{tot} \cdot \vec{H}_{In}\right] \tag{5.130}$$

Obviously, due to the nature of MIT coefficient, second term in (5.123) and (5.126), respectively, vanishes again.

Therefore, the usage of MIT RC in integral expressions in the case of curved wires of arbitrary shapes for the determination of induced currents along wires and related radiated fields significantly simplifies the formulation.

5.4.3.1 Numerical Solution Procedures

A set of the Pocklington integro-differential equations (5.122) has been solved by using the GB-IBEM described elsewhere, e.g. in [1].

Performing the Galerkin–Bubnov scheme of (GB-IBEM) in the frequency domain, the set of coupled integro-differential equations (5.122) is transformed into the following matrix equation [24]

$$\sum_{n=1}^{M}\sum_{i=1}^{N_n} [Z]_{ji}^e \{I\}_i^e = \{V\}_j^e \tag{5.131}$$

where the mutual impedance matrix is given by [24]

$$[Z]_{ij}^e = -\int_{-1}^{1}\int_{-1}^{1} \{D\}_j \{D'\}_i^T g_{0nm}\left(s_n, s_m'\right) \frac{ds_m'}{d\xi'} d\xi' \frac{ds_n}{d\xi} d\xi$$

$$+ k_1^2 \vec{e}_{s_n} \vec{e}_{s_m} \int_{-1}^{1}\int_{-1}^{1} \{f\}_j \{f'\}_i^T g_{0nm}\left(s_n, s_m'\right) \frac{ds_m'}{d\xi'} d\xi' \frac{ds_n}{d\xi} d\xi -$$

$$
- R_{TM} \int_{-1}^{1} \int_{-1}^{1} \{D\}_j \{D'\}_i^T g_{inm}\left(s_n, s_m^*\right) \frac{ds_m'}{d\xi'} d\xi' \frac{ds_n}{d\xi} d\xi
$$

$$
+ R^{tot} \cdot k_1^2 \vec{e}_{s_n} \vec{e}_{s_m^*} \int_{-1}^{1} \int_{-1}^{1} \{f\}_j \{f'\}_i^T g_{inm}\left(s_n, s_m^*\right) \frac{ds_m'}{d\xi'} d\xi' \frac{ds_n}{d\xi} d\xi
$$

$$
+ \frac{j}{4\pi\omega\varepsilon_0} \int_{-1}^{1} Z_T' \{f\}_j \{f'\}_j^T \frac{ds_n}{d\xi} d\xi
$$

$$(5.132)$$

while the voltage vector is [24]

$$
\{V\}_j^n = -j4\pi\omega\varepsilon_0 \int_{-1}^{1} E_{s_n}^{exc}(s_n) f_{jn}(s_n) \frac{ds_n}{d\xi} d\xi_n \tag{5.133}
$$

Once the current distribution is obtained, the radiated field can be obtained by applying a similar BEM formalism [24].

Thus, the total field is given by:

$$
\vec{E} = \sum_{k=1}^{N} \left[\vec{E}^e_{Sk} + R^{tot} \cdot \vec{E}^e_{Ik} \right] \tag{5.134}
$$

where the field components due to a wire segment radiation are given by

$$
\vec{E}^e_{Sk} = \frac{1}{j4\pi\omega\varepsilon_0} \sum_{i=1}^{n} \left[k^2 \int_{-1}^{1} \vec{e}_{ks'} I_{ik}^e f_i(\xi) g_{0k}\left(\vec{r}, \vec{r'}\right) \frac{ds_k'}{d\xi} d\xi \right.
$$

$$
\left. + \int_{-1}^{1} I_{ik}^e \frac{\partial f_i(\xi)}{\partial \xi} \nabla g_{0k}\left(\vec{r}, \vec{r'}\right) \frac{ds_k'}{d\xi} d\xi \right]
$$

$$(5.135)$$

$$
\vec{E}^e_I = \frac{1}{j4\pi\omega\varepsilon_0} \sum_{i=1}^{n} \left[k^2 \int_{-1}^{1} \vec{e}_{ks^*} I_{ik}^e f_i(\xi) g_{ik}\left(\vec{r}, \vec{r}^*\right) \frac{ds_k'}{d\xi} d\xi \right.
$$

$$
\left. - \int_{-1}^{1} I_{ik}^e \frac{\partial f_i(\xi)}{\partial \xi'} \nabla g_{ik}\left(\vec{r}, \vec{r}^*\right) \frac{ds_k'}{d\xi} d\xi \right]
$$

$$(5.136)$$

The total magnetic field is given by [24]

$$
\vec{H} = \sum_{k=1}^{N} \left[\vec{H}^e_{Sk} + R^{tot} \cdot \vec{H}^e_{Ik} \right] \tag{5.137}
$$

while the magnetic field components are given by [24]

$$\vec{H}^e_{Sk} = -\frac{1}{4\pi} \sum_{i=1}^{n} \int_{-1}^{1} I_{ik} f_i(\xi) \vec{e}_{Sk'} \times \nabla g_{0k}\left(\vec{r}, \vec{r}'\right) \frac{ds_k'}{d\xi} d\xi \qquad (5.138)$$

$$\vec{H}^e_{lk} = -\frac{1}{4\pi} \sum_{i=1}^{n} \int_{-1}^{1} I^e_{ik} f_i(\xi) \vec{e}_{ks^*} \times \nabla g_{ik}\left(\vec{r}, \vec{r}^*\right) \frac{ds_k'}{d\xi} d\xi \qquad (5.139)$$

More details on the applied numerical procedures, such as inclusion of lumped sources, could be found elsewhere, e.g. in [7, 29].

5.4.3.2 Computational Examples

Numerical results pertain to a simple PLC (power line communications) system (circuit) above a multilayer. Note that the excitation is delta-function generator [1, 2]. The discretization condition requires at least 20 samples per wavelength and $\Delta l > 2a$ (where Δl is a wire segment, and a is the radius), to avoid pancake elements.

It is worth noting that PLC technology aims to provide users with necessary communication means by using already existing and widely distributed power line networks and electrical installations in houses and buildings. However, one of the principal drawbacks of PLC technology is electromagnetic interference (EMI) problems, as overhead power lines at the PLC frequency range (1 to 30 MHz) behave as transmitting or receiving antennas, respectively [29].

Figure 5.53 shows the geometry of a simple PLC system above two-layer ground consisting of two conductors placed in parallel above each other at the distance d,

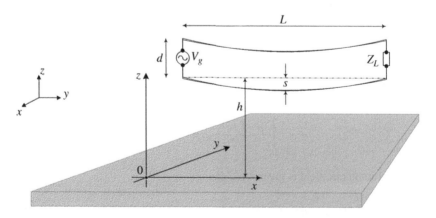

Figure 5.53 Simple PLC system above a two-layer ground.

suspended between two poles of equal height, thus having the shape of the catenary.

The geometry of a catenary is fully defined by such parameters as the distance between the points of suspension, L; the sag of the conductor, s; and the height of the suspension point, h as shown in Fig. 5.53.

The conductors are modeled as thin wire antennas driven by the voltage generator V_g at one end, and terminated by the load impedance Z_L at the other end.

The wires are oriented along x-axis (with distance between poles $L = 200$ m and the wire radius $a = 6.35$ mm) and suspended on the poles at heights $h_1 = 10$ m, and $h_2 = 11$ m, while the conductor sag is $s = 4$ m. The distance between the conductors is $d = 1$ m, and the ground parameters are $\varepsilon_r = 13$ and $\sigma = 5$ mS/m. The power of the applied voltage generator is 2.5 µW, (minimum power required for the PLC system operation), and operating frequency is 28 MHz. The value of the terminating load Z_L is 5 kΩ. The curved wire is discretized into 500 linear elements.

Two-layer ground consists of the vegetation cover above the homogeneous soil. The imperfectly conducting ground is characterized by relative permittivity $\varepsilon_r = 13$ and conductivity $\sigma = 5$ mS/m and it is considered to be constant for all examples.

The properties of the upper layer (vegetation) are varied as follows: $\varepsilon_r = 4, 10, 18,$ and 30, and $\sigma = 0.001, 0.01,$ and 0.1 S/m. The upper layer thickness is 10 cm, otherwise the thickness is varied from 1 cm to 0.5 m. However, the influence of the layer thickness is shown to be negligible.

Figure 5.54 shows the current distribution along the upper wire at the operating frequency $f = 28$ MHz in the case of homogeneous ground.

It is obvious that the current values do not differ significantly (less than 1%) for both approximations. This is expected since wires are located relatively high above

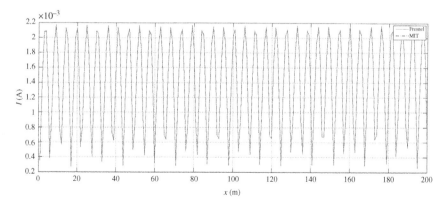

Figure 5.54 Current distribution along the upper wire at $f = 28$ MHz.

the ground. Thus, the effect of the reflected field on the current distribution is negligible.

Figures 5.55 and 5.56 show E_x and E_z radiated field components at $f = 28$ MHz, for various values of permittivity, and $\sigma = 1$ mS/m conductivity, respectively. The results obtained for the multilayer case are compared to the results obtained by using the homogeneous half-space assumption. The results are shown for E_x

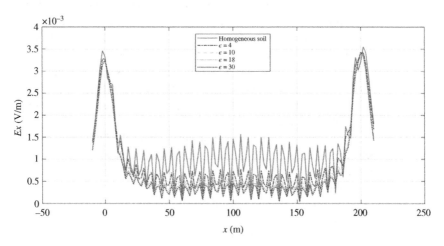

Figure 5.55 E_x component at 1 m above ground at $f = 28$ MHz and $\sigma = 0.001$ S/m.

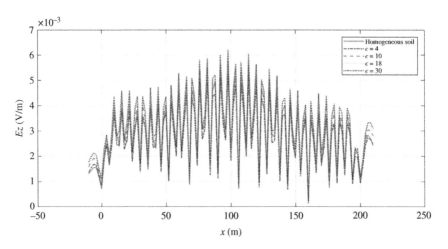

Figure 5.56 E_z component at 1 m above ground at $f = 28$ MHz and $\sigma = 0.001$ S/m.

and E_z components, respectively, obtained by using the RC arising from the MIT. The homogeneous ground scenario is simulated via the Fresnel RC approximation.

The last set of the simulation results is carried out with $\varepsilon_r = 4$, while the conductivity of the upper layer is varied. The corresponding field results are shown in Figs. 5.57 and 5.58.

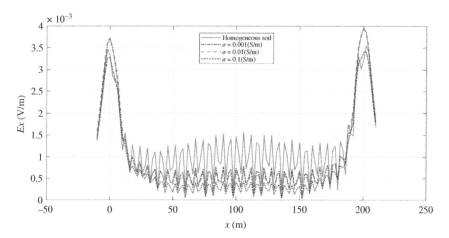

Figure 5.57 E_x component at 1 m above ground component at 1 m above ground at $f = 28$ MHz and $\varepsilon_r = 4$.

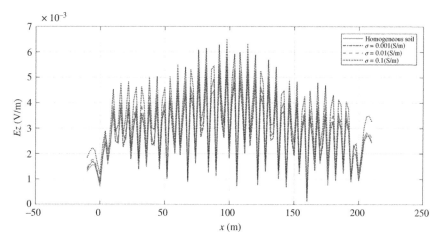

Figure 5.58 E_z component at 1 m above ground component at 1 m above ground at $f = 28$ MHz and $\varepsilon_r = 4$.

One can observe the discrepancy between the two-layer results and simplified homogeneous half-space case for different values of permittivity and conductivity, respectively. Larger deviations occur at the point where the wire is close to the interface (near the center of the conductor), which is expected due to the different approximations for the RCs being used.

Thus, E_x (tangential component) of radiated electric field is significantly different for the multi-layered soil compared to the homogeneous soil case (red line in Figs. 5.55 and 5.57). On the other hand, E_z (normal) component is almost the same as in the case of the homogeneous soil.

Only significant discrepancy of E_z values for multilayer case is observed for $\varepsilon_r = 30$ (Fig. 5.56) and $\sigma = 0.1$ S/m (Fig. 5.58).

On the basis of the present analysis, one may conclude that the calculated results for ILS system are in satisfactory agreement and the model used in this work is found to be useful for the engineering analysis of antenna system performances for the instrumental landing of the airplane.

5.4.4 Electromagnetic Field Coupling to Arbitrarily Shaped Aboveground Wires

The knowledge of an induced current distribution along considered curved wire configuration is a prerequisite to understanding the behavior of field coupling to overhead wires, such as power lines and communication cables [26]. The TL approach, often neglecting radiation effects, is an efficient approximation in engineering practice for a wide range of applications [2, 32, 33]. Nevertheless, some applications still require a more accurate approach related to the solution of a corresponding integral expression arising from scattering theory (full-wave models) [2].

The assessment of EMI induced along three-phase power line illuminated by the plane wave incident electric field by using via full-wave model has been presented in [33].

This section first deals with the frequency domain formulation based on the corresponding set of Pocklington integral equations for arbitrarily shaped overhead wires above a lossy ground, thus providing one to account for the conductor sag. The effects of a lossy half-space are taken into account via the rigorous Sommerfeld integral approach.

Next, GB-IBEM featuring linear and quadratic isoparametric elements is outlined.

Finally, some illustrative numerical results pertaining to some practical scenarios involving realistic power line configurations are presented.

5.4.4.1 Formulation via a Set of Coupled Integro-differential Equations

The currents $I_n(s')$ induced along the N_w curved wires located above a lossy ground (Fig. 5.59) due to a plane wave electric field E^{exc} (Fig. 5.60) are governed by the set of the Pocklington integro-differential equations [2].

The set of coupled Pocklington integro-differential equations is given by [33]

$$
\begin{aligned}
E^{exc}_{sm}(s) = -\frac{1}{j4\pi\omega\varepsilon_0}\sum_{n=1}^{N_w}
&\begin{bmatrix}
\displaystyle\int_{C_n'} I_n(s')\cdot\vec{s}\cdot\vec{s}'\cdot\left[k_0^2+\nabla\nabla\right]g_{0n}(s_m,s_n')ds' + \\[2mm]
\displaystyle\frac{k_g^2-k_0^2}{k_g^2+k_0^2}\int_{C_n'} I_n(s_n')\cdot\vec{s}\cdot\vec{s}^*\cdot\left[k_0^2+\nabla\nabla\right]g_{in}(s_m,s_n^*)ds' + \\[2mm]
\displaystyle\int_{C_n'} I_n(s')\cdot\vec{s}\cdot\vec{G}_s(s_m,s_n')ds'
\end{bmatrix} \\[2mm]
&+ Z_S\cdot I_m(s)
\end{aligned}
$$

$$(5.140)$$

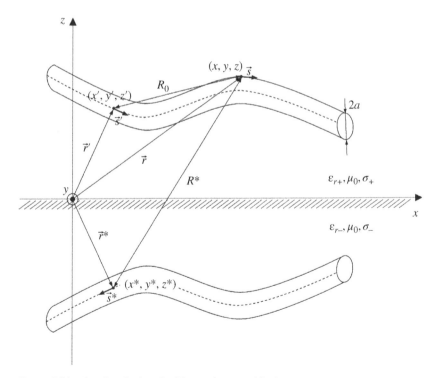

Figure 5.59 Overhead wire of arbitrary shape and its image.

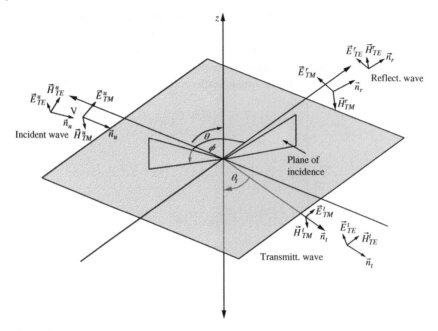

Figure 5.60 Incident, reflected, and transmitted waves.

where the corresponding Green functions are given by

$$g_{0mn}\left(s_m, s'_n\right) = \frac{e^{-jkR_{1mn}}}{R_{1mn}} \tag{5.141}$$

$$g_{imn}\left(s_m, s'_n\right) = \frac{e^{-jkR_{2mn}}}{R_{2mn}} \tag{5.142}$$

where R_{1mn} and R_{2mn} are distances from the source point (from real and image wire axis, respectively) to the observation point (at the real wire surface).

$$\vec{G}_s(s, s') = \left(\vec{e}_x \cdot \vec{s}'\right) \cdot \left(G_\rho^H \cdot \vec{e}_\rho + G_\phi^H \cdot \vec{e}_\phi + G_z^H \cdot \vec{e}_z\right)$$
$$+ \left(\vec{e}_z \cdot \vec{s}'\right) \cdot \left(G_\rho^V \cdot \vec{e}_\rho + G_z^V \cdot \vec{e}_z\right) \tag{5.143}$$

while the related wave numbers are

$$k_0^2 = \omega^2 \mu_0 \varepsilon_0 \tag{5.144}$$

$$k_g^2 = \omega^2 \mu_0 \varepsilon_{eff} = \omega^2 \mu_0 \left(\varepsilon_0 \varepsilon_{rg} - j\frac{\sigma_g}{\omega}\right) \tag{5.145}$$

where ε_{rg} and σ_g are relative permittivity and conductivity of the ground, respectively, and ω is the frequency of interest.

The kernel terms are as follows:

$$G_\rho^V = \frac{\partial^2}{\partial\rho\partial z} k_g^2 V^R \tag{5.146}$$

$$G_z^V = \left(\frac{\partial^2}{\partial z^2} + k_0^2\right) k_g^2 V^R \tag{5.147}$$

$$G_\rho^H = \cos\phi\left(\frac{\partial^2}{\partial\rho^2} k_0^2 V^R + k_0^2 U^R\right) \tag{5.148}$$

$$G_\phi^H = -\sin\phi\left(\frac{1}{\rho}\frac{\partial}{\partial\rho} k_0^2 V^R + k_0^2 U^R\right) \tag{5.149}$$

$$G_z^H = -j4\pi\omega\varepsilon_0 \cos\phi\, G_\rho^V \tag{5.150}$$

And the Sommerfeld integral terms are given by

$$U^R = \int_0^\infty D_1(\lambda)\, e^{-\gamma_0|z + z'|}\, J_0(\lambda\rho)\,\lambda\,d\lambda \tag{5.151}$$

$$V^R = \int_0^\infty D_2(\lambda)\, e^{-\gamma_0|z + z'|}\, J_0(\lambda\rho)\,\lambda\,d\lambda \tag{5.152}$$

$$D_1(\lambda) = \frac{2}{\gamma_0 + \gamma_g} - \frac{2k_0^2}{\gamma_0\left(k_0^2 + k_g^2\right)} \tag{5.153}$$

$$D_2(\lambda) = \frac{2}{k_g^2\gamma_0 + k_0^2\gamma_g} - \frac{2}{\gamma_0\left(k_0^2 + k_g^2\right)} \tag{5.154}$$

$$\gamma_0 = \sqrt{\lambda^2 - k_0^2}; \qquad \gamma_g = \sqrt{\lambda^2 - k_g^2} \tag{5.155}$$

where J_0 denotes the Bessel function.

The set of Pocklington IEs is handled via the GB-IBEM with isoparametric elements.

5.4.4.2 Numerical Solution of Coupled Pocklington Equations

The set of Pocklington IEs is handled via the GB-IBEM with isoparametric elements. The unknown current along the n-th wire segment is expressed by a sum of independent basis functions with unknown complex coefficients

$$I_n^e(s') = \sum_{i=1}^n I_{ni} f_{ni}(s') = \{f\}_n^T \{I\}_n \tag{5.156}$$

The use of isoparametric elements yields:

$$I_n^e(\zeta) = \sum_{i=1}^{n} I_{ni} f_{ni}(\zeta) = \{f\}_n^T \{I\}_n \tag{5.157}$$

Having performed the BEM discretization procedure, the set of coupled Pocklington equations is transformed into the following matrix equation

$$\sum_{n=1}^{N_w} \sum_{i=1}^{N_n} [Z]_{ji}^e \{I\}_i^e = \{V\}_j^e, m = 1, 2, ..., N_w; \quad j = 1, 2, ..., N_m \tag{5.158}$$

where the local matrix and right-side vector are given by

$$[Z]_{ij}^e = - \int_{-1}^{1} \int_{-1}^{1} \{D\}_j \{D'\}_i^T g_{0nm}(s_n, s_m') \frac{ds_m'}{d\xi'} d\xi' \frac{ds_n}{d\xi} d\xi$$

$$+ k_0^2 \vec{s}_n \cdot \vec{s}_m' \int_{-1}^{1} \int_{-1}^{1} \{f\}_j \{f'\}_i^T g_{0nm}(s_n, s_m') \frac{ds_m'}{d\xi'} d\xi' \frac{ds_n}{d\xi} d\xi -$$

$$- \frac{k_g^2 - k_0^2}{k_g^2 + k_0^2} \int_{-1}^{1} \int_{-1}^{1} \{D\}_j \{D'\}_i^T g_{inm}(s_n, s_m^*) \frac{ds_m'}{d\xi'} d\xi' \frac{ds_n}{d\xi} d\xi$$

$$+ \frac{k_g^2 - k_0^2}{k_g^2 + k_0^2} k_0^2 \vec{s}_n \cdot \vec{s}_m' \int_{-1}^{1} \int_{-1}^{1} \{f\}_j \{f'\}_i^T g_{inm}(s_n, s_m^*) \frac{ds_m'}{d\xi'} d\xi' \frac{ds_n}{d\xi} d\xi +$$

$$+ \vec{s}_m' \int_{-1}^{1} \int_{-1}^{1} \{f\}_j \{f'\}_i^T \vec{G}_{snm}(s_n, s_m^*) \frac{ds_m'}{d\xi'} d\xi' \frac{ds_n}{d\xi} d\xi$$

$$+ \frac{j}{4\pi\omega\varepsilon_0} \int_{-1}^{1} Z_T' \{f\}_j \{f'\}_j^T \frac{ds_n}{d\xi} d\xi \tag{5.159}$$

$$\{V\}_j^n = -j4\pi\omega\varepsilon_0 \int_{-1}^{1} E_{s_n}^{exc}(s_n) \{f\}_j \frac{ds_n}{d\xi} d\xi_n \tag{5.160}$$

Note that the physical meaning of (5.159) is the mutual impedance, while (5.160) represents the voltage.

5.4.4.3 Computational Example

The computational example deals with a plane wave ($f = 5$ MHz) coupling to three-phase power line, shown in Fig. 5.61, consisting of four wires (phase

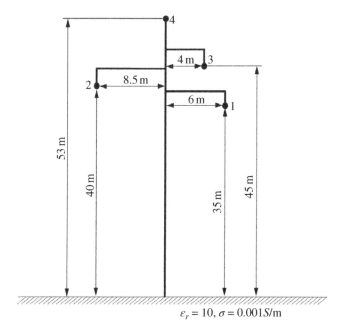

Figure 5.61 Three-phase power line system.

conductors + shield wire) above a lossy ground ($\varepsilon_r = 10$, $\sigma = 0.001$ S/m). The separation between towers is 300 m. The radius of perfectly conducting conductors is $a = 5$ mm.

More details about such a power line configuration can be found elsewhere, e.g. in [33].

Case number 1 deals with straight conductors, i.e. the power line system consisting of straight conductors illuminated by the plane wave incidence ($\alpha = 0°$, $\theta = 60°$, and $\varphi = 30°$) with amplitude $E_0 = 1$ V/m, and frequency $f = 5$ MHz is considered. Fig. 5.62a–d shows the real and imaginary parts of the current distribution along the wires calculated via GB-IBEM and compared to the results computed via the NEC software package [34].

An excellent agreement between the results obtained via GB-IBEM and NEC is considered.

Case number 2 deals with curved conductors; the wire sag (23 m for phase conductors, 13.5 m for shield wire) is taken into account. The power line system is illuminated by the plane wave incidence ($\alpha = 0°$, $\theta = 0°$, and $\varphi = 0$) with amplitude $E_0 = 1$ V/m and frequency $f = 5$ MHz. Figures to follow (Fig. 5.63a–d) show the current distribution along the wires calculated via GB-IBEM (linear and quadratic elements) and compared to the results obtained via the NEC software package.

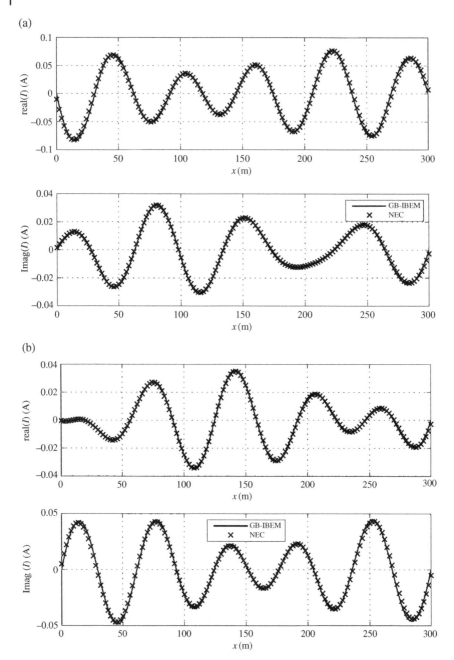

Figure 5.62 (a) Real and imaginary parts of the current distribution along conductor number 1, (b) Real and imaginary parts of the current distribution along conductor number 2, (c) Real and imaginary parts of the current distribution along conductor number 3, (d) Real and imaginary parts of the current distribution along the shielded wire.

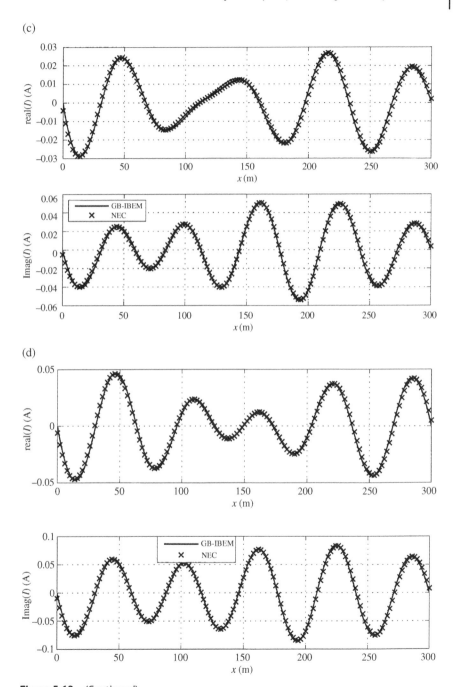

Figure 5.62 (Continued)

(a)

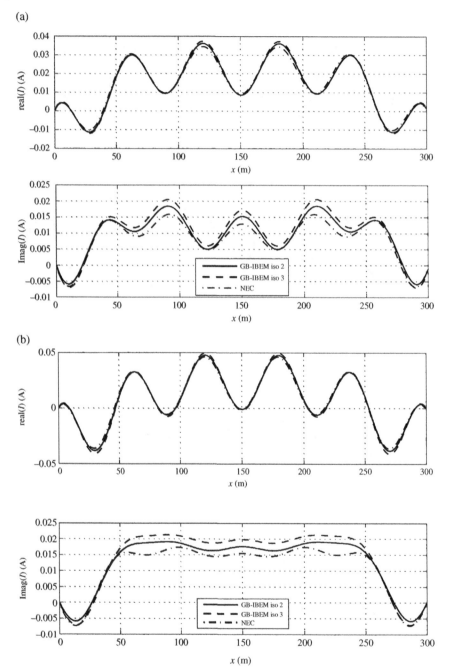

Figure 5.63 (a) Real and imaginary parts of the current distribution along conductor number 1, (b) Real and imaginary parts of the current distribution along conductor number 2, (c) Real and imaginary parts of the current distribution along conductor number 3, (d) Real and imaginary parts of the current distribution along the shielded wire.

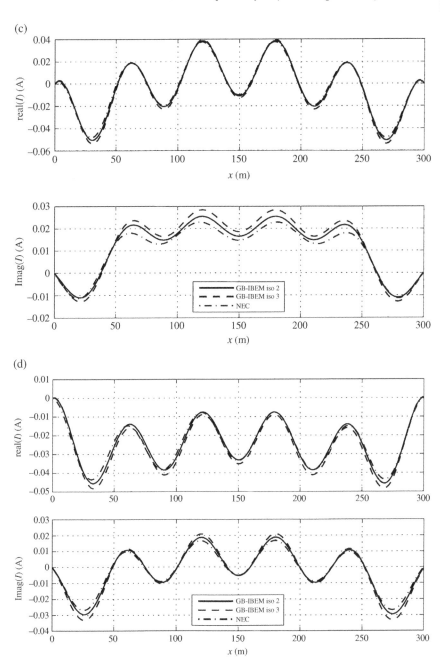

Figure 5.63 (Continued)

Some discrepancies are noticeable in the imaginary part of current distribution along the conductors.

Finally, Case number 3 deals with both straight and curved conductors, i.e. figures to follow (Fig. 5.64a–d) show the comparison of currents induced on the power line system for the case of straight and curved conductors, respectively. The excitation is a unit amplitude plane wave ($\alpha = 0°$, $\theta = 60°$, and $\varphi = 30°$) of frequency 5 MHz. The current distribution along the straight conductors is calculated by using linear interpolation only, while the current distribution along the curved wires is calculated by using linear and quadratic approximation, respectively.

The influence of the conductor sag is clearly demonstrated in Fig. 5.63a–d.

Therefore, the results obtained via different approaches agree satisfactorily. The influence of the conductor sag becomes clearly visible when the results obtained for straight and curved conductors are compared.

5.4.5 Buried Wires of Arbitrary Shape

Electromagnetic modeling of buried wire configurations has numerous applications, e.g. in the analysis of underground cables, or grounding systems [26, 35–37]. For most of the applications dealing with straight buried wires, the TL approach is used (e.g. [1, 26, 38–40]), primarily in the frequency domain, due to the simplicity of the formulation and computational efficiency. The TL approach, on the other hand, is mostly limited to straight wire structures and fails to account for the radiation effects at higher frequencies [26].

An alternative is to use the AT or full-wave approach, whose main disadvantage is the complexity of formulation and high computational cost [1, 37]. A possible way to reduce the computational cost by combining TL and AT approaches has been carried out in [36] for the case of straight buried wires.

An extension of the work reported in [36] to the analysis of curved wires buried in a lossy ground is available in [28] and is outlined in this subsection. The influence of a lossy ground is taken into account via the simplified RC arising from the MIT appearing within the related Green's function. The application of the generalized telegrapher's equations for buried wires is of particular importance in the transient analysis of grounding systems as the concept of scattered voltage is included within the formulation which is not the case in standard AT. Such a concept provides a straightforward determination of the transient voltage which is necessary for the evaluation of input impedance and step-voltage at the ground surface.

5.4.5.1 Formulation

The generalized telegrapher's equations for curved wire configurations can be derived by extending the simple case of a straight buried wire. A straight thin wire

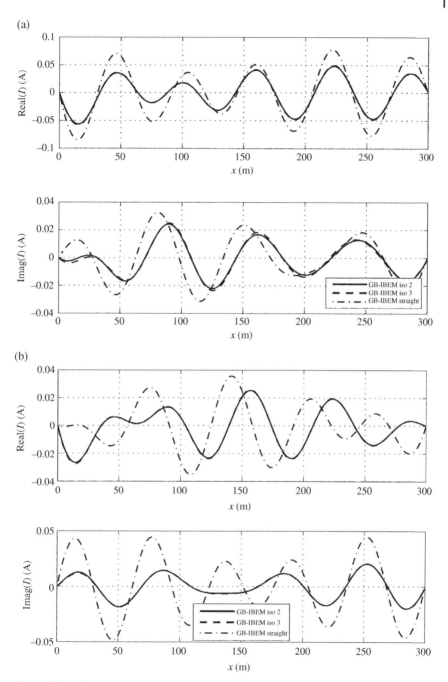

Figure 5.64 (a) Real and imaginary parts of the current distribution along conductor number 1, (b) Real and imaginary parts of the current distribution along conductor number 2, (c) Real and imaginary parts of the current distribution along conductor number 3, (d) Real and imaginary parts of the current distribution along the shielded wire.

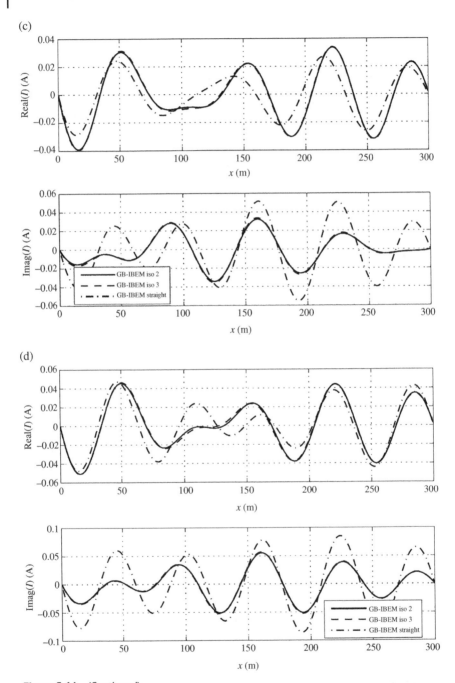

Figure 5.64 (Continued)

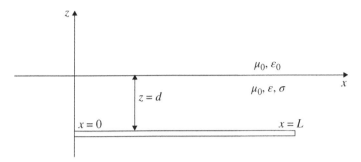

Figure 5.65 Straight thin wire buried in a lossy ground.

of length L and radius a buried in a lossy ground at depth d (see Fig. 5.65) is considered.

The buried wire is excited either by a plane wave (underground cable) or by an equivalent current source (grounding system). The current and voltage induced along the wire are obtained by solving the generalized telegrapher's equations derived in [36].

$$\int_0^L \frac{\partial I(x')}{\partial x'} g(x,x')dx' + j4\pi\omega\varepsilon_{eff} V^{sct}(x) = 0 \tag{5.161}$$

where $I(x')$ is the current distribution along the wire, and $g(x,x')$ stands for the corresponding Green's function.

The rigorous form of Green's function is expressed in terms of the Sommerfeld integrals whose evaluation is rather time-consuming [36].

An approximate approach is to use the Fresnel RC [36] leading to the following total Green's function

$$g(x,x') = g_0(x,x') - \Gamma_{ref} g_i(x,x') \tag{5.162}$$

where $g_0(x,x')$ is the free-space Green's function

$$g_0(x,x') = \frac{e^{-j\gamma R_o}}{R_o} \tag{5.163}$$

and $g_i(x,x')$ arises from the image theory and is given by

$$g_i(x,x') = \frac{e^{-j\gamma R_i}}{R_i} \tag{5.164}$$

In (5.163) and (5.164), R_o and R_i denote the corresponding distance from the source to the observation point, respectively, and the propagation constant of the lossy ground is given by

$$\gamma = \sqrt{j\omega\mu\sigma - \omega^2\mu\varepsilon} \tag{5.165}$$

The corresponding Fresnel reflection is of the form [1]

$$\Gamma_{ref} = \frac{\dfrac{\varepsilon_0}{\varepsilon_{eff}}\cos\theta - \sqrt{\dfrac{\varepsilon_0}{\varepsilon_{eff}} - \sin^2\theta}}{\dfrac{\varepsilon_0}{\varepsilon_{eff}}\cos\theta + \sqrt{\dfrac{\varepsilon_0}{\varepsilon_{eff}} - \sin^2\theta}} \tag{5.166}$$

which accounts for the presence of a lossy half-space.

The complex permittivity of the ground is given by

$$\varepsilon_{eff} = \varepsilon_r\varepsilon_0 - j\frac{\sigma}{\omega} \tag{5.167}$$

and argument θ is defined as

$$\theta = \operatorname{arctg}\frac{|x - x'|}{2d} \tag{5.168}$$

An alternative is to use the simplified RC arising from MIT:

$$\Gamma_{MIT} = \frac{\varepsilon_{eff} - \varepsilon_0}{\varepsilon_{eff} + \varepsilon_0} \tag{5.169}$$

The next step is to extend the formulation (5.161) to the case of an arbitrarily shaped wire.

The curved wire with the corresponding image is shown in Fig. 5.66.

Assuming the potential in the remote soil to be zero and by using straightforward mathematical manipulations, the following expression for the voltage between any point on the wire $(\vec{r})$ and remote soil is obtained:

$$\int_{C'} \frac{\partial I(s')}{\partial s'} \cdot g_0\left(\vec{r}, s'\right) ds' - \int_{C'} \frac{\partial I(s')}{\partial s^*} \cdot \Gamma_{MIT}g_i\left(\vec{r}, s^*\right) ds' +$$
$$+ j4\pi\omega\varepsilon_{eff}V^{sct}\left(\vec{r}\right) = 0 \tag{5.170}$$

in which the geometrical parameters are defined in Fig. 5.2.

Integral expression (5.170) represents the second telegrapher's equation for curved wires. Note that r stands for the observation point position, while s' and s^* are related to source variables over the buried wire and image in the air, respectively. Now the extension to the multiconductor case is straightforward, i.e. it follows

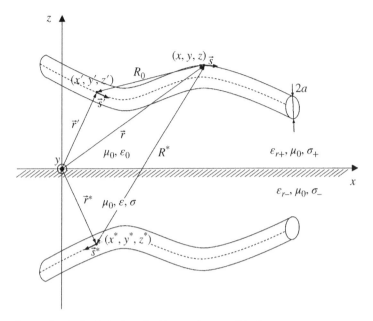

Figure 5.66 Buried wire of arbitrary shape and its image.

$$\sum_{k=1}^{N_W} \left\{ \int_{C'} \frac{\partial I_k(s')}{\partial s'} \cdot g_{0k}\left(\vec{r}, s'\right) ds' + \Gamma_{MIT} \int_{C'} \frac{\partial I_k(s')}{\partial s^*} \cdot g_{ik}\left(\vec{r}, s^*\right) ds' \right\} +$$

$$+ j4\pi\omega\varepsilon_{eff} V^{sct}\left(\vec{r}\right) = 0$$

$$(5.171)$$

where N_w denotes the total number of wires.

The first generalized telegrapher's equation for curved wires could be derived in a similar manner. Once the current distribution along the given wire configuration is determined by solving the corresponding integro-differential equation, it is possible to evaluate the scattered voltages (5.170) or (5.171), respectively.

5.4.5.2 Numerical Procedure

The current $I_n^e(\zeta)$ over the n-th segment is expressed via basis functions f_{ni}, and complex coefficients I_{ni}. Featuring the use of isoparametric elements, it follows [37]

$$I_n^e(\zeta) = \sum_{i=1}^{n} I_{ni} f_{ni}(\zeta) = \{f\}_n^T \{I\}_n \qquad (5.172)$$

where n is the number of local nodes per element.

The boundary element formalism applied to (5.171) yields

$$V\left(\vec{r}\right) = -\frac{1}{j4\pi\omega\varepsilon_{\mathit{eff}}}\sum_{i=1}^{N_w}\sum_{n=1}^{N_g}\sum_{k=1}^{nl}I_k^{e\,i}\int_{-1}^{1}\frac{\partial f_k^e(\zeta')}{\partial\zeta'}\cdot\left[g_0^i\left(\vec{r},s'\right)-\Gamma_{\mathit{MIT}}g_i^i\left(\vec{r},s^*\right)\right]d\zeta'$$

(5.173)

The transient voltage is obtained by applying the IFT to (5.173).

5.4.5.3 Computational Examples

The first set of numerical results deals with a grounding system composed of $60\,\mathrm{m}\times60\,\mathrm{m}$ grid $(10\,\mathrm{m}\times10\,\mathrm{m}$ square meshes), with conductor radius $a = 0.007\,\mathrm{m}$ (Fig. 5.67). The grid is buried at depth $d = 0.5\,\mathrm{m}$ in a lossy ground with conductivity $\sigma = 1\,\mathrm{mS/m}$ and relative permittivity $\varepsilon_r = 10$.

The grounding grid is excited at its center by a double exponential current source

$$i(t) = I\left(e^{-\alpha t} - e^{-\beta t}\right)$$

(5.174)

where $I = 1.0167\,\mathrm{kA}$, $\alpha = 0.0142\,\mathrm{\mu s^{-1}}$, and $\beta = 5.073\,\mathrm{\mu s^{-1}}$.

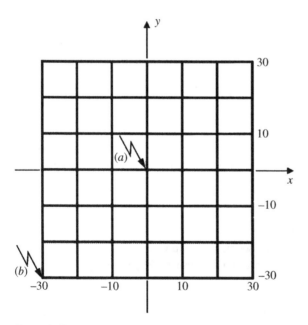

Figure 5.67 Geometry of a grounding grid.

Figure 5.68 shows the spatial distribution of the voltage induced along the grounding grid for the case of center injection at various time instances.

The obtained numerical results are in good agreement with the results published in [35].

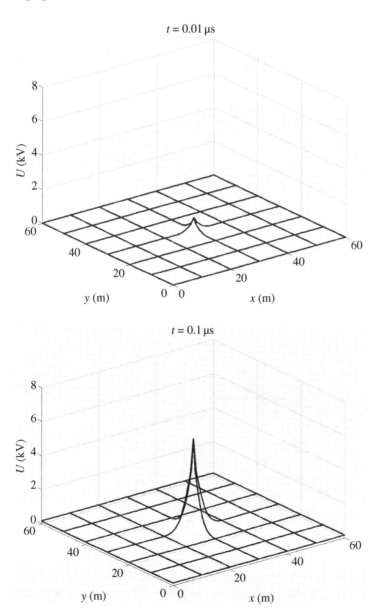

Figure 5.68 The voltage distribution along the grounding grid (center injection) at various instances of time.

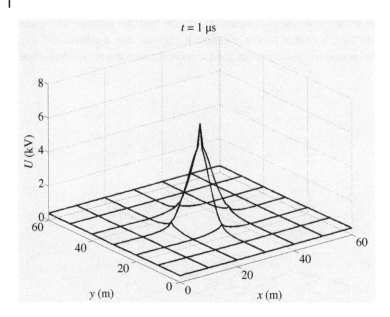

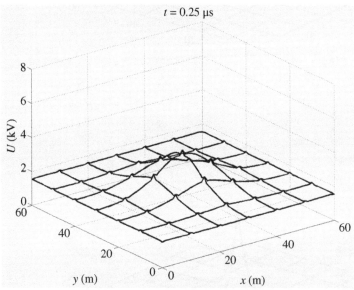

Figure 5.68 (Continued)

Next example deals with a typical grounding system of a wind turbine (WT), shown in Fig. 5.71. WT grounding system, composed of 2 copper rings with radii 3.25 and 6.8 m (buried at depths 5 and 55 cm), is placed in a homogeneous soil of conductivity $\sigma = 0.8$ mS/m and a relative dielectric constant $\varepsilon_r = 9$.

Note that a rectangular grounding grid of wire length 9.6 m buried at a depth of 2 m is located inside a bigger ring. Horizontal wires (Fe/Zn 30 mm × 4 mm) of lengths 28, 29, and 9 m are placed radially, as depicted in Fig. 5.69. The WT is excited by the lightning current given by the double exponential function (5.174) with: $I_0 = 1.1043$ A, $\alpha = 0.07924 \cdot 10^6$ s^{-1}, and $\beta = 4.0011 \cdot 10^6$ s^{-1}, representing a 1/10 μs pulse, as shown in Fig. 5.70.

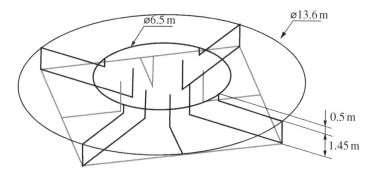

Figure 5.69 Typical configuration of WT grounding system.

Figure 5.70 WT excited by a lightning return-stroke pulse.

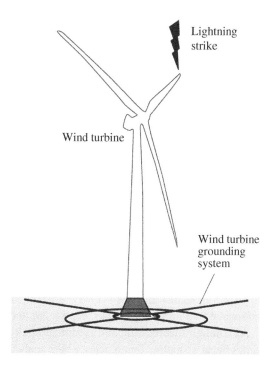

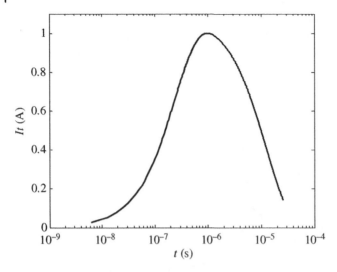

Figure 5.71 The lightning return-stroke current pulse.

The time-domain waveform of the lightning return-stroke current pulse is shown in Fig. 5.71.

Figure 5.72 shows the spatial distribution of the voltage at the ground surface above the WT grounding system.

The model of buried wires of arbitrary shape could be improved using the rigorous Sommerfeld integral approach to account for the presence of a lossy half-space, or a multilayer, respectively.

5.5 Complex Power of Arbitrarily Shaped Thin Wire Radiating Above a Lossy Half-Space

Modeling of Wireless Power Transfer (WPT) systems includes the analysis of power generated and received by the antennas in the near field. Most commonly used measure of efficiency of such systems is power transfer efficiency (PTE) defined as a ratio between received and transmitted power.

Though the analysis of WPT systems can be carried out by using relatively simple approaches, circuit theory, or TL models, respectively, rigorous studies use AT approach in some scenarios [41–43]. A trade-off between different approaches is addressed in a number of books, e.g. [1, 26] and papers, e.g. [36, 44–48].

AT, though theoretically and computationally demanding, accounts for radiation effects at higher frequencies when analyzing finite length wires, which is

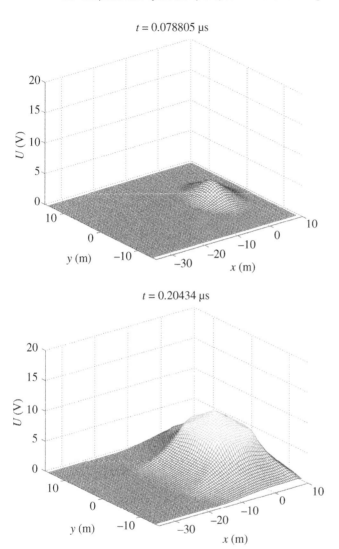

Figure 5.72 The voltage distribution induced along the WT grounding system (center injection) at various time instants.

not possible to provide by using simplified TL approximation, or oversimplified circuit theory approximation. Particular difficulties with TL approach arise when wires radiate in the presence of a lossy half-space.

AT models generally require numerical methods for the solution, although there are some examples where the analytical approach has been successfully applied [49, 50].

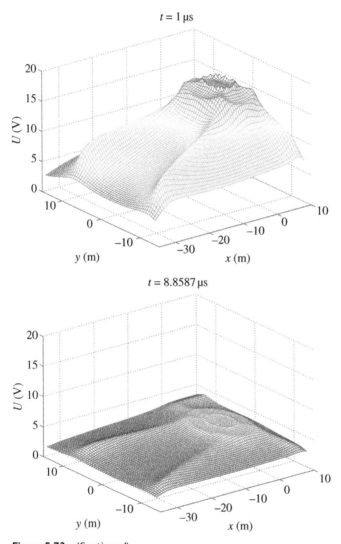

Figure 5.72 (Continued)

The received complex power along the straight horizontal wire scatterer in the presence of a lossy ground has been recently reported in [51]. The use of generalized telegrapher's equations provides to calculate induced current and the scattered voltage along the scatterer, as derived in [44]. Knowing the current and voltage distributions along the scatterer, the active, reactive, and apparent power, respectively, arising from the standard circuit theory definitions are calculated [51], which could be useful in the analysis of PTE for the WPT applications.

An efficient alternative approach for the calculation of complex power generated by the transmitting antenna above a lossy ground has been reported in [14] providing the assessment of the antenna's active, reactive, and apparent power. Such an approach does not require the evaluation of generated electric and magnetic fields and carry out subsequent integration of corresponding Poynting vector over a closed surface around the wire. Once obtaining the antenna current by numerically solving the Pocklington integro-differential equation via GB-IBEM [1], the complex power of the antenna versus frequency can be computed by solving the integral over the inner product of tangential component of the electric field and current distribution along the wire arising from the Poynting integral theorem. The numerical procedure for the calculation of complex power uses the vectors and matrices already constructed within the numerical procedure for the calculation of current distribution along the wire which is rather attractive advantage of GB-IBEM. Such an approach could be also used to compute input impedance spectrum as well [1, 2].

5.5.1 Theoretical Background

A thin wire antenna of an arbitrary shape of finite length C_w and radius a located above a lossy ground and driven by an equivalent voltage generator (Fig. 5.73) is considered.

The apparent complex power P_a generated in the volume of interest V containing sources for time-harmonic fields can be derived from the Poynting integral theorem and is given by

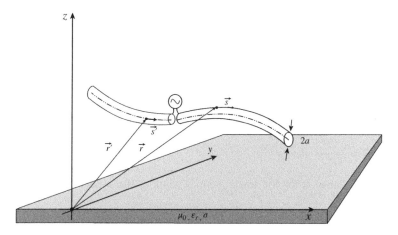

Figure 5.73 Center-fed arbitrarily shaped antenna above a lossy half-space.

$$P_a = -\frac{1}{2} \int_V \vec{E} \cdot \vec{J}^* \, dV \tag{5.175}$$

where E stands for the field generated by the source and J is the distribution of corresponding current density along the source.

If thin wires are of interest, the problem becomes one-dimensional, i.e. it follows

$$\vec{J}^* \, dV = I^* S \, d\vec{s} \tag{5.176}$$

where S stands for wire cross-section and I^* is the complex conjugate of the antenna current, and (5.175) simplifies into

$$P_a = -\frac{1}{2} \int_{C_W} \vec{E} \cdot I^* \, d\vec{s} \tag{5.177}$$

where C_w represents the geometry of curved wires.

Note that the real part of (5.177) represents the active power, while the imaginary part is reactive power.

The Pocklington integro-differential equation for the wire of arbitrary shape above a lossy half-space can be written as follows

$$E_s^{exc}(s) = \frac{j}{4\pi\omega\varepsilon_0} \int_{C_w} \left\{ \left[k^2 \vec{e}_s \vec{e}_{s'} - \frac{\partial^2}{\partial s \partial s'} \right] g_0(s,s') + \right. $$
$$\left. R_{TM} \left[k^2 \vec{e}_s \vec{e}_{s''} - \frac{\partial^2}{\partial s \partial s''} \right] g_i(s,s'') \right\} I(s') I^*(s) ds' \tag{5.178}$$

where $I(s')$ is the induced current along the wire, s' and s are the source and observation point, respectively, $\vec{e}_{s'}$ is the vector tangential to the curved wire, and $\vec{e}_s$ is related to the vector tangential to the observation point at the wire surface.

Furthermore, $g_0(s,s')$ stands for the lossless medium Green function

$$g_0(s,s') = \frac{e^{-jkR}}{R} \tag{5.179}$$

and R is the distance from the source point to the observation point, respectively, while $g_i(s, s'')$ arises from the image theory and is given by

$$g_i(s,s'') = \frac{e^{-jkR''}}{R''} \tag{5.180}$$

and R'' is the distance from the image source point to the observation point, respectively.

The propagation constant of the lossless homogeneous medium is

$$k = \omega\sqrt{\mu_0 \varepsilon_0} \tag{5.181}$$

The influence of a lower medium is taken into account via the Fresnel RC for TM polarization

$$R'_{TM} = \frac{\underline{n} \cos \theta' - \sqrt{\underline{n} - \sin^2 \theta'}}{\underline{n} \cos \theta' + \sqrt{\underline{n} - \sin^2 \theta'}} \tag{5.182}$$

where θ' is the angle of incidence and n is given by

$$\underline{n} = \frac{\varepsilon_{eff}}{\varepsilon_0}, \quad \varepsilon_{eff} = \varepsilon_r \varepsilon_0 - j \frac{\sigma}{\omega} \tag{5.183}$$

Once the current distribution is known, the scattered field can be obtained by forcing the continuity condition

$$\vec{e}_s \cdot \left(\vec{E}^{exc} + \vec{E}^{sct} \right) = 0 \tag{5.184}$$

where $\vec{E}^{sct}$ stands for the scattered field.

Inserting (5.178) into (5.177), one obtains

$$P_a = \frac{j}{8\pi\omega\varepsilon_0} \int\limits_{C_w} \int\limits_{C_w} \left\{ \left[k^2 \vec{e}_s \vec{e}_{s'} - \frac{\partial^2}{\partial s \partial s'} \right] g_0(s, s') + \right.$$

$$\left. R_{TM} \left[k^2 \vec{e}_s \vec{e}_{s''} - \frac{\partial^2}{\partial s \partial s''} \right] g_i(s, s'') \right\} I(s') I^*(s) ds' ds'' \tag{5.185}$$

Utilizing the integration by parts, it is possible to avoid quasi-singularity problems with the kernel, i.e. (5.185) becomes

$$P_a = \frac{j}{8\pi\omega\varepsilon_0} \left[- \int\limits_{C_w} \int\limits_{C_w} \frac{\partial I(s')}{\partial s'} \cdot \frac{\partial I^*(s)}{\partial s} g_0(s, s') ds' ds + \right.$$

$$k^2 \int\limits_{C_w} \int\limits_{C_w} \vec{e}_s \vec{e}_{s'} I(s') I^*(s) ds' ds g_0(s, s') ds' ds -$$

$$R_{TM} \int\limits_{C_w} \int\limits_{C_w} \frac{\partial I(s'')}{\partial s''} \cdot \frac{\partial I^*(s)}{\partial s} g_i(s, s'') ds'' ds +$$

$$\left. R_{TM} \int\limits_{C_w} \int\limits_{C_w} \vec{e}_s \vec{e}_{s''} I(s'') I^*(s) g_i(s, s'') ds'' ds \right] \tag{5.186}$$

Applying the GB-IBEM featuring isoparametric elements, integral expression (5.186) is transformed into matrix equation

$$P_a = \frac{j}{8\pi\omega\varepsilon_0}\{I\}^T[Z]\{I^*\} \tag{5.187}$$

where $\{I\}$ and $\{I\}^T$ are the vector of current and its transpose vector, respectively, while $[Z]$ is impedance matrix stemming from the numerical procedure for the current distribution determination. Global impedance matrix is assembled from the generalized mutual impedance matrices for the *j-th* and *i-th* isoparametric elements given by

$$
\begin{aligned}
[Z]_{ji}^e = & \int_{-1}^{1}\int_{-1}^{1}\Bigg[-\{D\}_j \cdot \{D'\}_i^T g_0(s,s') + \\
& + k^2 \vec{e}_s \vec{e}_{s'} \{f\}_j \cdot \{f'\}_i^T ds' ds g_0(s,s') \Bigg] \frac{ds'}{d\xi'} d\xi' \frac{ds}{d\xi} d\xi + \\
& R_{TM} \int_{-1}^{1}\int_{-1}^{1}\Bigg[-\{D\}_j \cdot \{D'\}_i^T g_i(s,s'') + \\
& + k^2 \vec{e}_s \vec{e}_{s''} \{f\}_j \cdot \{f'\}_i^T g_i(s,s'') \Bigg] \frac{ds'}{d\xi'} d\xi' \frac{ds}{d\xi} d\xi
\end{aligned}
\tag{5.188}
$$

where vectors $\{f\}$ and $\{f'\}$ contain shape functions while vectors $\{D\}_i^n$ and $\{D\}_j^m$ contain shape function derivatives.

Note that vectors $\{I\}^T$ and vector of complex conjugate values of current $\{I\}^*$ and mutual impedance matrix $[Z]$ are already available from the calculation procedure for the current distribution along the wire.

Input impedance of the antenna can be defined as follows [1]

$$Z_{in} = \frac{2P_a}{|I_0|^2} = -\frac{1}{|I_0|^2} \int_C \vec{E} \cdot \vec{I}^* \, ds \tag{5.189}$$

where I_0 stands for the input current.

Taking into account matrix equation (5.187), the expression for computation of input impedance becomes

$$Z_{in} = -\frac{1}{j4\pi\omega\varepsilon_0|I_0|^2}\{I\}^T[Z]\{I^*\} \tag{5.190}$$

More details on the calculation of the input impedance of a dipole antenna can be found in [5]. It is worth stressing that the proposed procedure is highly efficient as it practically requires no additional computation, only instantaneous manipulation with already existing matrices and vectors from the numerical modeling of the Pocklington equation which is, in a sense of computational cost, a matter of few seconds.

5.5.2 Numerical Results

Computational example deals with center-fed dipole antenna above a lossy ground (Figure 5.74).

First example deals with the free space scenario.

Figures 5.75–5.77 show frequency spectra of the active, reactive, and apparent power, respectively, for the wire lengths $L = 0.25$ m, $L = 0.5$ m, and $L = 1$ m. Input impedance is shown in Fig. 5.78.

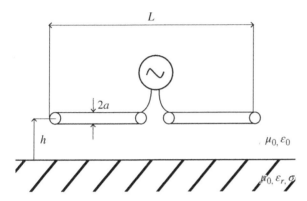

Figure 5.74 Center-fed dipole antenna above a lossy half-space.

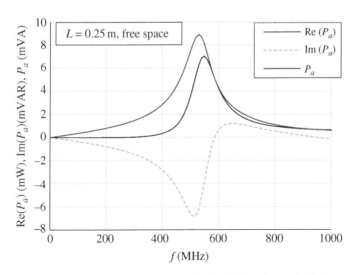

Figure 5.75 Power versus frequency for the 0.25 m-long wire in free space.

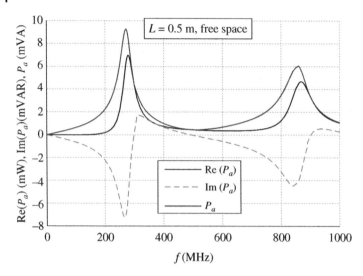

Figure 5.76 Power versus frequency for the 0.5 m-long wire in free space.

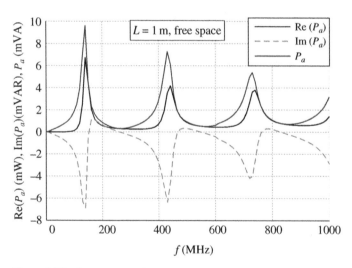

Figure 5.77 Power versus frequency for the 1 m-long wire in free space.

As can be observed in Fig. 5.79, peaks of the active power coincide with minima of the input impedance and vice versa. Figs. 5.79–5.82 show the results for the active, reactive, and apparent powers, respectively, of the wire placed at height $h = 0.25$ m above lossy ground with conductivity $\sigma = 0.001$ S/m and relative permittivity $\varepsilon_r = 10$.

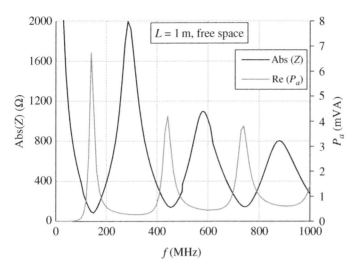

Figure 5.78 Input impedance and active power for the 1 m-long wire in free space.

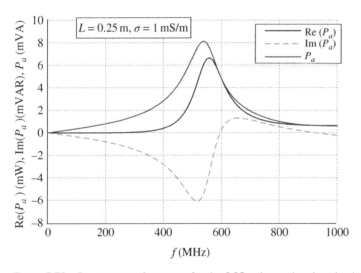

Figure 5.79 Power versus frequency for the 0.25 m-long wire placed at height $h = 0.25$ m above a lossy ground ($\sigma = 0.001$ S/m, $\varepsilon_r = 10$).

As could be seen from the numerical results, the power amplitude is slightly changed due to the presence of the lossy ground when compared to the case of free space. Furthermore, influence of the ground is more obvious at the first maximum which is expected as the wire is electrically closer to the ground at lower frequencies.

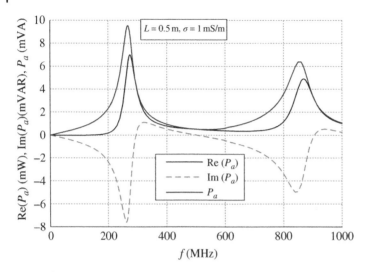

Figure 5.80 Power versus frequency for the 0.5 m-long wire placed at height $h = 0.25$ m above a lossy ground ($\sigma = 0.001$ S/m, $\varepsilon_r = 10$).

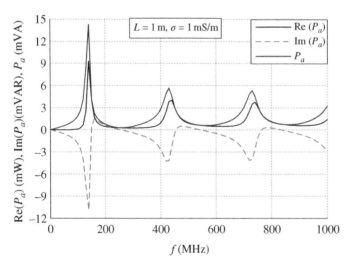

Figure 5.81 Power versus frequency for the 1 m-long wire placed at height $h = 0.25$ m above a lossy ground ($\sigma = 0.001$ S/m, $\varepsilon_r = 10$).

Figures 5.82 and 5.83 show frequency span of the active and reactive powers, respectively, for the case of $L = 0.25$ m and $h = 0.25$ m for the different values of the ground conductivity. Furthermore, Fig. 5.84 shows detail from the curves shown in Fig. 5.82 around the first maximum.

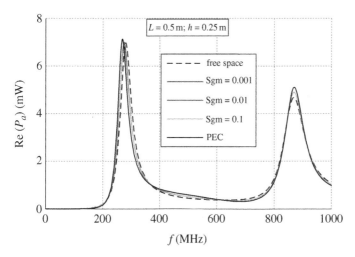

Figure 5.82 Active power versus frequency for the 0.5 m-long wire placed at height $h = 0.25$ m above a lossy ground for various values of the ground conductivity ($\varepsilon_r = 10$).

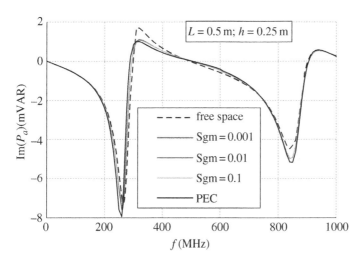

Figure 5.83 Reactive power versus frequency for the 0.5 m-long wire placed at height $h = 0.25$ m above a lossy ground for various values of the ground conductivity ($\varepsilon_r = 10$).

It could be noticed that presence of the ground causes a slight shift of the maximum toward lower frequencies (about 6 MHz) while in the case of the perfect ground, this shift is even more appreciable (around 12 MHz). The same behavior could be observed for the minima of the input impedance (Fig. 5.85).

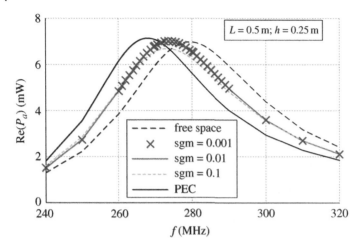

Figure 5.84 First maximum of the active power versus frequency for the 0.5 m-long wire placed at height $h = 0.25$ m above a lossy ground for various values of the ground conductivity ($\varepsilon_r = 10$).

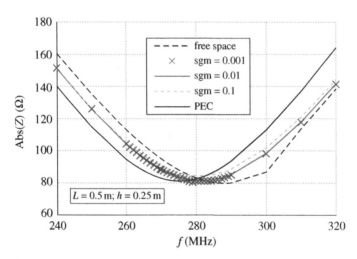

Figure 5.85 First minimum of the input impedance versus frequency for the 0.5 m-long wire placed at height $h = 0.25$ m above a lossy ground for various values of the ground conductivity ($\varepsilon_r = 10$).

The same analysis has been carried out for the wire located closer to the ground at the height of $h = 0.1$ m. Figs. 5.86 and 5.87 show the results for the apparent and active powers, respectively. In this case, the frequency shift is also visible for the second maximum. The amplitude of the apparent power at the first maximum is

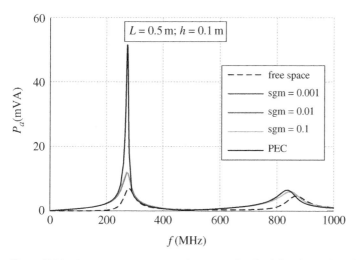

Figure 5.86 Apparent power versus frequency for the 0.5 m-long wire placed at height $h = 0.1$ m above a lossy ground for various values of the ground conductivity ($\varepsilon_r = 10$).

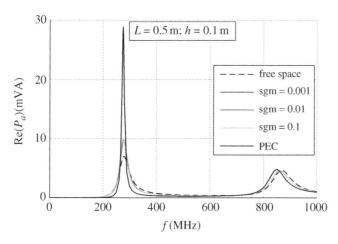

Figure 5.87 Active power versus frequency for the 0.5 m-long wire placed at height $h = 0.1$ m above a lossy ground for various values of the ground conductivity ($\varepsilon_r = 10$).

significantly higher for the perfect ground case. It is worth noting that for the cases of the real ground, curves remain almost the same regardless of the conductivity values.

Figure 5.88 shows the apparent power for the same wire located 0.1 m above the ground but for different values of the ground permittivity. The frequency shift of

(a)

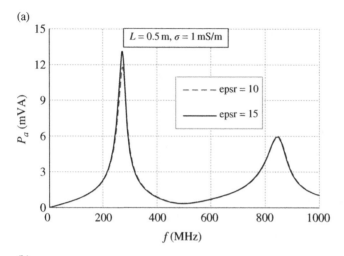

(b)

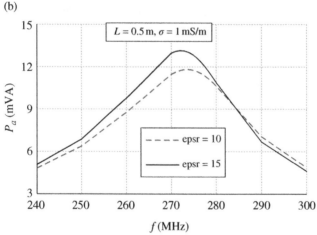

Figure 5.88 (a) Apparent power versus frequency for the 0.5 m-long wire placed at height $h = 0.1$ m above a lossy ground for various values of the ground permittivity ($\sigma = 0.001$ S/m); (b) detailed view.

the maximum is minimal and the amplitude change due to height variation is comparable to the influence of the ground conductivity.

Finally, Figs. 5.89 and 5.90 show the results obtained for the apparent and active power in the case of the wire above a perfectly conducting ground for different heights.

As is expected, the closer the wire positioned to the ground, the higher is the impact to the peak of the apparent power.

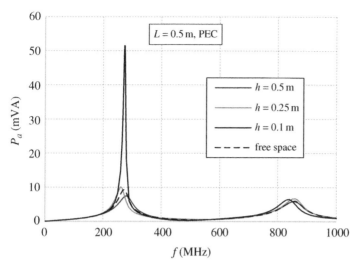

Figure 5.89 Apparent power versus frequency for the 0.5 m-long wire placed at different heights above a perfect ground.

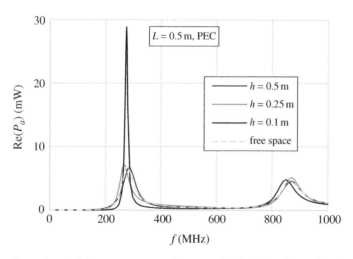

Figure 5.90 Active power versus frequency for the 0.5 m-long wire placed at different heights above a perfect ground.

References

1 D. Poljak, Advanced Modeling in Computational Electromagnetic Compatibility, New York: John Wiley and Sons, 2007.
2 D. Poljak and K. El Khamlichi Drissi, Computational Methods in Electromagnetic Compatibility, New York: John Wiley and Sons, 2018.
3 E. K. Miller, A. J. Poggio, G. J. Burke and E. S. Selden, "Analysis of wire antennas in the presence of a conducting half-space. Part I. The vertical antenna in free space," *Canadian Journal of Physiscs*, vol. 50, pp 879–888, 1972.
4 E. K. Miller, A. J. Poggio, G. J. Burke and E. S. Selden, "Analysis of wire antennas in the presence of a conducting half-space. Part II. The horizontal antenna in free space," *Canadian Journal of Physiscs*, 50, pp 2614–2627, 1972.
5 T. Takashima, T. Nakae and R. Ishibashi, "Calculation of complex fields in conducting media," *IEEE Transactions on Electrical Insulation*, vol. EI-15, no. 1, pp. 1–7, 1980.
6 D. Poljak and M. Rancic, On the Frequency Domain Analysis of Straight Thin Wire Radiating Above a Lossy Half-Space, Pocklington Equation versus Hallén equation revisited: 80th Anniversary of the Hallén Integral Equation, Split: Proc. SoftCOM, 2018.
7 T. K. Sarkar, "Analysis of arbitrarily oriented thin wire antennas over a plane imperfect ground," *Archiv fur elektronik und ubertragungstechnik*, 31, pp. 449–457, 1977.
8 K. K. Mei, "On the integral equation of thin wire antennas," *IEEE Tranactions on AP*, vol. 13, pp. 59–62, 1965.
9 P. Parhami and R. Mittra, "Wire antennas over a lossy half-space," *IEEE Transactions on AP*, 28, pp. 397–403, 1980.
10 D. Poljak and M. Rancic, Frequency Domain Modeling of a Dipole Antenna Buried in Lossy Half-Space Pocklington Equation versus Hallén Equation Revisited: 80th Anniversary of the Hallén Integral Equation, Vesteros: Proc. CAMA, 2018.
11 K. A. Michalski and J. R. Mosig, "Multilayered media Green's functions in integral equation formulations," *IEEE Transactions on Antennas and Propagation*, vol. 45, no 3, pp. 508–519, 1997.
12 P. Ylä-Oijala and M. Taskinen, "Efficient formulation of closed-form Green's functions for general electric and magnetic sources in multilayered media," *IEEE Transactions on Antennas and Propagation*, vol. 51, no. 8, pp. 2106–2115, 2003.
13 M. Yuan, T. K. Sarkar and M. Salazar-Palma, "A direct discrete complex image method from the closed-form Green's functions in multilayered media," *IEEE Transactions on Microwave Theory and Techniques*, vol. 54, no 3, pp. 1025–1032, 2006.
14 R. R. Boix, A. L. Fructos and F. Mesa, "Closed-form uniform asymptotic expansions of Green's functions in layered media," *IEEE Transactions on Antennas and Propagation*, vol. 58, no. 9. pp. 2934–2945, 2010.

15 D. Poljak, C. Y Tham, A. McCowen, and V. Roje, "Transient analysis of two coupled horizontal wires over a real ground," *IEEE Proceedings – Microwave, Antennas and Propagation*, vol. 147, pp. 87–94, 2000.

16 M. H. Haddad, M. Ghaffari-Miab and R. Faraji-Dana, "Transient analysis of thin-wire structures above a multilayer medium using complex-time Green's functions," *IET Microwaves, Antennas and Propagation*. vol. 4, no. 11, pp. 1937–1947, 2010.

17 N. Hojjat, S. Safavi-Naeini, R. Faraji-Dana and Y.L. Chow, "Fast computation of the nonsymmetrical components of the Green's function for multilayer media using complex images," *IEEE Proceedings - Microwaves, Antennas and Propagation*, vol. 145, no. 4, pp. 285–288, 1998.

18 D. Poljak et al., A simple analysis of dipole antenna radiation above a multilayered medium, *2017 9th International Workshop on Advanced Ground Penetrating Radar (IWAGPR)*, Edinburgh, University of Edinburgh, 2017.

19 K. El Poljak K. Khamlichi Drissi and S. Sesnic Kerroum, "Comparison of analytical and boundary element modeling of electromagnetic field coupling to overhead and buried wires," *Engineering Analysis with Boundary Elements*, vol. 35, no. 3, pp. 555–563, 2011.

20 A. Susnjara, V. Doric and D. Poljak, Electric Field Radiated in the Air by a Dipole Antenna Placed above a Two-Layered Lossy Half Space: Comparison of Plane Wave Approximation with the Modified Image Theory Approach, Split, Croatia: Proc. SoftCOM, 2018.

21 A. Susnjara, V. Doric and D. Poljak, Electric Field Radiated By a Dipole Antenna and Transmitted Into a Two-Layered Lossy Half Space: Comparison of Plane Wave Approximation with the Modified Image Theory Approach, Split, Croatia: Proc. Splitech, 2018.

22 V. Arnautovski-Toseva, K. El Khamlichi Drissi and K. Kerroum, "Comparison of image and transmission line models of energized horizontal wire above two-layer soil," *Automatika*, vol. 53, no. 1, pp. 38–48, 2012.

23 D. Poljak, M. Birkić and V. Dorić, Analysis of LPDA Radiation above a Multilayer, *26th International Conference on Software, Telecommunications and Computer Networks, SoftCOM 2018,* Supetar, Croatia, 2018.

24 D. Poljak, V. Dorić and M. Birkić, "Analysis of curved thin wires radiating over a layered medium and some engineering applications," *Journal of Electromagnetic Waves and Applications*, vol. 35, 2020.

25 J. A. Stratton, Electromagnetic Theory, New York: McGraw-Hill Book Company, pp. 511–512, 1941.

26 F. Tesche, M. Ianoz and F. Carlsson, EMC Analysis Methods and Computational Models. New York (NY): John Wiley and Sons; 1997.

27 R. R. Boix, A. L. Fructos, F. Mesa, "Closed-form uniform asymptotic expansions of Green's functions in layered media," *IEEE Transactions on Antennas and Propagation*, vol. 58, no. 9, pp. 2934–2945, 2010.

28 D. Poljak, D. Cavka and F. Rachidi, Farhad Generalized Telegrapher's Equations for Buried Curved Wires. *Proceedings of the 2nd URSI Atlantic Radio Science Meeting*, Gran Canaria, Španjolska, 2018. E01-05, 4

29 V. Doric and D. Poljak. EMC Analysis of the PLC System Based on the Antenna Theory, Dubrovnik, Croatia: ICECom, 2010.

30 D. Poljak and K. El Khamlichi Drissi, "Electromagnetic field coupling to overhead wire configurations:antenna model versus transmission line approach," *International Journal of Antennas and Propagation*, vol. 2012, pp. 730145-1–730145-18, 2012.

31 D. Poljak, V. Doric and E. Lerinc, Boundary Element Modeling of Curved Wire Configurations. BEM/MRM 39. New Forest: WIT Press, 2015.

32 G. E. Bridges and L. Shafai, "Plane wave coupling to multiple conductor transmission lines above a lossy earth," *IEEE Transactions on Electromagnetic Compatability*, vol. EMC-31, pp. 21–33, 1989.

33 D. Poljak and D. Cavka, BEM Analysis of Plane Wave Coupling to Three-Phase Power Line, Southampton: BEM, 2018.

34 G. J. Burke, A. J. Poggio, I. C. Logan and J. W. Rockway, "Numerical electromagnetics code—A program for antenna system analysis," in *Proceedings of EMC 1979 Symposium*, Rotterdam, The Netherlands, pp. 89–94.

35 L. Grcev, "Computer analysis of transient voltages in large grounding systems," *IEEE Transactions on Power Delivery*, vol. 11, no. 2, pp. 815–823, 1996.

36 D. Poljak, V. Doric, F. Rachidi, K. El Khamlichi Drissi, K. Kerroum, S. Tkachenko and S. Sesnic, "Generalized form of telegrapher's equations for the electromagnetic field coupling to buried wirtes of finite length," *IEEE Transactions on EMC*, vol. 51, no. 2, pp. 331–337, 2009.

37 D. Poljak, K. El-Khmlichi Drissi and B. Nekhoul, "Electromagnetic field coupling to arbitrary wire configurations buried in a lossy ground: a review of antenna model and transmission line approach," *International J.ournal of Computational Methods and Experimental Measurements*, vol. 1, no. 2, pp. 142–163, 2013.

38 E. Petrache, F. Rachidi M. Paolone, C. A. Nucci V. A. Rakov and M. A. Uman, "Lightning-induced disturbances in buried cables", Part I: theory, *IEEE Transactions on EMC*, vol. 47, no. 3, pp. 498–508, 2005.

39 M. Paolone, E. Petrache, F. Rachidi, C. A. Nucci, V. Rakov, M. Uman, D. Jordan, K. Rambo, J. Jerauld, M. Nyffeler and J. Schoene, "Lightning-induced disturbances in buried cables. Part II: Experiment and model validation," *IEEE Transactions on on EMC*, vol. 47, no. 3, pp. 509–520, 2005.

40 E. Petrache, M. Paolone, F. Rachidi, C. A. Nucci, V. Rakov, M. Uman, D. Jordan, K. Rambo, J. Jerauld, M. Nyffeler and J. Schoene, "Lightning-induced currents in buried coaxial cables: a frequency-domain approach and its validation using rocket-triggered lightning," *Journal of Electrostatics*, vol. 65, pp. 322–328, 2007, doi:https://doi.org/10.1016/j.elstat.2006.09.015.

41 S. Y. Ron Hui, "Past, present and future trends of non-radiative wireless power transfer," *CPSS Transactions on Power Electronics and Applications*, vol. 1, no. 1, pp. 83–91, 2016.

42 Q. Chen et al., "Antenna characterization for wireless power-transmission system using near-field coupling," *IEEE Antennas and Propagation Magazine*, vol. 54, no. 4, pp. 108–116, 2012.

43 M. Skiljo, Z. Blazevic and D. Poljak, "Interaction between human and near-field of wireless power transfer system," *Progress in Electromagnetics Research C*, vol. 67, pp. 1–10, 2016.

44 D. Poljak, F. Rachidi and S. V. Tkachenko "Generalized form of telegrapher's equations for the electromagnetic field coupling to finite-length lines above a lossy ground," *IEEE Transactions on Electromagnetic Compatibility*, vol. 49, no. 3, pp. 689–697, 2007.

45 D. Poljak et al., "Time-domain generalized telegrapher's equations for the electromagnetic field coupling to finite length wires above a lossy ground," *IEEE Transactions on Electromagnetic Compatibility,*vol. 54, no. 1, pp. 218–224, 2012.

46 D. Poljak et al. "Time domain generalized telegrapher's equations for the electromagnetic field coupling to finite-length wires buried in a lossy half-space," *Electric Power Systems Research*, vol. 160, pp. 199–204, 2018.

47 R. Olsen, J. Young and D. Chang, "Electromagnetic wave propagation on a thin wire above earth," *IEEE Transaction on Antennas and Propagation*, vol. 48, no. 9, pp. 1413–1419, 2000.

48 R. Olsen and D. Chang, "Analysis of semi-infinite and finite thin-wire antennas above a dissipative earth," *Radio Science*, vol. 11, no. 11, pp. 867–874, 1976, doi: https://doi.org/10.1029/RS011i011p00867.

49 J. Nitsch and S. Tkachenko, "Complex-valued transmission-line parameters and their relation to the radiation resistance," *IEEE Transaction on Electromagnetic Compatibility*, vol. EMC - 46, no. 3, pp. 477–487, 2004.

50 J. B. Nitsch and S. V. Tkachenko, "Global and modal parameters in the generalized transmission line theory and their physical meaning," *Radio Science Bulletin*, vol. 312, pp.21–31, 2005.

51 D. Poljak and V. Doric, "Calculation of complex power generated by a transmitting thin wire antenna radiating above a lossy half-space," *Journal of Communications Software and Systems*, vol. 18, no 2, pp. 175–181, 2022.

6

Wire Configurations – Time Domain Analysis

Transient analysis of thin wire configurations in the presence of a lossy half-space is of considerable interest in various areas of computational electromagnetics (CEM), such as electromagnetic compatibility (EMC), or ground penetrating radar (GPR). To obtain time domain (TD) response for a given radiating structure, one may use indirect approach featuring the formulation in the frequency domain (FD), while the corresponding TD response is obtained by applying some frequency to TD transformation technique, typically inverse fast Fourier transform (IFFT) [1, 2]. Though relatively simple, when compared to direct TD modeling, the indirect FD approach suffers from efficiency drawbacks, and potential problems arising from the frequency to TD transformation particularly occurring in the analysis of wires radiating in the presence of a lossy half-space when highly resonant structures are analyzed [3–5]. Such problems could be overcome by using a direct TD approach [6]. Direct TD approach, on the other hand, implies more demanding formulations and often suffers from instability problems, i.e. the appearance of the nonphysical late time oscillations in the numerical results [7]. The origin of these oscillations is usually discretization of the second-order space-time differential operator [8, 9].

Direct time-domain analysis of the thin wire structures can be carried out by using different methods [10, 11], and the Galerkin–Bubnov Indirect Boundary Element Method (GB-IBEM) [11], which, although complicated, has proven to provide.

Original space-time Hallen integral equation formulation [11, 12] has been developed for a single wire located horizontally above real ground, and the influence of the ground has been taken into account by means of the space-time reflection coefficient [13] obtained by using the inverse Laplace transform of Fresnel reflection coefficient. Direct numerical solution of space-time integral equation is efficient for the cases where the finite ground conductivity could be

Deterministic and Stochastic Modeling in Computational Electromagnetics: Integral and Differential Equation Approaches, First Edition. Dragan Poljak and Anna Šušnjara.
© 2024 The Institute of Electrical and Electronics Engineers, Inc.
Published 2024 by John Wiley & Sons, Inc.

ignored, i.e. for the case of dielectric half-space, or for the case of perfect (PEC) ground. Namely, by adopting these approximations, the time dependence in the reflection coefficient function vanishes, and the resulting matrix system is significantly simplified leading to attractive possibility of separating spatial from the temporal part [6], which enables one to efficiently perform marching-on-in-time (MoT) scheme for the solution of space-time integral equation.

However, for the cases of finite conductivities, the space-time dependent reflection coefficient is required yielding to the tedious evaluation of certain convolution integrals, more complicated integral equation formulation, and tremendous increase in computational cost.

This chapter deals with the analysis of transient behavior of thin wires featuring the use of several TD energy measures, such as specific absorption (SA), measures for the energy stored in electric and magnetic fields, and root mean square (rms) value of transient response. First, the basic equations for TD analysis of thin wires are outlined and then the TD energy measures are discussed. Numerical procedures are outlined here, while full mathematical details can be found elsewhere, e.g. in [11, 12].

6.1 Single Wire Above a Lossy Ground

The geometry of interest pertaining to straight wire of length L and radius a horizontally located at height h above a lossy ground with relative permittivity ε_r and conductivity σ is shown in Fig. 6.1.

Derivation of space-time Hallen integral equation for a single wire above a lossy ground is available elsewhere, e.g. in [11]. The Hallen equation for the unknown current $I(x, t)$ induced along the horizontal wire is given by

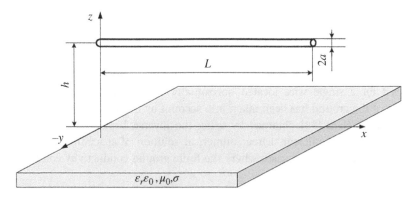

Figure 6.1 Geometry of the problem.

$$\int_0^L \frac{{}^L I\left(x', t - \frac{R}{c}\right)}{4\pi R} dx' - \int_{-\infty}^t \int_0^L r(\theta, \tau) \frac{I\left(x', t - \frac{R^*}{c} - \tau\right)}{4\pi R^*} dx' d\tau$$

$$= \frac{1}{2Z_0} \int_0^L E_x^{inc}\left(x', t - \frac{|x - x'|}{c}\right) dx' + F_0\left(t - \frac{x}{c}\right) + F_L\left(t - \frac{L - x}{c}\right) \tag{6.1}$$

where distance from the source (and its image) axis to the observation point at the wire surface R and R^* are given by

$$R = \sqrt{(x - x')^2 + a^2} \; ; \; R^* = \sqrt{(x - x')^2 + 4h^2} \tag{6.2}$$

Note that E_x^{inc} is the tangential component of the incident electric field, while F_0 and F_L are unknown functions taking into account reflection at wire-free ends defined via auxiliary functions K_i, K_L as follows [11, 14]:

$$F_0(t) = \sum_{n=0}^{\infty} K_0\left(t - \frac{2nL}{c}\right) - \sum_{n=0}^{\infty} K_L\left(t - \frac{(2n + 1)L}{c}\right) \tag{6.3a}$$

$$F_L(t) = \sum_{n=0}^{\infty} K_L\left(t - \frac{2nL}{c}\right) - \sum_{n=0}^{\infty} K_0\left(t - \frac{(2n + 1)L}{c}\right) \tag{6.3b}$$

with auxiliary functions K_0, K_L defined as follows [11]:

$$K_0(t) = \int_0^L \frac{{}^L I\left(x', t - \frac{R_0}{c}\right)}{4\pi R_0} dx' - \int_{-\infty}^t \int_0^L r(\theta, \tau) \frac{I\left(x', t - \frac{R_0^*}{c} - \tau\right)}{4\pi R_0^*} dx' d\tau$$

$$- \frac{1}{2Z_0} \int_0^L E_x^{inc}\left(x', t - \frac{x'}{c}\right) dx'$$

$$\tag{6.4a}$$

$$K_L(t) = \int_0^L \frac{{}^L I\left(x', t - \frac{R_L}{c}\right)}{4\pi R_L} dx' - \int_{-\infty}^t \int_0^L r(\theta, \tau) \frac{I\left(x', t - \frac{R_L^*}{c} - \tau\right)}{4\pi R_L^*} dx' d\tau$$

$$- \frac{1}{2Z_0} \int_0^L E_x^{inc}\left(x', t - \frac{L - x'}{c}\right) dx'$$

$$\tag{6.4b}$$

where R_0^* and R_L^* are the distances between source point x' on the image wire and observation points on the wire ends:

$$R_0^* = R^*|_{x=0} = \sqrt{x'^2 + 4h^2} \tag{6.5a}$$

$$R_L^* = R^*|_{x=L} = \sqrt{(L-x')^2 + 4h^2} \tag{6.5b}$$

Unknown functions F_0 and F_L can be determined by assuming the zero current at the wire ends [11, 14].

It is worth noting that the incident field is due to the voltage source in the case of the antenna mode, while if scattering mode is of interest, the incident field is replaced by the excitation field being composed of the incident field and the field reflected from the two-media interface.

In particular, if the geometry of interest pertains to the antenna (radiation) mode, i.e. to the horizontal center-fed thin wire antenna, shown in Fig. 6.2, then integral Eq. (6.1) is adapted by expressing the incident field via the equivalent voltage source.

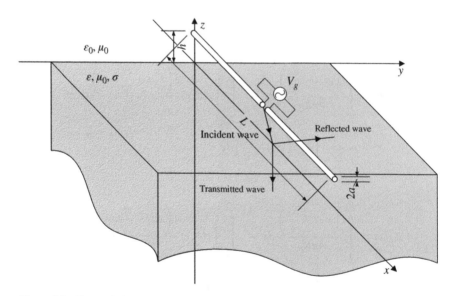

Figure 6.2 Transmitting dipole antenna.

Thus, in the case of a transmitting dipole-antenna, the voltage generator, which differs from zero only in the feed-gap area: $x_g - \Delta l_g < x < x_g + \Delta l_g$ and the incident field can be written in terms of Dirac impulse notation

$$E_x^{inc}(x,t) = V_g(t)\delta(x - x_g) \tag{6.6}$$

where V_g denotes the voltage generator and $x_g = L/2$.

For sufficiently small gap dimensions (delta-function generator), it follows

$$\int_{x_g - \Delta l_g}^{x_g + \Delta l_g} E_x^{inc}\left(x', t - \frac{|x - x'|}{c}\right) dx' = V_g\left(x', t - \frac{|x - x_g|}{c}\right) \tag{6.7}$$

and the resulting Hallen integral equation for the radiation mode is of the form

$$\int_0^L \frac{I(x', t - R_a/c)}{4\pi R_a} dx' - \int_{-\infty}^t \int_0^L r(\theta, \tau) \frac{I(x', t - R_d{}^*/c - \tau)}{4\pi R_d{}^*} dx' d\tau$$
$$= F_0\left(t - \frac{x}{c}\right) + F_L\left(t - \frac{L-x}{c}\right) + \frac{1}{2Z_0} V_g\left(x', t - \frac{|x - x_g|}{c}\right) \tag{6.8}$$

while (6.4a) and (6.4b) become:

$$K_0(t) = \int_0^L \frac{I(x', t - R_{0a}/c)}{4\pi R_{0a}} dx' - \int_{-\infty}^t \int_0^L r(\theta', \tau) \frac{I(x', t - R_{0d}^*/c - \tau)}{4\pi R_{0d}^*} dx' d\tau$$
$$- \frac{1}{2Z_0} V_g\left(x', t - \frac{|x - x_g|}{c}\right)$$
$$\tag{6.9}$$

$$K_L(t) = \int_0^L \frac{I(x', t - R_{La}/c)}{4\pi R_{La}} dx' - \int_{-\infty}^t \int_0^L r(\theta', \tau) \frac{I(x', t - R_{Ld}^*/c - \tau)}{4\pi R_{Ld}^*} dx' d\tau$$
$$- \frac{1}{2Z_0} V_g\left(x', t - \frac{|x - x_g|}{c}\right)$$
$$\tag{6.10}$$

while R_{0a} and R_{La} are the distances from the wire ends to the source point, and R_{0d}^*, R_{Ld}^* are the distances from the image wire ends to the image source point.

Space-time integral Eqs. (6.1) and (6.8), respectively, can be solved by forcing the zero current at the free ends of the wire and with the initial conditions requiring the wire not to be excited before the certain instant $t = t_0$.

Combining (6.1) to (6.5a) and (6.5b), one obtains

$$
\int_0^L \frac{I\left(x',t-\dfrac{R}{c}\right)}{4\pi R}dx' - \int_{-\infty}^{t}\int_0^L r(\theta,\tau)\frac{I\left(x',t-\dfrac{R^*}{c}-\tau\right)}{4\pi R^*}dx'\,d\tau = \frac{1}{2Z_0}\int_0^L E_x^{inc}\left(x',t-\frac{|x-x'|}{c}\right)dx'
$$

$$
+\int_0^L\sum_{n=0}^{\infty}\frac{I\left(x',t-\dfrac{R_0}{c}-\dfrac{2nL}{c}-\dfrac{x}{c}\right)}{4\pi R_0}dx'
$$

$$
-\int_{-\infty}^{t}\int_0^L\sum_{n=0}^{\infty}r(\theta,\tau)\frac{I\left(x',t-\dfrac{R_0^*}{c}-\dfrac{2nL}{c}-\dfrac{x}{c}\right)}{4\pi R_0^*}dx'\,d\tau
$$

$$
+\int_{-\infty}^{t}\int_0^L\sum_{n=0}^{\infty}r(\theta,\tau)\frac{I\left(x',t-\dfrac{R_L^*}{c}-\dfrac{(2n+1)L}{c}-\dfrac{x}{c}\right)}{4\pi R_L^*}dx'\,d\tau
$$

$$
+\frac{1}{2Z_0}\int_0^L\sum_{n=0}^{\infty}E_x^{inc}\left(x',t-\frac{L-x'}{c}-\frac{(2n+1)L}{c}-\frac{x}{c}\right)dx'
$$

$$
+\int_0^L\sum_{n=0}^{\infty}\frac{I\left(x',t-\dfrac{R_L}{c}-\dfrac{2nL}{c}-\dfrac{L-x}{c}\right)}{4\pi R_L}dx'
$$

$$
-\int_{-\infty}^{t}\int_0^L\sum_{n=0}^{\infty}r(\theta,\tau)\frac{I\left(x',t-\dfrac{R_L^*}{c}-\dfrac{2nL}{c}-\dfrac{L-x}{c}\right)}{4\pi R_L^*}dx'\,d\tau
$$

$$
-\frac{1}{2Z_0}\int_0^L\sum_{n=0}^{\infty}E_x^{inc}\left(x',t-\frac{L-x'}{c}-\frac{2nL}{c}-\frac{L-x}{c}\right)dx'
$$

$$
-\int_0^L\sum_{n=0}^{\infty}\frac{I\left(x',t-\dfrac{R_0}{c}-\dfrac{(2n+1)L}{c}-\dfrac{L-x}{c}\right)}{4\pi R_0}dx'
$$

$$
+\int_{-\infty}^{t}\int_0^L\sum_{n=0}^{\infty}r(\theta,\tau)\frac{I\left(x',t-\dfrac{R_0^*}{c}-\dfrac{(2n+1)L}{c}-\dfrac{L-x}{c}\right)}{4\pi R_0^*}dx'\,d\tau
$$

$$
+\frac{1}{2Z_0}\int_0^L\sum_{n=0}^{\infty}E_x^{inc}\left(x',t-\frac{x'}{c}-\frac{(2n+1)L}{c}-\frac{L-x}{c}\right)dx'
$$

(6.11)

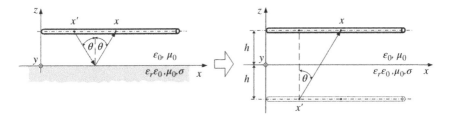

Figure 6.3 Source and the image wire.

Comparing (6.11) to (6.1), it can be seen that (6.11) now contains the current along the wire as the only unknown.

The reflecting effect of the ground is taken into account via the image theory, where the finitely conducting ground is replaced by the image wire, as depicted in Fig. 6.3.

The reflected part of the radiated/scattered electric field from the imperfect ground is taken into account via the space-time dependent reflection coefficient function for TM-polarization, proposed in [13] and given by

$$r(\theta, t) = K\delta(t) + \frac{4\beta}{1-\beta^2} \frac{e^{-at}}{t} \sum_{n=1}^{\infty} (-1)^{n+1} nK^n I_n(at) \tag{6.12}$$

where:

$$\tau = \frac{\sigma}{\varepsilon_0 \varepsilon_r}; \quad \beta = \frac{\sqrt{\varepsilon_r - \sin^2\theta}}{\varepsilon_r \cos\theta}; \quad \gamma = \frac{\tau}{1 - \frac{\sin^2\theta}{\varepsilon_r}}$$

$$\theta = arctg\frac{|x-x'|}{2h}; \quad K = \frac{1-\beta}{1+\beta}; \quad \alpha = \frac{\tau}{2} \tag{6.13}$$

and I_n is the modified Bessel function of the first order, n-th degree.

The main problem in the formulation (6.1) to (6.13) is the treatment of convolution integral in (6.1). For the case of very low, or rather high conductivity, one can get rid of the convolution integral, while this simplification is not possible in scenarios with finite ground conductivity.

6.1.1 Case of Perfectly Conducting (PEC) Ground and Dielectric Half-Space

The problem of solving the tedious convolution required by second term of the left-hand side in (6.1) can be avoided if the ground conductivity is high enough, i.e. the ground can be treated as perfectly conducting (PEC), i.e. $\sigma \to \infty$ when the reflection coefficient function simply becomes 1. On the contrary, for negligible values of conductivity, the ground can be treated as a pure dielectric medium, i.e. $\sigma \to 0$ and the reflection coefficient function becomes only spatially dependent, thus simplifying into following expression

$$r(\theta, t) = \frac{1-\beta}{1+\beta} \delta(t) \tag{6.14}$$

By adopting these approximations, the convolution integral (6.1) vanishes, and integral Eq. (6.1) simplifies into:

$$\int_0^L \frac{I\left(x', t - \frac{R}{c}\right)}{4\pi R} dx' - \int_0^L r(\theta) \frac{I\left(x', t - \frac{R^*}{c}\right)}{4\pi R^*} dx' = \frac{1}{2Z_0} \int_0^L E_x^{inc}\left(x', t - \frac{|x - x'|}{c}\right) dx'$$

$$+ F_0\left(t - \frac{x}{c}\right) + F_L\left(t - \frac{L-x}{c}\right)$$

$$\tag{6.15}$$

where

$$r(\theta) = \frac{1-\beta}{1+\beta} \tag{6.16}$$

More details in this formulation could be found elsewhere, e.g. in [11, 12].

6.1.2 Modified Reflection Coefficient for the Case of an Imperfect Ground

The use of reflection coefficient (6.12) proposed in [13] requires unacceptably long computational time [15]. The TD reflection coefficient is usually derived from the Fresnel reflection coefficient function using some transformation technique. As an exact, closed form is not achievable, the use of different additional approximations is necessary. Depending on the frequency of TD transformation technique used, as well as type of approximation adopted, various forms of TD reflection coefficients varying in both accuracy and analytical complexity can be derived.

Transformation methods used are typically based on analytical Laplace [13, 16] or Fourier [17, 18] transform and implemented approximations often determine the accuracy of the function. If relatively rough approximations are adopted, the resulting function will typically be simpler and less accurate or appropriate only for a narrow range of problems [19–22]. Conversely, the lesser degree of approximation will typically yield a rather complex but more accurate form of TD reflection coefficient variant [13, 17, 18]. Thus, the reflection coefficient presented in [21] is intended to be simple and computationally efficient, but at the same time, sufficiently accurate.

If lossy ground is modeled, the reflection coefficient function becomes space-time dependent, and the resulting convolution integrals have to be numerically evaluated and often exploded in unacceptably long computational times. In that case, it is of outmost importance to use the reflection coefficient function which should be as simple as possible with regards to evaluation of the convolution integrals.

An efficient approach to derive the TD reflection coefficient via Gaver-Stehfest algorithm (one of the simplest algorithms for numerical Laplace transform, shown to be very well suited for application to transformation of the Fresnel reflection coefficient from frequency to TD) [23] has been presented in [24]. The transformation results in a rather simple expression, avoiding the use of the Bessel functions, which may be convenient for numerical evaluation, as well as additional analytical manipulation. At the same time, the accuracy of the resulting RC function is not significantly reduced [24].

In [23], first a simplified FD reflection coefficient for a TM polarization is derived from the Fresnel reflection coefficient which is then transformed to the TD using the Gaver-Stehfest algorithm. Accuracy of the resulting function has been also discussed.

The procedure described in this chapter is based on the work presented in [13, 16], and [24].

If a plane wave is incident from the free space to the half plane with conductivity σ and relative permittivity ε_r, as depicted in Fig. 6.4, then the Fresnel reflection coefficient for TM polarization is given by

$$R(\theta, s) = \frac{\varepsilon_r\left(1 + \dfrac{\sigma}{s\varepsilon_0\varepsilon_r}\right)\cos\theta - \sqrt{\varepsilon_r\left(1 + \dfrac{\sigma}{s\varepsilon_0\varepsilon_r}\right) - \sin^2\theta}}{\varepsilon_r\left(1 + \dfrac{\sigma}{s\varepsilon_0\varepsilon_r}\right)\cos\theta + \sqrt{\varepsilon_r\left(1 + \dfrac{\sigma}{s\varepsilon_0\varepsilon_r}\right) - \sin^2\theta}} \tag{6.17}$$

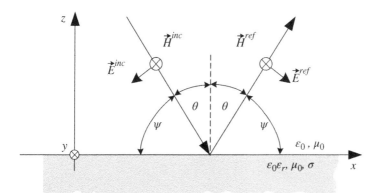

Figure 6.4 TM plane wave incident on an interface.

where the complex frequency is

$$s = j\omega = j2\pi f \tag{6.18}$$

Now (6.17) can be rewritten as follows

$$R(\theta, s) = \frac{(s+\tau)\dfrac{\varepsilon_r \cos\theta}{s} - \sqrt{\varepsilon_r\left(1 + \dfrac{\tau}{s}\right) - \sin^2\theta}}{(s+\tau)\dfrac{\varepsilon_r \cos\theta}{s} + \sqrt{\varepsilon_r\left(1 + \dfrac{\tau}{s}\right) - \sin^2\theta}} \tag{6.19}$$

where the time constant is

$$\tau = \frac{\sigma}{\varepsilon_0 \varepsilon_r} \tag{6.20}$$

Furthermore, it follows

$$R(\theta, s) = \frac{(s+\tau)\dfrac{\varepsilon_r \cos\theta}{s} - \sqrt{\varepsilon_r - \sin^2\theta}\sqrt{1 + \dfrac{\gamma}{s}}}{(s+\tau)\dfrac{\varepsilon_r \cos\theta}{s} + \sqrt{\varepsilon_r - \sin^2\theta}\sqrt{1 + \dfrac{\gamma}{s}}} \tag{6.21}$$

where

$$\gamma = \frac{\tau}{1 - \dfrac{\sin^2\theta}{\varepsilon_r}} \tag{6.22}$$

and one obtains

$$R(\theta, s) = \frac{s + \tau - \beta\sqrt{s(s+\gamma)}}{s + \tau + \beta\sqrt{s(s+\gamma)}} \tag{6.23}$$

where

$$\beta = \frac{\sqrt{\varepsilon_r - \sin^2\theta}}{\varepsilon_r \cos\theta} \tag{6.24}$$

In case the ε_r is not too small, and the incident angle θ not too large, i.e. approximately:

$$\varepsilon_r \geq 10, \quad \theta < 80° \tag{6.25}$$

then it follows

$$\frac{\sin^2\theta}{\varepsilon_r} \ll 1 \tag{6.26}$$

and one may assume

$$\gamma \cong \tau \tag{6.27}$$

Now (6.23) simplifies into

$$R(\theta, s) = \frac{\sqrt{s+\tau} - \beta\sqrt{s}}{\sqrt{s+\tau} + \beta\sqrt{s}} \tag{6.28}$$

Introducing the auxiliary function:

$$A(s) = \frac{1}{\sqrt{1 + \dfrac{\tau}{s}}} \tag{6.29}$$

(6.28) can be written as

$$R(s, \theta) = \frac{1 - \beta A(s)}{1 + \beta A(s)} \tag{6.30}$$

Now, separating the frequency and space-dependent variables, (6.30) can be written in following general form:

$$R(\theta, s) = R_0(\theta) + R''(\theta, s) \tag{6.31}$$

where $R_0(\theta)$ is defined as

$$R_0(\theta) = \frac{1 - \beta}{1 + \beta} \tag{6.32}$$

and space-time dependent function $R''(\theta)$ is given by

$$R''(\theta, s) = \frac{2\beta}{1 + \beta} \frac{1 - A(s)}{1 + \beta A(s)} \tag{6.33}$$

Next step is a transformation of (6.33) into the TD, i.e. one has

$$r(\theta, t) = R_0(\theta)\delta(t) + L^{-1}[R''(\theta, s)] = r'(\theta, t) + r''(\theta, t) \tag{6.34}$$

Laplace transform of $r''(\theta, t)$ can be obtained using numerical inverse Laplace transform. One of the simplest numerical inversion methods, Gaver-Stehfest algorithm is used [25].

If $F(s)$ is a general FD function, then, in accordance with the Gaver-Stehfest algorithm, its TD transform $f(t)$ is given by [23, 25]

$$f(t) = \frac{\ln 2}{t} \sum_{n=1}^{N} V_n F\left(\frac{n \ln 2}{t}\right) \tag{6.35}$$

where coefficients $V_1, V_2, \ldots, V_n$ are

$$V_n = (-1)^{\frac{N}{2}+1} \sum_{k=\frac{n+1}{2}}^{\min\left(n, \frac{N}{2}\right)} \frac{k^{\frac{N}{2}+1}(2k)!}{\left(\dfrac{N}{2} - k\right)! k!} \tag{6.36}$$

Although (6.34) is not very simple, the coefficients are calculated only once.

Therefore, the complexity of the coefficient calculation is effectively irrelevant with regard to the computational efficiency of the method using the reflection coefficient function. The number of summands N must be chosen as an even number approximately between 8 and 16.

Now, one has

$$r''(\theta, t) = \frac{2\ln 2}{t} \frac{\beta}{1+\beta} \sum_{n_1=1}^{N_1} V_{n_1} \frac{1 - A\left(\frac{n_1 \ln 2}{t}\right)}{1 + \beta A\left(\frac{n_1 \ln 2}{t}\right)} \tag{6.37}$$

which can be rewritten as follows:

$$r''(\theta, t) = \frac{2\ln 2}{t} \frac{\beta}{1+\beta} \sum_{n=1}^{N} V_n \frac{\sqrt{1 + \frac{K}{n}t} - 1}{\sqrt{1 + \frac{K}{n}t} + \beta} \tag{6.38}$$

where

$$K = \frac{\sigma}{\ln 2 \, \varepsilon_0 \varepsilon_r} \tag{6.39}$$

Finally, (6.34) now can be written as

$$r(\theta, t) = \frac{1-\beta}{1+\beta}\delta(t) + \frac{2\ln 2}{t} \frac{\beta}{1+\beta} \sum_{n=1}^{N} V_n \frac{\sqrt{1 + \frac{K}{n}t} - 1}{\sqrt{1 + \frac{K}{n}t} + \beta} \tag{6.40}$$

Generally, in the Gaver-Stehfest algorithm, the value of N is recommended to be chosen as an even number so that $8 < N < 16$ [25]: However, in this particular application, satisfactory accuracy can be achieved for $2 < N < 8$, and higher values of N do not appreciably contribute to the increase of the result accuracy [24].

The reflection coefficient is often subjected to both spatial and temporal integration. Thus, it can be rather convenient to rewrite the function so that the space and time integrals can be solved independently.

Applying the binomial expansion to the summand denominator in Eq. (6.38) leads to the following expression

$$\frac{1}{1 + \beta A\left(\frac{n_1 \ln 2}{t}\right)} = \sum_{m=0}^{M} \left[-\beta A\left(\frac{n_1 \ln 2}{t}\right)\right]^m \tag{6.41}$$

Furthermore, a condition

$$\beta A\left(\frac{n_1 \ln 2}{t}\right) < 1 \tag{6.42}$$

should be satisfied for $t = 0$, and therefore all other time instants, and it follows

$$\sin \theta < \sqrt{\frac{\varepsilon_r}{\varepsilon_r + 1}} \tag{6.43}$$

Condition (6.43) is in alignment with approximation condition given by (6.26). Taking (6.41) into account, (6.37) becomes

$$r''(\theta, t) = D \sum_{n=1}^{N} V_n \left[1 - A \left(\frac{n \ln 2}{t} \right) \right] \sum_{m=0}^{M} \left[-\beta A \left(\frac{n \ln 2}{t} \right) \right]^m \tag{6.44}$$

where D is

$$D = \frac{2 \ln 2}{t} \frac{\beta}{1 + \beta} \tag{6.45}$$

Finally, inserting (6.44) into (6.34) gives

$$r(\theta, t) = r'(\theta, t) + 2 \ln 2 \sum_{m=0}^{M} r_m^*(\theta) \sum_{n=1}^{N} V_n r_{n,m}^{**}(t) \tag{6.46}$$

where

$$r'(\theta, t) = \frac{1 - \beta}{1 + \beta} \delta(t) \tag{6.47}$$

$$r_m^*(\theta) = -\frac{(-\beta)^{m+1}}{1 + \beta} \tag{6.48}$$

$$r_{n,m}^{**}(t) = \frac{\sqrt{1 + \dfrac{K}{n} t} - 1}{t \left(\sqrt{1 + \dfrac{K}{n} t} \right)^{m+1}} \tag{6.49}$$

It is worth emphasizing that the resulting space-time dependent reflection coefficient (6.46) does not require the use of Bessel functions, and can be used either in the form (6.40), or somewhat less accurate (6.46). However, if (6.46) is used, the reflection coefficient is given via linear combination of only space and only time-dependent functions, which is very useful when such reflection coefficient is used within convolution integral (6.1). Using this form of function, on the other hand, further limits the parameters of the configuration, according to a convergence condition (6.43) to obtain valid results.

The validity of TD reflection coefficient (6.46) is numerically verified in [24] via example used in [13]. The plane wave is incident on the ground with the angle $\theta = 45°$, where the ground parameters are: $\varepsilon_r = 10$, $\sigma = 10 \, \text{mS/m}$.

The waveform of the incident electric field is given by double-exponential function

$$E^{inc}(t) = E_0\left(e^{-\alpha t} - e^{-\beta t}\right) \tag{6.50}$$

where $E_0 = 52.5$ kV/m, $\alpha = 4 \cdot 10^6\, s^{-1}$, $\beta = 4.76 \cdot 10^8\, s^{-1}$.

The reflected field can be calculated using space-time reflection coefficient via following convolution integral [13]

$$E^{ref}(t) = \int_0^t r(\theta, \tau)E^{inc}(t - \tau)d\tau \tag{6.51}$$

where $r(\theta, t)$ is given by either (6.40) or (6.46).

Accuracy of the proposed reflection coefficient (6.44) can be illustrated by comparing the reflected field obtained by numerically evaluating convolution integral (6.51) with the reference waveform obtained from the FD response via IFFT. Fig. 6.5 depicts the comparison of the reflected field calculated for several values of N.

As expected, a higher number of summands N leads to more accurate results. It can be seen that appreciable difference between the results obtained via space-time

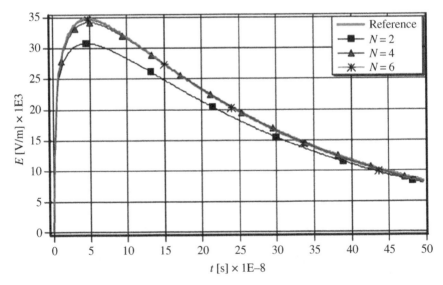

Figure 6.5 Reflected electric field for different values of N compared with referent waveform obtained via IFFT. *Source:* Adapted from Antonijevic and Poljak [24].

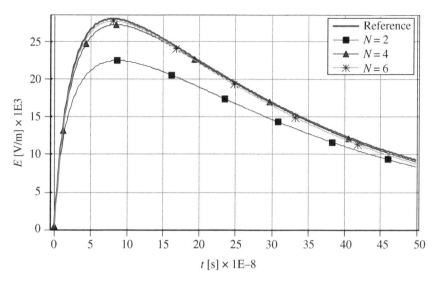

Figure 6.6 Reflected electric field for different values of N compared with referent waveform obtained via IFFT for $\theta = 70°$. *Source:* Adapted from Antonijevic and Poljak [24].

reflection coefficient (6.46) and the reference results, obtained via IFFT, appears for very low value of N, i.e. for $N = 2$. For the next even number, $N = 4$, the accuracy is significantly improved.

Reflected field $\theta = 70°$, obtained in this case using (6.40), is shown in Fig. 6.6.

Finally, decrease in accuracy as the function of the angle of incidence is presented in Fig. 6.7.

It is therefore obvious that the use of (6.40) is limited to angles of incidence less than approximately 80°. However, this limitation is not expected to be of considerable importance for practical applications because the reflection coefficient is typically intended for use in reflection coefficient approximation (RCA) [26, 27]. Since RCA assumes that the observed structure is far enough from the interface, and since the incident wave can be considered a plane wave, the angles of incidence that are usually used in RCA are not expected to be too large.

Therefore, the resulting reflection coefficient proposed in [24] and presented in this chapter is rather simple, does not require the evaluation of Bessel functions, and achieves a good accuracy within a relatively few terms in series used, making it convenient for efficient computation. If an additional approximation is introduced, by simple binomial expansion, the resulting reflection coefficient can be represented in terms of a linear combination of only space and only time-dependent functions, which might be useful in some applications.

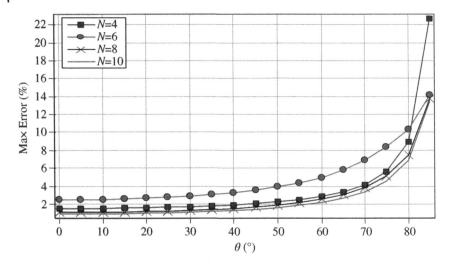

Figure 6.7 Maximum error (in percentage) between reference results and the results obtained via (6.38) for various values of N and θ. *Source:* Adapted from Antonijevic and Poljak [24].

6.2 Numerical Solution of Hallen Equation via the Galerkin–Bubnov Indirect Boundary Element Method (GB-IBEM)

The Hallen equation (6.1) is numerically solved via the Galerkin–Bubnov Indirect Boundary Element Method (GB-IBEM) [11]. An attractive benefit of using the Hallen equation instead of the Pocklington equation [12] is to avoid the second-order space-time differential operator. It was shown that discretization of this operator within a given numerical scheme may be the origin of certain numerical instabilities. The use of the Hallen equation implies that no derivatives exist, which proved to yield highly stable and rather accurate solutions, as suggested by many numerical tests on various configurations that have been carried out over the years using this method. On the other hand, the cost of this type of formulation is significantly increased complexity, which evidently is the most prominent drawback of the GB-IBEM.

What follows is an outline of GB-IBEM. Full mathematical details of the method could be found elsewhere, e.g. in [11].

Space-time discretization is carried out using the weighted residual approach [11].

First, space discretization is applied to Hallen equation (6.1), i.e. the numerical procedure starts with a local approximation for the current on wire segment, given by

$$I(x',t') = \{f\}^T\{I\} \tag{6.52}$$

where $\{I\}$ denotes space-time dependent solution vector, and $\{f\}^T$ denotes vector containing spatial shape functions:

$$f_r(x') = \frac{x_{r+1}-x'}{x_{r+1}-x_r}, f_{r+1}(x') = \frac{x'-x_r}{x_{r+1}-x_r} \tag{6.53}$$

where r is the global node index of the observed wire segment.

Having performed some mathematical manipulations, the following local system of linear equations is obtained

$$[A]\{I\}_i\Big|_{t-\frac{R}{c}} - [A^*]\{I\}_i\Big|_{t-\frac{R^*}{c}} - \{\hat{A}\}\Big|_{t-\frac{R^*}{c}} = [B]\{E\}\Big|_{t-\frac{|x-x'|}{c}}$$

$$+ [C]\left\{\sum_{n=0}^{\infty} I^n\right\}\Bigg|_{i\Big|_{t-\frac{R_0}{c}-\frac{2nL}{c}-\frac{x}{c}}} \quad -[C^*]\left\{\sum_{n=0}^{\infty} I^n\right\}\Bigg|_{i\Big|_{t-\frac{R_0^*}{c}-\frac{2nL}{c}-\frac{x}{c}}}$$

$$- [B]\left\{\sum_{n=0}^{\infty} E^n\right\}\Bigg|_{t-\frac{x'}{c}-\frac{2nL}{c}-\frac{x}{c}} \quad -[D]\left\{\sum_{n=0}^{\infty} I^n\right\}\Bigg|_{i\Big|_{t-\frac{R_L}{c}-\frac{(2n+1)L}{c}-\frac{x}{c}}}$$

$$+ [D^*]\left\{\sum_{n=0}^{\infty} I^n\right\}\Bigg|_{i\Big|_{t-\frac{R_L^*}{c}-\frac{(2n+1)L}{c}-\frac{x}{c}}} \quad +[B]\left\{\sum_{n=0}^{\infty} E^n\right\}\Bigg|_{t-\frac{L-x'}{c}-\frac{(2n+1)L}{c}-\frac{x}{c}}$$

$$+ [D]\left\{\sum_{n=0}^{\infty} I^n\right\}\Bigg|_{i\Big|_{t-\frac{R_L}{c}-\frac{2nL}{c}-\frac{L-x}{c}}} \quad -[D^*]\left\{\sum_{n=0}^{\infty} I^n\right\}\Bigg|_{i\Big|_{t-\frac{R_L^*}{c}-\frac{2nL}{c}-\frac{L-x}{c}}}$$

$$- [B]\left\{\sum_{n=0}^{\infty} E^n\right\}\Bigg|_{t-\frac{L-x'}{c}-\frac{2nL}{c}-\frac{L-x}{c}} \quad -[C]\left\{\sum_{n=0}^{\infty} I^n\right\}\Bigg|_{i\Big|_{t-\frac{R_0}{c}-\frac{(2n+1)L}{c}-\frac{L-x}{c}}}$$

$$+ [C^*]\left\{\sum_{n=0}^{\infty} I^n\right\}\Bigg|_{i\Big|_{t-\frac{R_0^*}{c}-\frac{(2n+1)L}{c}-\frac{L-x}{c}}} \quad +[B]\left\{\sum_{n=0}^{\infty} E^n\right\}\Bigg|_{t-\frac{x'}{c}-\frac{(2n+1)L}{c}-\frac{L-x}{c}}$$

$$- \left\{\sum_{n=0}^{\infty} \hat{C}^n\right\}\Bigg|_{t-\frac{R_0^*}{c}-\frac{2nL}{c}-\frac{x}{c}} \quad +\left\{\sum_{n=0}^{\infty} \hat{D}^n\right\}\Bigg|_{t-\frac{R_L^*}{c}-\frac{(2n+1)L}{c}-\frac{x}{c}}$$

$$- \left\{\sum_{n=0}^{\infty} \hat{C}^n\right\}\Bigg|_{t-\frac{R_0^*}{c}-\frac{(2n+1)L}{c}-\frac{L-x}{c}} \quad +\left\{\sum_{n=0}^{\infty} \hat{D}^n\right\}\Bigg|_{t-\frac{R_L^*}{c}-\frac{2nL}{c}-\frac{L-x}{c}}$$

$$\tag{6.54}$$

where space-dependent local matrices for observation element j and source element i are:

$$[A] = \int_{\Delta l_j} \int_{\Delta l_i} \{f\}_j \{f\}_i^T \frac{1}{4\pi R} dx' dx; \quad [B] = \frac{1}{2Z_0} \int_{\Delta l_j} \int_{\Delta l_i} \{f\}_j \{f\}_i^T dx' dx$$

$$[C] = \int_{\Delta l_j} \int_{\Delta l_i} \{f\}_j \{f\}_i^T \frac{1}{4\pi R_0} dx' dx; \quad [D] = \int_{\Delta l_j} \int_{\Delta l_i} \{f\}_j \{f\}_i^T \frac{1}{4\pi R_L} dx' dx$$

$$(6.55)$$

$$[A^*] = \int_{\Delta l_j} \int_{\Delta l_i} \{f\}_j \{f\}_i^T \frac{K}{4\pi R^*} dx' dx; \quad [C^*] = \int_{\Delta l_j} \int_{\Delta l_i} \{f\}_j \{f\}_i^T \frac{K}{4\pi R_0^*} dx' dx$$

$$[D^*] = \int_{\Delta l_j} \int_{\Delta l_i} \{f\}_j \{f\}_i^T \frac{K}{4\pi R_L^*} dx' dx.$$

Distance between observation (i-th) and source (j-th) point on the actual wire is R, distances between source point and wire ends are R_0 and R_L, while asterisked terms pertain to the distances for source points on the image wire. Shape function vector $\{f\}$ contains shape functions which are chosen as simple linear Lagrange polynomials, and K is defined in [15], and represents the value of reflection coefficient function if ground conductivity is neglected, i.e. if $r''(\theta, \tau) = 0$.

Local space and time-dependent vectors in [15] are:

$$\{\hat{A}\} = \int_0^{t - \frac{R^*}{c}} \int_{\Delta l_j} \int_{\Delta l_i} \{f\}_j \{f\}_i^T H_1 dx' dx \{I(\tau)\}_i d\tau$$

$$\{\hat{C}^n\} = \int_0^{t - \frac{R_0^*}{c} - \frac{2nL}{c} - \frac{x}{c}} \int_{\Delta l_j} \int_{\Delta l_i} \{f\}_j \{f\}_i^T H_2 dx' dx \{I(\tau)\}_i d\tau \qquad (6.56)$$

$$\{\hat{D}^n\} = \int_0^{t - \frac{R_L^*}{c} - \frac{(2n+1)L}{c} - \frac{x}{c}} \int_{\Delta l_j} \int_{\Delta l_i} \{f\}_j \{f\}_i^T H_3 dx' dx \{I(\tau)\}_i d\tau$$

where auxiliary functions H are given by:

$$H_1 = \frac{r''\left(\theta, t - \dfrac{R^*}{c} - \tau\right)}{4\pi R^*}$$

$$H_2 = \frac{r''\left(\theta, t - \dfrac{R_0^*}{c} - \dfrac{2nL}{c} - \dfrac{x}{c} - \tau\right)}{4\pi R_0^*}$$

$$H_3 = \frac{r''\left(\theta, t - \dfrac{R_L^*}{c} - \dfrac{(2n+1)L}{c} - \dfrac{x}{c} - \tau\right)}{4\pi R_0^*}$$

(6.57)

Assembling the local matrices and vectors into the global matrices and vectors results in the following general form of the global system [11, 12]

$$[A]\{I\}\Big|_{t-\frac{R}{c}} = \{g\}\Big|_{\substack{\text{previous time} \\ \text{instants}}} + \{\hat{g}\}\Big|_{\substack{\text{previous time} \\ \text{instants}}}$$

(6.58)

Performing the time sampling, and separating the members relating to the current at present instance, a recurrent formula, forming the basis of MoT procedure, is obtained

$$I_j\Big|_{t_k} = \frac{\displaystyle\sum_{i=1}^{N^s} a_{ji}I_j\Big|_{t_k - \frac{R}{c}} - g_j\Big|_{\substack{\text{previous time} \\ \text{instants}}} - \hat{g}_j\Big|_{\substack{\text{previous time} \\ \text{instants}}}}{a_{jj}} \; ; i \neq j.$$

(6.59)

Vector $\{\hat{g}\}$ is assembled from local vectors representing the time-dependent contribution of ground conductivity, and is time dependent. Thus, unlike $\{g\}$, it has to be re-evaluated for each time step during the marching-on-in-time (MoT) procedure, which significantly increases computational time of the overall procedure, when compared to cases where ground conductivity is neglected.

Furthermore, numerical evaluations of convolution integrals in (6.56) impose even greater computational burden since $r''(\theta, \tau)$ contains infinite sum of Bessel functions. This makes evaluation of the vector $\{\hat{g}\}$ drastically more complex than vector $\{g\}$.

It is worth noting [15] that the overall increase in computational time is several orders of magnitude, clearly rendering GB-IBEM practically infeasible in case a real ground conductivity is to be considered.

6.2.1 Computational Examples

First illustrative computational example deals with scattering from a horizontal thin wire above a lossy ground deals with wire of length $L = 1$ m, radius $a = 2$ mm, placed at certain height h above a lossy ($\varepsilon_r = 10$, $\sigma = 0.01$ S/m). The wire is illuminated by the normally incident tangential electromagnetic pulse (EMP) plane wave (6.50) with $E_0 = 1$ V/m, $\alpha = 4 \cdot 10^7$ s^{-1}, $\beta = 6 \cdot 10^8$ s^{-1}.

Figure 6.8 depicts the transient response of the center of the wire located at $h = 0.25$ m, $h = 0.5$ m and $h = 1$ m calculated via the improved TD version of the GB-IBEM. The obtained numerical results are compared to the results obtained by using NEC2 combined with IFFT.

A rather satisfactory agreement of the numerical results obtained via different approaches can be observed in Fig. 6.8. The influence of the height h variation on the transient response is obvious.

A trade-off between the dielectric and conducting half-space, respectively, regarding the calculation time is given in detail in [15].

Finally, Fig. 6.9 shows the transient current induced at the center of the wire for different ground conductivities if the wire is located at heights $h = 0.25$ m and $h = 0.5$ m.

The influence of the ground conductivity is obviously most visible in the range from about 0.1–1 S/m. Outside this range, approximating the ground as either pure dielectric or perfect conductor (PEC) yields results of plausible accuracy. Comparing the transient currents for different heights, it can be clearly observed that the greater the influence of the ground conductivity, the closer the wire is to the ground. Also, the early time response looks similar for all cases as the reflected wave has not reached the observation point for such early time intervals.

The last example in this chapter deals with antenna mode for the straight wire of length $L = 1$ m, radius $a = 2$ mm, horizontally located above a lossy ground ($\sigma = 1$ mS/m, $\varepsilon_r = 10$) at height $h = 0.8$ m. The antenna is excited in its center via the Gaussian pulse voltage source

$$V_g(t) = V_0 e^{-g^2(t-t_0)^2} \tag{6.60}$$

with $V_0 = 1$ V, $g = 2 \cdot 10^9$ s^{-1}, $t_0 = 2$ ns.

Figure 6.10 shows the transient current induced at the antenna driving point obtained via TD GB-IBEM approach.

The obtained results via GB-IBEM seem to be in good agreement with the results computed via NEC2/IFFT approach.

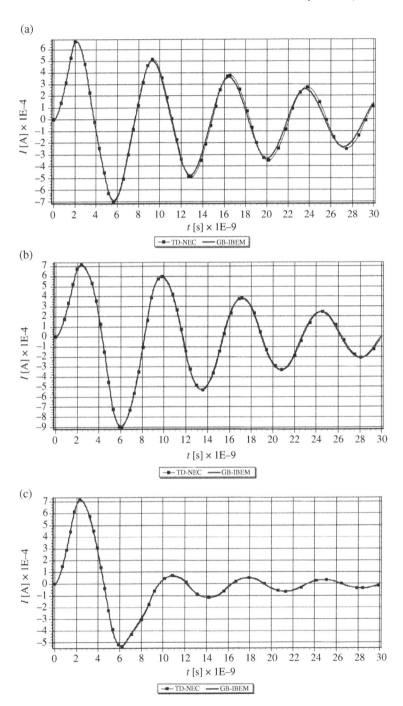

Figure 6.8 Transient current at the wire center (L = 1 m, a = 2 mm, σ = 10 mS/m, ε_r = 10). (a) h = 0.25 m. (b) h = 0.5 m. (c) h = 1 m.

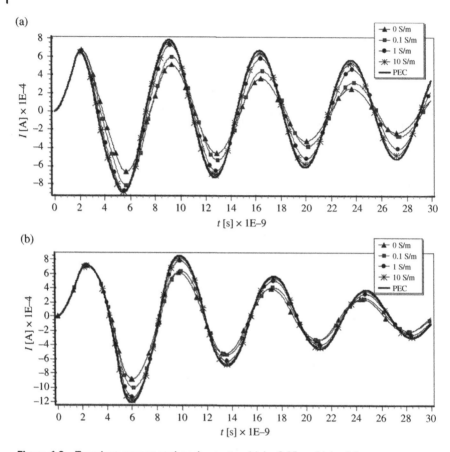

Figure 6.9 Transient current at the wire center. (a) $h = 0.25$ m. (b) $h = 0.5$ m.

6.3 Application to Ground-Penetrating Radar

GPR is a device using propagation of electromagnetic waves to probe the subsurface [28] with a number of applications in civil engineering, archeology, earth sciences, underground engineering, mine detection, cultural heritage, forensics, etc. One of the crucial parts of any GPR system is the antenna moving along the ground surface and transmitting/receiving electromagnetic waves with velocities determined by the medium properties. The waves propagating from the antenna are partly reflected from the ground, while a significant portion of the energy is emitted into a lossy lower half-space in the form of the transmitted wave. The waves scattered from the target in a lossy half-space are propagated back and

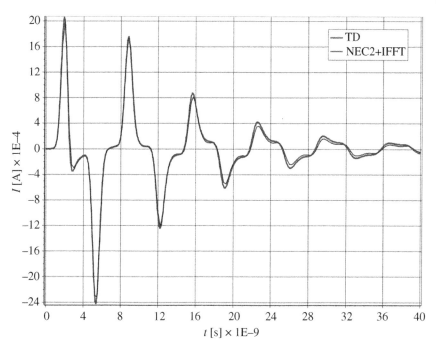

Figure 6.10 Transient current at the wire center (L = 1 m, a = 2 mm, h = 0.8 m, σ = 0.001 S/m, ε_r = 10).

detected by the GPR antenna. The antenna type depends on the particular application, but the most commonly used antennas are dipole antennas and bowtie antennas which can also be represented by an equivalent dipole antenna [28–30]. The high-frequency (HF) antennas are smaller in size and provide higher resolution but, at the same time, suffer from a lower penetration depth. On the other hand, the low-frequency (LF) antennas are physically larger and provide a deeper penetration into soil with a cost of resolution [28–31]. The knowledge of the energy transmitted into the lossy medium aims to improve an antenna design and to provide deeper insight into the understanding of the target-reflected waves.

Many numerical techniques in either frequency or TD, respectively, have been used to provide an accurate analysis of the transient GPR dipole radiation aiming to determine the behavior of the field reflected from the air–earth interface and the field transmitted into the ground.

Thus, the frequency domain (FD) analysis of GPR dipole antenna based on the Pocklington integro-differential equation approach has been reported elsewhere, e.g. in [11]. The main advantage of FD approach is relative simplicity of the formulation and consequently low computational cost if a study at single frequency is of interest.

Direct TD modeling, on the other hand, is convenient when broad frequency spectrum is considered, e.g. [31]. It is worth emphasizing that finite difference TD (FDTD) approach is the most commonly used technique for TD modeling [32–35]. Nevertheless, direct TD modeling based on the integral equation approach is much more convenient for wire configurations and the analysis of GPR antennas [36], while it represents appreciably less convenient tools where non-homogeneous half-space problems are of interest. A trade-off between the finite-difference (FDTD), finite-integration technique (FIT), and integral-equation (TDIE) approaches has been carried out in some papers. e.g. in [30].

Stochastic study modeling GPR antenna parameters as well as intrinsic uncertainties in electric properties of the soil coupling the stochastic collocation (SC) with the TD integral equation has been reported in [37] to demonstrate the statistical behavior of GPR antenna transient response.

The review of deterministic frequency and TD methods for the calculation of the fields reflected from the interface and of the fields propagated into lossy media due to horizontal GPR dipole antenna radiating over a lossy half-space has been given in [38].

This section deals with the analysis of reflected/transmitted fields due to the GPR dipole antenna (Fig. 6.11). The wire is located at height h above ground and is assumed to be perfectly conducting (PEC).

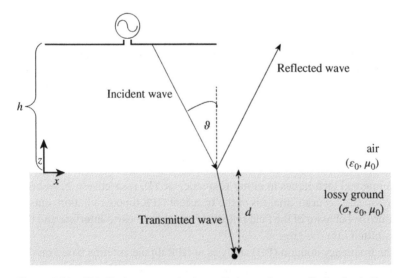

Figure 6.11 GPR dipole antenna horizontally located over a dissipative half-space.

Subsections to follow deal with the TD modeling of the transient field reflected from the interface, and the transient field transmitted into the lossy ground.

6.3.1 Transient Field due to Dipole Radiation Reflected from the Air–Earth Interface

This subsection deals with the direct TD calculation of a transient electric field in the upper half-space generated by a dipole antenna above a lossy ground [38, 39]. The transient current along the wire is governed by the Hallen integral Eq. (6.8) which is numerically solved by means of GB-IBEM. Once the current distribution along the dipole is known, the related radiated field is computed from certain integral formulas.

The electric field in the upper half-space radiated by a thin wire is given by following integral expression [38, 39]

$$E_x(r,t) = \frac{\mu_0}{4\pi} \left(\int_0^L \frac{\partial}{\partial t} \frac{I(x',t')}{R_1} dx' - \int_{-\infty}^t \Gamma_{ref}(\tau) \int_0^L \frac{\partial}{\partial t} \frac{I\left(x', t - \frac{R_1^*}{c} - \tau\right)}{R_1^*} dx' d\tau \right)$$

(6.61)

where R_1^* is

$$R_1^* = \sqrt{(x - x\prime)^2 + (2h + z)^2}$$

(6.62)

The reflecting properties of a conducting half-space are taken into account via the space-time reflection coefficient arising from the Modified Image Theory (MIT) [40, 41]

$$\Gamma_{ref}^{MIT}(t) = \frac{\tau_1}{\tau_2}\delta(t) + \frac{1}{\tau_2}\left(1 - \frac{\tau_1}{\tau_2}\right)e^{-\frac{t}{\tau_2}}$$

(6.63)

where

$$\tau_1 = \frac{\varepsilon_0(\varepsilon_r - 1)}{\sigma}$$

(6.64)

$$\tau_2 = \frac{\varepsilon_0(\varepsilon_r + 1)}{\sigma}$$

(6.65)

To simplify the computation procedure, the reflection coefficient is rewritten in a way to separate the dielectric from the conducting part, respectively

$$\Gamma_{MIT}^{ref}(t) = \Gamma_{ref,D}^{MIT} + \Gamma_{ref,V}^{MIT}$$

(6.66)

where:

$$\Gamma_{ref,D}^{MIT} = \; = \frac{\tau_1}{\tau_2}\delta(t) \tag{6.67}$$

$$\Gamma_{ref,V}^{MIT} = \frac{1}{\tau_2}\left(1 - \frac{\tau_1}{\tau_2}\right)e^{-\frac{t}{\tau_2}} \tag{6.68}$$

Now, the total field above a lossy medium can be written as follows

$$E_x^{tot}(r,t) = \frac{\mu_0}{4\pi}\int_0^L \frac{\partial}{\partial t}\frac{I(x',t')}{R_1}dx' - \frac{\mu_0}{4\pi}\int_0^L \frac{\tau_1}{\tau_2}\frac{\partial I\left(x',t-\frac{R_1^*}{c}-\tau\right)}{\partial t}\frac{1}{R_1^*}dx'$$

$$- \frac{\mu_0}{4\pi}\int_{-\infty}^t\int_0^L \frac{\partial}{\partial t}\Gamma_{MIT}^{ref,L}(t-\tau)\frac{I\left(x',t-\frac{R_1^*}{c}-\tau\right)}{R_1^*}dx'd\tau \tag{6.69}$$

A part of the reflection coefficient by which conducting properties of the soil are taken into account can be differentiated analytically, i.e. one has

$$\frac{\partial}{\partial t}\Gamma_{MIT}^{ref,V}(t-\tau) = \frac{\partial}{\partial t}\left[\frac{1}{\tau_2}\left(1 - \frac{\tau_1}{\tau_2}\right)e^{-\frac{t}{\tau_2}}\right] = -\frac{1}{\tau_2}\left(1 - \frac{\tau_1}{\tau_2}\right)e^{-\frac{t}{\tau_2}} \tag{6.70}$$

and the final expression for the total field in the air becomes

$$E_x^{tot}(r,t) = -\frac{\mu_0}{4\pi}\left[\int_0^L \frac{\partial}{\partial t}\frac{I(x',t')}{R_1}dx' + \int_0^L \frac{\tau_1}{\tau_2}\frac{\partial I\left(x',t-\frac{R_1^*}{c}-\tau\right)}{\partial t}\frac{1}{R_1^*}dx'\right.$$

$$\left. + \int_{-\infty}^t\int_0^L\left[-\frac{1}{\tau_2}\left(1 - \frac{\tau_1}{\tau_2}\right)e^{-\frac{t}{\tau_2}}\right]I\left(x',t-\frac{R_1^*}{v}-\tau\right)\frac{1}{R_1^*}d\tau\right] \tag{6.71}$$

Formula (6.71) is evaluated using the BEM formalism.

6.3.1.1 Numerical Evaluation Procedure

Provided the transient antenna current is known, integral (6.63) can be evaluated using the boundary element formalism. The details regarding the determination of the current distribution along the dipole by solving the Hallen equation could be found elsewhere, e.g. in [11, 12]. According to the BEM procedure, the solution for the current $I(x',t')$ on a wire segment can be written as follows

$$I(x',t') = \sum_{i=1}^{N_g} I_i(t')f_i(x') \tag{6.72}$$

or, in the matrix notation, one has:

$$I(x', t') = \{f\}^T \{I(t')\} \tag{6.73}$$

where f stands for the set of shape functions.

The total field is then calculated as follows

$$E_x^{tot} = \sum_{i=1}^{M} \left(E_{x,i}^{inc} + E_{x,i}^{ref,D} + E_{x,i}^{ref,V} \right) \tag{6.74}$$

where:

$$E_{x,i}^{inc}(r,t) = -\frac{\mu_0}{4\pi} \frac{\partial}{\partial t} \{I\}|_{t'} \int_{\Delta l_i} \{f'\}_i^T \frac{1}{R_1} dx' \tag{6.75}$$

$$E_{x,i}^{ref,D}(r,t) = \frac{\mu_0}{4\pi} \frac{\tau_1}{\tau_2} \frac{\partial}{\partial t} \{I\}|_{t_1^*} \int_{\Delta l_i} \{f'\}_i^T \frac{1}{R^*} dx' \tag{6.76}$$

$$E_{x,i}^{ref,L}(r,t) = \frac{\mu_0}{4\pi} \int_{-\infty}^{t} \Gamma_{MIT}^{ref,diff}(t-\tau)\{I(\tau_1)\}|_{t_1^*} \int_{\Delta l_i} \{f'\}^T \frac{1}{R_1^*} dx' d\tau \tag{6.77}$$

and M denotes the total number of wire segments.

The shape functions are given by

$$f_r(x') = \frac{x_{r+1} - x'}{x_{r+1} - x_r} \tag{6.78}$$

$$f_{r+1}(x') = \frac{x' - x_r}{x_{r+1} - x_r} \tag{6.79}$$

Having completed spatial discretization procedure, the time-dependent unknown coefficients for the solution of current I_i^k are expressed as follows

$$I_i(t') = I\left(t - \frac{R}{c}\right) = \sum_{k=1}^{N_t} I_i^k(t') T^k(t') \tag{6.80}$$

where T^k stands for time-dependent shape functions:

$$T^k(t') = \frac{t_{k+1} - t'}{t_{k+1} - t_k} \tag{6.81a}$$

$$T^{k+1}(t') = \frac{t' - t_k}{t_{k+1} - t_k} \tag{6.81b}$$

Now the total radiated field can be obtained by inserting following expressions

$$E_{x,i}^{inc}(r,t) = -\frac{\mu_0}{4\pi} \frac{I_{i1}^{m+1}-I_{i1}^m}{\Delta t} \int_{x_{i1}}^{x_{i2}} \frac{x_{i2}-x'}{\Delta x} \frac{1}{R_1} dx' - \frac{\mu_0}{4\pi} \frac{I_{i2}^{m+1}-I_{i2}^m}{\Delta t} \int_{x_{i1}}^{x_{i2}} \frac{x'-x_{i1}}{\Delta x} \frac{1}{R_1} dx'$$

(6.82)

$$E_{x,i}^{ref,D}(r,t) = \frac{\mu_0}{4\pi} \frac{I_{i1}^{m+1}-I_{i1}^m}{\Delta t} \frac{\tau_1}{\tau_2} \int_{x_{i1}}^{x_{i2}} \frac{x_{i2}-x'}{\Delta x} \frac{1}{R_1^*} dx' + \frac{\mu_0}{4\pi} \frac{I_{i2}^{m+1}-I_{i2}^m}{\Delta t} \frac{\tau_1}{\tau_2} \int_{x_{i1}}^{x_{i2}} \frac{x'-x_{i2}}{\Delta x} \frac{1}{R_1^*} dx'$$

(6.83)

$$E_{x,i}^{ref,V}(r,t) = \frac{\mu_0}{4\pi} \int_{t_k}^{t_{k+1}} \Gamma_{MIT}^{ref,diff}(t-\tau) \left[\left(I_{i1}^m \frac{\tau^{m+1}-\tau}{\Delta \tau} + I_{i1}^{m+1} \frac{\tau-\tau^m}{\Delta \tau} \right) \int_{x_{i1}}^{x_{i2}} \frac{x_{i2}-x'}{\Delta x} \frac{1}{R_1^*} dx' \right.$$

$$\left. + \left(I_{i2}^m \frac{\tau^{m+1}-\tau}{\Delta \tau} + I_{i2}^{m+1} \frac{\tau-\tau^m}{\Delta \tau} \right) \int_{x_{i1}}^{x_{i2}} \frac{x'-x_{i1}}{\Delta x} \frac{1}{R_1^*} dx' \right] d\tau$$

(6.84)

Integrals in (6.82)–(6.84) are solved numerically by means of Gaussian quadrature.

More mathematical details on the TD calculation of the field above a lossy half-space can be found in [38, 39].

6.3.1.2 Numerical Results

Numerical results are obtained for the field above a lossy ground radiated by a dipole antenna of length $L = 1$ m, $a = 6.74$ mm, located at a certain height above ground h. Antenna is excited at its center by the Gaussian pulse voltage source (6.52) with parameters: $V_0 = 1$ V, $g = 1.5 \times 10^9 \text{s}^{-1}$, $t_0 = 1.43$ ns.

Figure 6.12 shows the spatial-time dependence of the radiated electric field above a lossy half-space ($\varepsilon_r = 10$, $\sigma = 1$ mS/m) for the dipole placed at height $h = 0.5$ m.

The results shown in Figs. 6.13–6.15 are related to the field component in the broadside direction.

Figure 6.13 shows the total field above a lossy ground ($\varepsilon_r = 10$, $\sigma = 1$ mS/m) for different heights: $h = 1.5$ m, $h = 2$ m, and $h = 4$ m.

Figure 6.14 shows the total field above a lossy ground ($\varepsilon_r = 10$, $\sigma = 10$ mS/m) for different heights: $h = 1.5$ m, $h = 2$ m, and $h = 4$ m. The results obtained via direct TD analysis are compared against the FD approach and are found to be in rather satisfactory agreement.

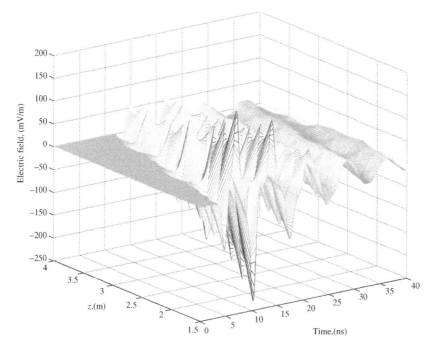

Figure 6.12 Spatial-time dependent electric field above a lossy ground (ε_r = 10, σ = 10 mS/m).

The higher the ground conductivity, the lesser is the field amplitude. Also, by increasing the dipole height, the influence of the reflected field component decreases. Figure 6.15 shows the field above a lossy ground for various values of ground permittivity and conductivity, respectively, computed at different altitudes above the interface.

It is visible that the field amplitude increases as the observation point is higher, but the field values are less sensible to the increase in permittivity.

6.3.2 Transient Field Transmitted into a Lossy Ground Due to Dipole Radiation

One of the crucial parameters in the study of GPR antenna operation is the transient field radiated by the antenna and transmitted into the lossy half-space.

The information on the amount of irradiated electromagnetic energy transferred into the lower medium provides an accurate antenna design and interpretation of the target-reflected wave, e.g. [28–30]. Taking into account that the assessment of the transient field transmitted into the lossy ground can be performed in either

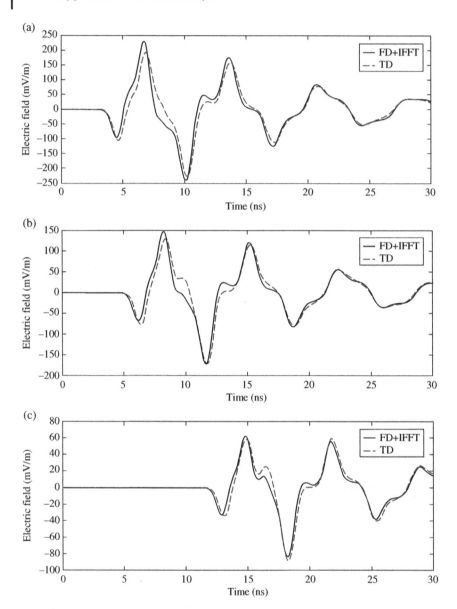

Figure 6.13 Radiated field above a lossy half-space (ε_r = 10, σ = 1 mS/m). (a) h = 1.5 m. (b) h = 2.0 m. (c) h = 4.0 m.

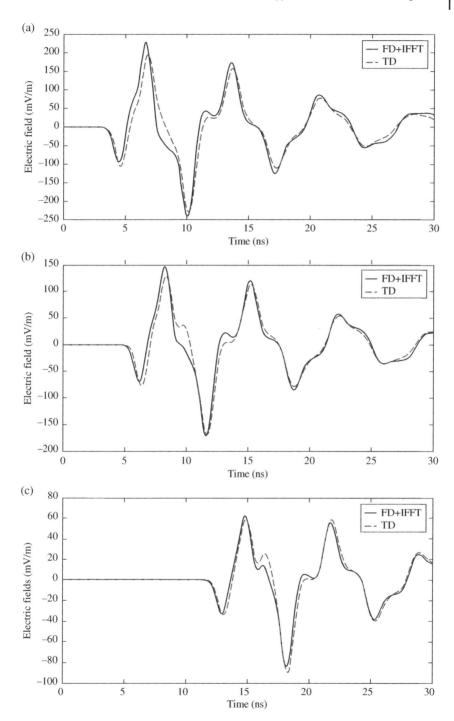

Figure 6.14 Radiated field above a lossy half-space (ε_r = 10, σ = 10 mS/m). (a) h = 1.5 m. (b) h = 2.0 m. (c) h = 4.0 m.

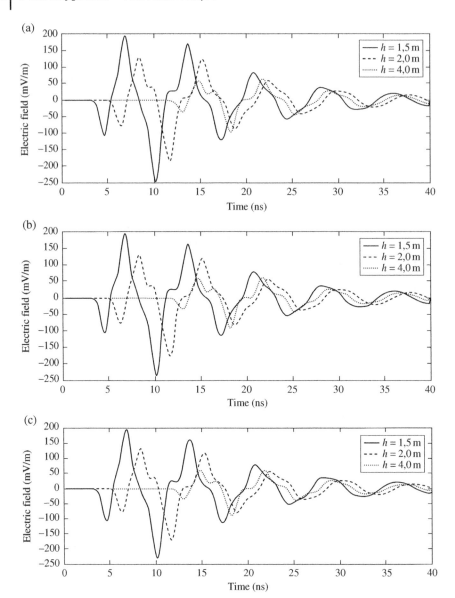

Figure 6.15 Radiated field above a lossy half-space ($\varepsilon_r = 10$, $\sigma = 10$ mS/m). (a) ($\varepsilon_r = 5$, $\sigma = 1$ mS/m). (b) ($\varepsilon_r = 8$, $\sigma = 1$ mS/m). (c) ($\varepsilon_r = 10$, $\sigma = 1$ mS/m).

frequency or TD, respectively, there are some advantages in TD modeling of GPR antennas [31]. Though most of the direct TD GPR antenna models are related to the use of the FDTD method, e.g. [32–34], integral equation approaches, such as the Boundary Element Method (BEM), offer some computational advantages, e.g. avoidance of stair-casing approximation errors [35].

Contrary to the widely used FDTD approach, an alternative approach to the assessment of transient electric field transmitted into the ground due to the GPR dipole antenna by means of the BEM has been reported in [31].

Direct time-domain analysis of a dipole antenna radiating over a dissipative half-space is based on the space-time Hallen integral Eq. (6.1), which is numerically solved via GB-IBEM [11, 12].

Furthermore, a deterministic and deterministic-stochastic model for the transient analysis of GPR dipole antenna radiating above a dielectric half-space is reported in [36] and [37], respectively.

The work reported in [33] extends the analysis reported in [36] and [37] to the more realistic case of a lossy half-space. The analysis of the electric field transmitted into a lower lossy half-space is based on more demanding formulation and numerical solution procedures, respectively, than the corresponding analysis of the field transmitted into a dielectric medium. The transient field transmitted into a lossy half-space is evaluated by numerically solving the corresponding field integrals by means of the boundary element formalism. Note that the influence of the air-lossy ground is taken into account via the simplified space-time transmission coefficient arising from the MIT [21, 31].

The geometry of interest is the horizontal straight center-fed thin wire antenna, as shown in Fig. 6.11. The antenna is positioned at height h above a lossy medium and is excited via voltage source.

The transient response of the dipole antenna driven by an equivalent voltage generator is obtained by solving the space-time Hallen integral equation for the unknown axial current $I(x,t)$ along the wire (6.8). The Hallen equation (6.8) is solved via space-time version of GB-IBEM [11, 12].

Knowing the current distribution along the dipole provides further assessment of the field transmitted into the lower medium.

The transient field transmitted in a lossy half-space is given by [31]

$$E_x^{tr}(r,t) = \frac{\mu_0}{4\pi} \int\limits_{-\infty}^{t} \int\limits_{0}^{L} \Gamma_{tr}^{MIT}(\tau) \frac{\partial I(x', t - R''/v - \tau)}{\partial t} \frac{e^{-\frac{1}{\tau_g} \frac{R''}{v}}}{R''} dx' d\tau \qquad (6.85)$$

where v is velocity of wave propagation in the lower medium and R'' is the distance from the dipole antenna to the observation point located in the lower medium

$$R'' = \sqrt{(x-x')^2 + (z+h)^2} \qquad (6.86)$$

The influence of the two-media interface is taken into account via the simplified space-time transmission coefficient arising from the MIT [21, 31]

$$\Gamma_{tr}(t) = \frac{\tau_3}{\tau_2}\delta(t) + \frac{1}{\tau_2}\left(2 - \frac{\tau_3}{\tau_2}\right)e^{-\frac{t}{\tau_2}} \tag{6.87}$$

where:

$$\tau_2 = \frac{\varepsilon_r + 1}{\sigma}\varepsilon_0, \tau_3 = \frac{2\varepsilon}{\sigma} \tag{6.88}$$

For convenience, expression for the transmitted field can be written as follows

$$E_x^{tr}(r,t) = -\frac{\mu}{4\pi}\frac{\partial}{\partial t}\int_{-\infty}^{t}\int_{0}^{L}\Gamma_{tr}(t-\tau)I\left(x',\tau-\frac{R''}{v}\right)\frac{e^{-\frac{1}{\tau_3}\frac{R''}{v}}}{R''}dx'd\tau \tag{6.89}$$

Furthermore, it is convenient to write a transmitted field in terms of dielectric part and conductive part

$$E_x^{tr} = E_{xd}^{tr} + E_{xg}^{tr} \tag{6.90}$$

Inserting (6.56) into (6.58), utilizing the Dirac impulse property, one obtains

$$E_{xd}^{tr}(r,t) = -\frac{\mu_0}{4\pi}\int_{0}^{L}\frac{\tau_3}{\tau_2}\frac{\partial}{\partial t}I\left(x',t-\frac{R''}{v}\right)\frac{e^{-\frac{1}{\tau_3}\frac{R''}{v}}}{R''}dx' \tag{6.91}$$

$$E_{xg}^{tr}(r,t) = -\frac{\mu}{4\pi}\int_{-\infty}^{t}\int_{0}^{L}\frac{\partial}{\partial t}\Gamma_{tr}^{ll}(t-\tau)I\left(x',\tau-\frac{R''}{v}\right)\frac{e^{-\frac{1}{\tau_3}\frac{R''}{v}}}{R''}dx'd\tau \tag{6.92}$$

where Γ_{tr}^{ll} is the time dependent part of the transmission coefficient (6.55)

$$\Gamma_{tr}^{ll}(t-\tau) = \frac{1}{\tau_2}\left(2 - \frac{\tau_3}{\tau_2}\right)e^{-\frac{t-\tau}{\tau_2}} \tag{6.93}$$

It should be pointed out that when performing the analysis of the lossless medium, as in [36], the transmitted field is expressed only with relation (6.83), which is significantly less demanding formulation, thus providing an appreciably simple and fast numerical solution procedure.

Furthermore, (6.91) can be written as

$$E_{xg}^{tr} = -\frac{\mu}{4\pi}\int_{-\infty}^{t}\int_{0}^{L}\Gamma_{tr}^{d}(t-\tau)I\left(x',\tau-\frac{R''}{v}\right)\frac{e^{-\frac{1}{\tau_3}\frac{R''}{v}}}{R''}dx'd\tau \tag{6.94}$$

$$\Gamma_{tr}^d(t-\tau) = \frac{\partial}{\partial t}\Gamma_{tr}^{II}(t-\tau) = \frac{\partial}{\partial t}\left[\frac{1}{\tau_2}\left(2 - \frac{\tau_3}{\tau_2}\right)e^{-\frac{t-\tau}{\tau_2}}\right] = -\frac{1}{\tau_2^2}\left(2 - \frac{\tau_3}{\tau_2}\right)e^{-\frac{t-\tau}{\tau_2}}$$

$$(6.95)$$

The field integrals (6.86) and (6.87) are handled via the boundary element formalism [38, 39].

6.3.2.1 Numerical Evaluation of the Transmitted Field

The local expansion for the space-time dependent current can be written as follows

$$I\left(x', t - \frac{R''}{v}\right) = \sum_{i=1}^{N_g} I_i\left(t - \frac{R''}{v}\right) f_i(x')$$

$$(6.96)$$

which can also be written in the matrix notation

$$I(x', t') = \{f\}^T\{I(t')\}$$

$$(6.97)$$

Note that linear approximation was shown to be sufficient for straight-wire problems [11]. For the case of linear approximation, the spatial shape functions are of the form

$$f_r(x') = \frac{x_{r+1} - x'}{x_{r+1} - x_r}, \; f_{r+1}(x') = \frac{x' - x_r}{x_{r+1} - x_r}$$

$$(6.98)$$

where $r = 1,2$.

The space discretization yields

$$E_x^{tr} = \sum_{i=1}^{M}\left(E_{xd,i}^{tr} + E_{xg,i}^{tr}\right)$$

$$(6.99)$$

where the corresponding field integrals are given by

$$E_{xd,i}^{tr}(r, t) = \frac{\mu_0}{4\pi}\frac{\tau_3}{\tau_2}\left[\int_{\Delta l_i}\frac{\partial}{\partial t}\{I\}|_{t''=t-\frac{R''}{v}}\{f'\}^T\frac{e^{-\frac{1}{\tau_3}\frac{R''}{v}}}{R''}dx'\right]$$

$$(6.100)$$

$$E_{xg,i}^{tr} = -\frac{\mu}{4\pi}\int_{-\infty}^{t}\Gamma_{tr}^d(t-\tau)\sum_{i=1}^{N_g}I_i(\tau)\int_{\Delta x_i}\{f'\}_i^T\frac{e^{-\frac{1}{\tau_3}\frac{R''}{v}}}{R''}dx'd\tau$$

$$(6.101)$$

Once the space discretization is carried out, the temporal discretization starts by assuming the time dependence for the current as follows

$$I_i\left(t - \frac{R''}{v}\right) = \sum_{k=1}^{N_t}I_i^k(t'')T^k\left(t\frac{1}{2}\right), t'' = t - \frac{R''}{v}$$

$$(6.102)$$

where I_i^k are the unknown coefficients and T^k are the TD shape functions given by

$$T^k(t'') = \frac{t_{k+1} - t''}{\Delta t''}, \quad T^{k+1}(t'') = \frac{t'' - t_k}{\Delta t''}, \quad \Delta t'' = t_{k+1} - t_k \tag{6.103}$$

where $\Delta t''$ denotes the time increment.

Having completed the temporal discretization, one obtains

$$E_{x,i}^{tr}(r,t) = \frac{\mu_0}{4\pi} \frac{\tau_3}{\tau_2} \left[\int_{\Delta l_i} \frac{I_{i1}^{m+1} - I_{i1}^m}{\Delta t} f_{i1}(x') \frac{e^{-\frac{1}{\tau_3}\frac{R''}{v}}}{R''} dx' + \int_{\Delta l_i} \frac{I_{i2}^{m+1} - I_{i2}^m}{\Delta t} f_{i2}(x') \frac{e^{-\frac{1}{\tau_3}\frac{R''}{v}}}{R''} dx' \right] \tag{6.104}$$

$$E_{xg,i}^{tr} = -\frac{\mu}{4\pi} \int_{t_k}^{t_{k+1}} \Gamma_{tr}^d(t_k - \tau) \left[\left(I_{i1}^m \frac{\tau^{m+1} - \tau}{\Delta \tau} + I_{i1}^{m+1} \frac{\tau - \tau^m}{\Delta \tau} \right) \int_{x_{i1}}^{x_{i2}} \frac{x_{i2} - x'}{\Delta x} \frac{e^{-\frac{1}{\tau_3}\frac{R''}{v}}}{R''} dx' \right.$$

$$\left. + \left(I_{i2}^m \frac{\tau^{m+1} - \tau}{\Delta \tau} + I_{i2}^{m+1} \frac{\tau - \tau^m}{\Delta \tau} \right) \int_{x_{i1}}^{x_{i2}} \frac{x' - x_{i1}}{\Delta x} \frac{e^{-\frac{1}{\tau_3}\frac{R''}{v}}}{R''} dx' \right] d\tau \tag{6.105}$$

Assembling the local spatial and temporal contributions, the transmitted field is obtained.

6.3.2.2 Numerical Results

Computational example is related to the horizontal dipole antenna radiating above a lossless and lossy half-space, respectively. The wire of length $L = 1$ m, radius $a = 6.74$ mm, is located horizontally above the interface at height $h = 0.1$ m. The relative permittivity of the medium is $\varepsilon_r = 10$. The dipole is excited by the Gaussian pulse voltage source (6.52) with parameters: $V_0 = 1$ V, $g = 1.5 \times 10^9$ s^{-1}, $t_0 = 1.43$ ns.

Transient electric field, radiated by the GPR dipole antenna and transmitted into the dielectric half-space for different depths from the interface is shown in Fig. 6.16. Numerical results obtained via direct TD approach are compared to the results obtained via FD approach and inverse FFT (FD+IFFT).

The numerical results obtained via different approaches agree satisfactorily. Multiple peaks correspond to the current waveform induced along the antenna and diminishing amplitude is due to time delay. Transient field transmitted into a lossy medium with different values of conductivity is shown in Figs. 6.17 and 6.18. Note that the homogeneous medium with conductivity of 1 and 10 mS/m is considered, respectively.

Comparing the results obtained for different soil conductivities, it can be noticed that the waveform of the transient electric field remains almost identical, actually

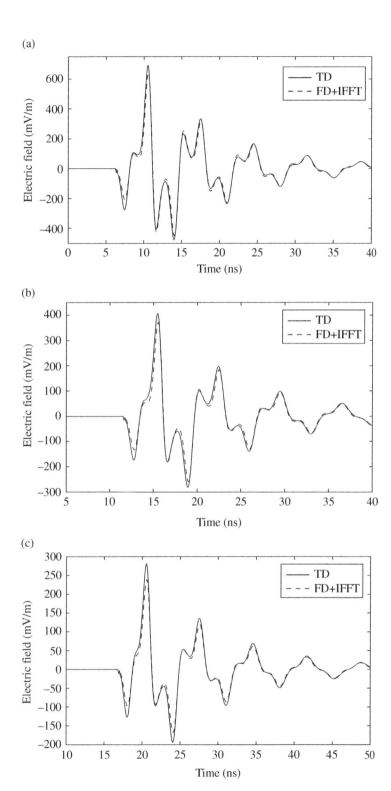

Figure 6.16 Transmitted field into the lossless half-space. (a) $d = 0.5$ m. (b) $d = 1.0$ m. (c) $d = 1.5$ m.

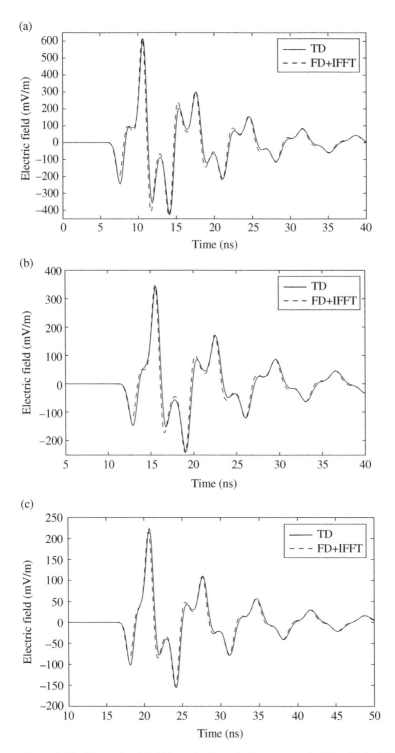

Figure 6.17 Transmitted field into the lossy half-space, $\sigma = 1$ mS/m. (a) $d = 0.5$ m. (b) $d = 1.0$ m. (c) $d = 1.5$ m.

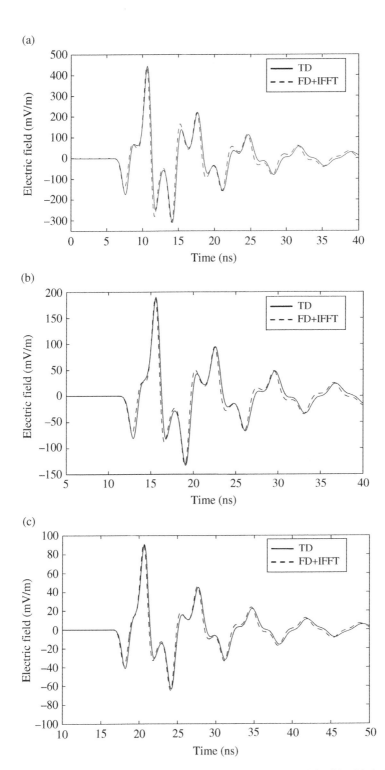

Figure 6.18 Transmitted field into the lossy half-space, $\sigma = 10$ mS/m. (a) $d = 0.5$ m. (b) $d = 1.0$ m. (c) $d = 1.5$ m.

independent of the variation of conductivity. Basically, this is expected behavior as the soil conductivity dominantly influences the energy dissipation (resulting in amplitude decrease). In addition, the drop-off in the amplitude is more prominent for the medium with higher conductivity since the energy dissipation is higher. It can be concluded that the variations in amplitude due to larger depth of the observation point in the case of lossless medium are primarily due to the time delay caused by the propagation distance, whereas in the case of the lossy medium, both propagation distance and energy dissipation play a prominent role.

The agreement between the results obtained via different approaches verifies the validity of a direct TD approach.

6.4 Simplified Calculation of Specific Absorption in Human Tissue

Human exposure to transient radiation can be also analyzed using the TD modeling of thin wires. One of the quantities pertaining to the study of transient exposures is SA. Simplified calculation of the SA in planar representation of the human muscle tissue when exposed to transient radiation from dipole antenna has been reported in [40].

Generally, direct TD modeling of the human body exposed to transient radiation is of continuous interest to the bioelectromagnetics community. However, this topic has been analyzed to an appreciably lesser extent compared to the FD studies of continuous wave exposures. TD analysis can be carried out by means of realistic, anatomically based models [41–43] or by means of simplified body representations, such as human equivalent antenna model [44]. The coupling of transient electromagnetic fields to anatomically based body representations is mostly being analyzed via the finite-difference time-domain (FDTD) method, e.g. [41].

Note that the human equivalent antenna has been particularly designed for experimental dosimetry, and is valid within the frequency range from 50 Hz to 110 MHz [45]. The dimensions of the human equivalent antenna are within the thin wire approximation and the effective frequency bandwidth of transient waveforms corresponds to the frequency range of the human equivalent antenna. The TD human equivalent antenna model has been based on the Hallen integral equation and on the dimensions available from [46]. A solution of this integral equation using the TD Galerkin–Bubnov scheme of the Boundary Element Method (GB-BEM) has been presented in [47] and more recently documented in detail in [48].

The evaluation of SA induced in planar model of the human muscle tissue when exposed to transient radiation from dipole antenna excited at its center by Gaussian pulse voltage source reported in [49] is based on numerical handling of Hallen

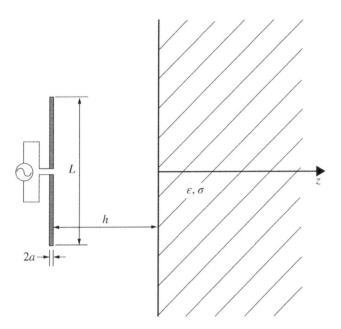

Figure 6.19 Dipole in front of planar muscle representation.

space-time integral Eq. (6.8) and the related part radiated field being transmitted into the human tissue (6.77) obtained by a corresponding space-time integral expression. The part of the radiated field is transmitted into the human tissue (6.77). The human tissue is represented as an unbounded finitely conducting half-space [40]. According to the standard definition [50], SA is evaluated as the time integral of the SA rate (SAR), i.e. by temporal integration of the squared value of the field strength in the tissue. It is worth emphasizing that SAR and SA are calculated per unit length and integration over control volume is required to obtain standard dosimetric quantities.

The geometry of interest pertaining to the dipole antenna in front of the planar model of the human muscle is presented in Fig. 6.19.

Once the space-time current and related field transmitted in the planar tissue model are obtained by numerically handling relations (6.8) and (6.77).

6.4.1 Calculation of Specific Absorption

Transient response of the human body exposed to dipole antenna radiation is quantified by SA which is derived from basic equations in electromagnetics.

According to the well-known general definition, the total energy absorbed by the resistive material can be obtained by temporally integrating the instantaneous power $p_{rad}(t)$

$$W_{tot} = \int_0^{T_0} p_{rad}(t)dt \tag{6.106}$$

where T_0 is the interval of interest.

The power dissipated in the human body is given by a volume integral over power density $\overline{P}_d$

$$p_{rad}(t) = \int_V \overline{P}_d dV \tag{6.107}$$

where $\overline{P}_d$ is defined as

$$\overline{P}_d = \sigma \left| \vec{E}\left(\vec{r},t\right) \right|^2 = \frac{\left| \vec{J}\left(\vec{r},t\right) \right|^2}{\sigma} \tag{6.108}$$

while $\vec{E}\left(\vec{r},t\right)$ and $\vec{J}\left(\vec{r},t\right)$ represent the electric field and current density induced inside the body, respectively.

Combining the Eqs. (6.106)–(6.108) yields

$$W_{tot} = \int_0^{T_0} \int_V \sigma \left| \vec{E}\left(\vec{r},t\right) \right|^2 dV dt \tag{6.109}$$

SA is defined as a quotient of the incremental energy dW absorbed by an incremental mass dm contained in the volume dV of density ρ

$$SA = \frac{dW}{dm} = \frac{dW}{\rho dV} \tag{6.110}$$

Substituting expression (6.101) into (6.102) leads to the expression

$$SA = \int_0^{T_0} \frac{1}{\rho} \frac{d}{dV} \left(\int_V \sigma \left| \vec{E}\left(\vec{r},t\right) \right|^2 dV \right) dt \tag{6.111}$$

and then it follows

$$SA = \int_0^{T_0} \frac{1}{\rho} \sigma \left| \vec{E}\left(\vec{r},t\right) \right|^2 dt \tag{6.112}$$

For the case of open time interval, SA becomes space-time dependent

$$SA\left(\vec{r},t\right) = \int_0^t \frac{1}{\rho}\sigma\left|\vec{E}\left(\vec{r},t\right)\right|^2 dt \tag{6.113}$$

Furthermore, taking into account definition (6.110), temperature increase can be expressed in terms of SA as follows

$$\Delta T = \frac{SA}{C} \tag{6.114}$$

where C stands for the specific heat capacity of the tissue.

6.4.2 Numerical Results

Computational example deals with the dipole in front of the human tissue as indicated in Fig. 6.1 and excited at its center by the Gaussian pulse (6.52) with parameters: $V_0 = 1\,\text{V}$, $g = 2.5 \cdot 10^9\,\text{s}^{-1}$, $t_0 = 1\,\text{ns}$. Length of dipole is $L = 5\,\text{cm}$, radius is $a = 1\,\text{mm}$, and the height above the interface is $h = 5\,\text{cm}$.

Muscle properties at 1 GHz are used [51]; relative dielectric permittivity is $\varepsilon_r = 54.8$, and conductivity is $\sigma = 0.978\,\text{S/m}$. Furthermore, the heat capacity is $C = 3421\,\text{J/kg/°C}$, while the density is $\rho = 1090\,\text{kg/m}^3$.

Figure 6.20 shows the antenna current induced at the center of the dipole versus time for a given excitation pulse.

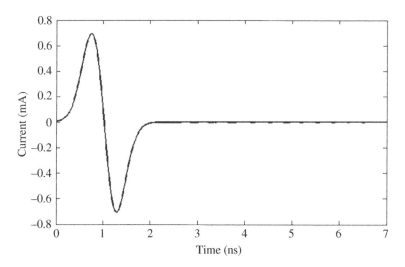

Figure 6.20 Transient current at the center of the dipole.

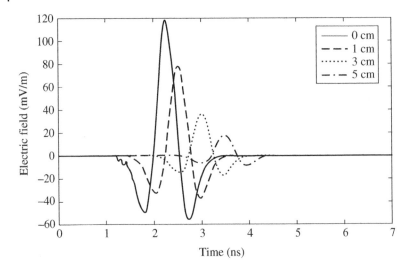

Figure 6.21 Transmitted electric field at various depths inside the muscle.

Figure 6.21 shows the electric field transmitted into the tissue at different depths where expected behavior of the electric field is observed. Namely, the peak value of the transmitted electric field strength decreases by 83% at the depth of 5 cm compared with the value at the surface.

The SA versus depth computed as a time integral of SA rate (SAR) over a whole considered time interval is presented in Fig. 6.22.

As the electric field decreases with depth, SA also shows similar behavior rapidly dropping off inside the muscle tissue.

SA versus time computed for different depths is shown in Fig. 6.23. SA achieves its maximum value after 3–4 ns (depending on depth) and afterward remains constant.

Finally, Fig. 6.24 deals with the temperature elevation inside the planar muscle model, which shows a waveform similar to SA.

Next computational example deals with the same dipole excited at its center by the Gaussian pulse, but for somewhat different tissue parameters, $\varepsilon_r = 10$ and $\sigma = 1\,\text{S/m}$.

Figure 6.25 shows the antenna current induced at the center of the dipole versus time for a given excitation pulse.

Furthermore, Fig. 6.26 shows the electric field transmitted into the tissue at different depths where behavior of the electric field is observed.

It is obvious that the peak value decreases rapidly. Figure 6.27 shows *SA* versus depth computed as a time integral of *SAR* over a whole time interval.

SA rapidly decreases inside the muscle tissue. Figure 6.28 shows *SA* versus time computed for different depths.

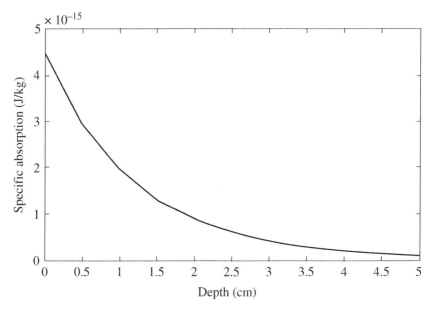

Figure 6.22 SA inside the muscle versus depth.

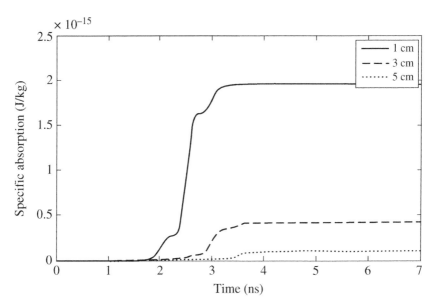

Figure 6.23 SA at various depths inside the muscle versus time.

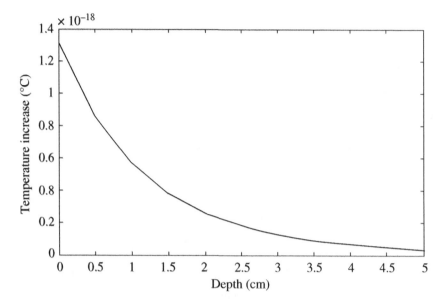

Figure 6.24 Temperature increase inside the muscle.

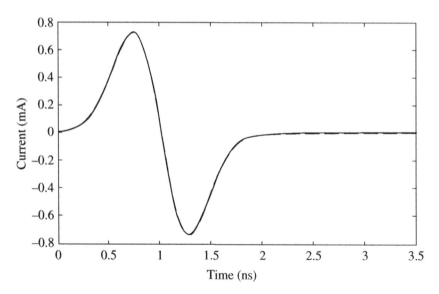

Figure 6.25 Current at the center of the dipole.

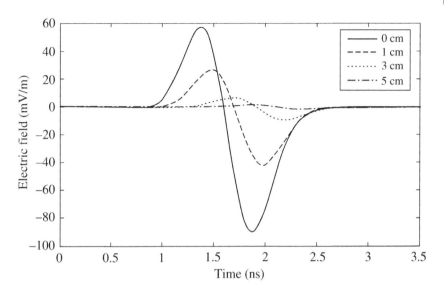

Figure 6.26 Transmitted electric field at various depths inside the muscle.

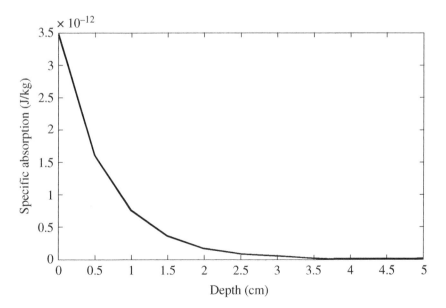

Figure 6.27 Specific absorption inside the muscle.

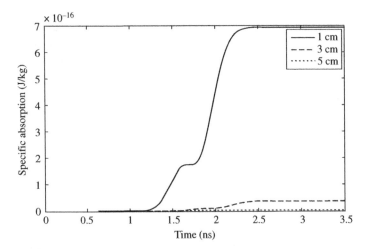

Figure 6.28 Specific absorption at various depths inside the muscle.

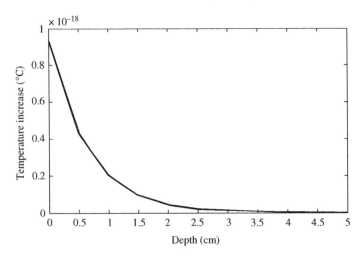

Figure 6.29 Temperature increase inside the muscle.

SA achieves its maximum value after approx. 2.5 ns and afterward remains constant. Figure 6.29 shows the temperature elevation inside the planar muscle model, which behaves similar to *SA*.

It is worth mentioning that all results obtained for unitary Gaussian pulse could be scaled to any realistic excitation type. Moreover, as Gaussian pulse represents numerical equivalent of a delta function, the related transient response to such an excitation is then, according to the signal theory, the impulse response.

6.5 Time Domain Energy Measures

TD energy measures proposed in [49] are very useful tools to estimate the nature of radiation of thin wire antennas aiming to provide an assessment of the energy stored in the antenna electric and magnetic field, respectively.

The approach in [49] proposed for the wires in free space is improved and extended to the case of dipole antenna above lossy ground [52–54].

To study the transient behavior of dipole antenna above a homogeneous lossy half space in terms of energy measures, the first step is to determine the transient current along the wire by solving the space-time dependent Hallen integral Eq. (6.8) using the numerical approach presented in [11, 12]. Furthermore, the transient charge along the wire is computed by means of one-dimensional continuity equation. Finally, the TD energy measures have been computed as spatial integral of squared current and charge, respectively, as proposed in [49]. More recently, a modified approach to compute energy measures by integrating the squared charge has been proposed in [53]. This modification has been adopted in this paper to calculate the measure of energy stored in the electric field as well.

Provided the transient current is determined by solving the Hallen equation and the charge density is computed from the one-dimensional continuity equation, the energy measures are obtained by evaluating following integrals [49]:

$$W_{tot}(t) = W_i(t) + W_q(t) = \frac{\mu}{4\pi} \int_0^L i^2(x',t)dx' + \frac{1}{4\pi\varepsilon} \int_0^L q^2(x',t)dx' \quad (6.115)$$

where $W_i(t)$ and $W_q(t)$ are energy measures due to the field sources in terms of integrals over squared current and charge, respectively.

The original formulation of $W_q(t)$ [49] is modified aiming to represent charge behavior at the wire ends [53], as follows

$$W_q^{MOD}(t) = \frac{1}{4\pi\varepsilon} \left[\int_{\Delta L}^{L-\Delta L} q^2(x',t)dx' + \frac{1}{3} \left(\int_0^{\Delta L} q^2(x',t)dx' + \int_{L-\Delta L}^L q^2(x',t)dx' \right) \right]$$

$$= W_q(t) - \frac{2}{3} \cdot \frac{1}{4\pi\varepsilon} \left(\int_0^{\Delta L} q^2(x',t)dx' + \int_{L-\Delta L}^L q^2(x',t)dx' \right) = W_q(t) - W_q^{ENDS}(t)$$

$$(6.116)$$

where ΔL represents the length of the wire segment.

Finally, one may also calculate TD power measure, simply given by

$$P_{tot}(t) = \frac{dW_{tot}(t)}{dt} \tag{6.117}$$

Mathematical details on the Hallen integral equation formulation and related numerical solution procedures could be found elsewhere, e.g. in [11].

Dipole antenna of length $L = 1$ m, radius $a = 2$ mm, horizontally located above ground ($\sigma = 0.001$ S/m, $\varepsilon_r = 10$) at height $h = 0.1$ m is considered.

Figure 6.30 shows GPR dipole antenna of radius a, and length L, horizontally located at height h above a lossy ground along x-axis and center-fed by a Gaussian voltage source (6.60).

The Gaussian pulse parameters are, as follows: $V_0 = 1$ V, $g = 2 \cdot 10^9$ s^{-1}, and $T_0 = 2$ ns. Figure 6.31 shows the transient driving point current obtained by using TD and frequency domain (FD) variant of the Galerkin–Bubnov Indirect Boundary Element Method (GB-IBEM) [11, 12] and compared to the NEC featuring the RCA and the rigorous Sommerfeld integral approach. The transient response from the frequency responses is obtained by using inverse IFFT.

Regardless of some discrepancies in late time instants, the results obtained via different approaches seem to be in satisfactory agreement.

Figures 6.32 and 6.33 show the related TD energy and power measures, respectively.

Finally, Figs. 6.34 and 6.35 show the total energy for different heights and ground conductivities.

It could be noticed that total energy measure decreases slower for lower heights and higher conductivities.

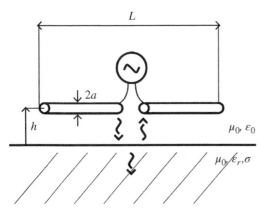

Figure 6.30 Straight thin wire above a lossy half-space.

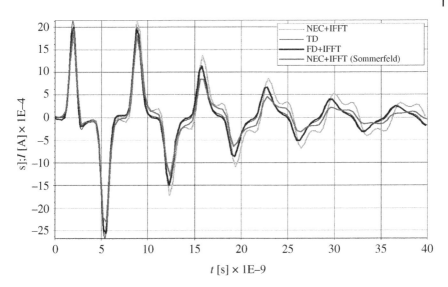

Figure 6.31 Transient current at the wire center (L = 1 m, a = 2 mm, h = 0.1 m, σ = 0.001 S/m, ε_r = 10).

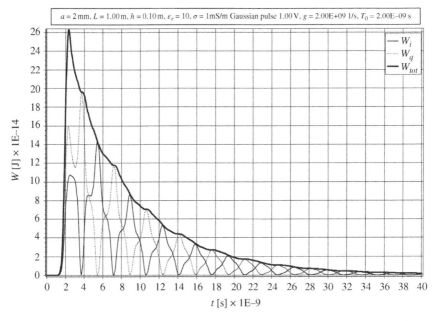

Figure 6.32 Energy measures W_i, W_q, and W_{tot} (L = 1 m, a = 2 mm, h = 0.1 m, σ = 0.001 S/m, ε_r = 10).

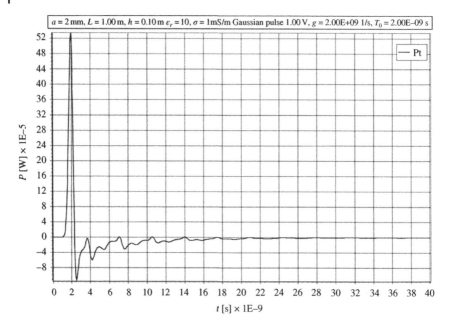

Figure 6.33 Power measure P_{tot} (L = 1 m, a = 2 mm, h = 0.1 m, σ = 0.001 S/m, ε_r = 10).

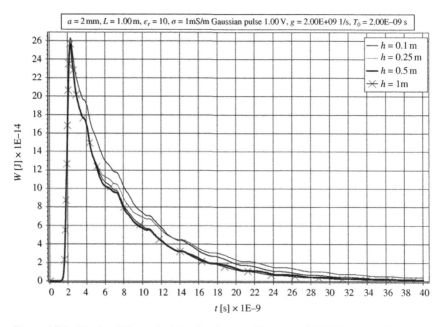

Figure 6.34 W_{tot} for different heights (L = 1 m, a = 2 mm, σ = 0.001 S/m, ε_r = 10).

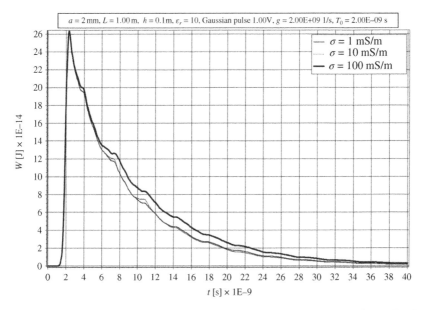

Figure 6.35 W_{tot} for different values of ground conductivity (L = 1 m, a = 2 mm, h = 0.1 m, ε_r = 10).

Finally, it is interesting to compare the total energy measure for the case of the dipole in free space, above a PEC ground, and above a dissipative half space ($\sigma = 0.001$ S/m, $\varepsilon_r = 10$), as it is depicted in Fig. 6.36.

It is clearly visible that the energy loss is the most rapid in the case of dipole in free space. This loss is slower for the case of the wire above a lossy ground as the

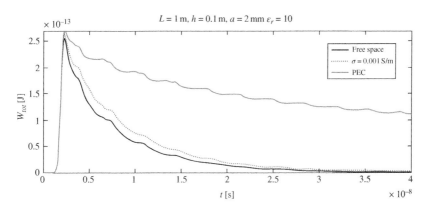

Figure 6.36 W_{tot} for the wire above free space, PEC ground and dissipative half-space (L = 1 m, a = 2 mm, h = 0.1 m, σ = 0.001 S/m, ε_r = 10).

part of the energy is stored in the reflected field. Finally, the loss is the smallest in the case of the dipole above a PEC ground and the reflection effect is the most pronounced.

6.6 Time Domain Analysis of Multiple Straight Wires above a Half-Space by Means of Various Time Domain Measures

This section deals with a direct TD analysis of multiple overhead straight wire configurations using various TD measures based on the energy stored in the near field. The space-time dependent currents along the wires are obtained by numerically solving the set of coupled Hallen integral equations. Once the current distribution is determined, it is possible to calculate the charge distribution along the wire by using the one-dimensional version of the continuity equation. Knowing the space-time distribution of current and charge along the straight wires, TD energy measures involving spatial integrals of squared values of transient current and charge, pertaining to the energy stored in the electric and magnetic field, respectively, of dipole antenna, are evaluated. Furthermore, total TD power measure defined as time derivative of the total energy measure is computed. Finally, according to definition stemming from the circuit theory, the distribution of root-mean-square (RMS) values of transient current is calculated. Some illustrative computational examples for transient current, TD energy, and power measures, respectively, and RMS distribution along the wires operating in both scattering and antenna mode are presented.

In general, transient analysis of arbitrary configurations of straight wires above real ground finds many applications in various areas of EMC and antennas and propagation such as antenna arrays, short-pulse radar, lightning protection systems, wide-band radio communications, electromagnetic interference (EMI) coupling to printed circuit boards (PCB), non-uniform transmission lines, power cables, GPR, etc., e.g. [11, 14, 28, 30, 49]. The information about the amount and distribution of energy stored in the irradiated electromagnetic field by a radiating structure is crucial to understanding the behavior and properties of various wire configurations. TD energy measures have been proposed in [49] and subsequently used in [52], for two coupled wires above a dielectric half-space. Quite recently, the analysis has been extended to a single wire above a lossy ground in [54]. In addition to TD energy measures proposed in [49], TD power measure has been presented in [44] and, finally, a distribution root-man square value of current has been used in [46, 55].

More recently, a modified approach to compute energy measures by integrating the squared charge has been proposed in [53]. This modification has been adopted in [54] to improve the calculation of the measure of energy stored in the electric field.

Note that a particular advantage of TD energy measures [49] is to deal with sources (in terms of current and charge density) instead of electric and magnetic fields, which are consequence of these sources. Thus, once the current and charge distribution, respectively, are known, it is possible to determine TD energy, power measures, and distribution of RMS values of sources without necessity to compute electric and magnetic fields.

Direct TD analysis of the parallel wires located above a dielectric half-space at arbitrary heights has been reported in [13, 14]. The mathematical model is based on a set of the coupled space-time integral equations of the Hallen type which are solved by means of the Galerkin–Bubnov scheme of the Indirect Boundary Element Method (GB-IBEM) [2].

The effect of a two-media configuration is taken into account by the corresponding space-time reflection coefficient appearing within the integral-equation kernels. Calculations are carried out for the case of a lossless dielectric half-space, thus significantly reducing computational cost of the procedure. In [15, 56, 57], the TD results have been compared to the results computed via NEC2 + inverse Fourier transform and the results obtained via different approaches agree satisfactorily.

Finally, the work undertaken in [58] extends the analysis reported in [28, 49], and [15, 44, 46, 55, 56] and pertains to the evaluation of the energy measures, power measures, and RMS of the arbitrary straight wires above dielectric half-space in the antenna and scattering mode, respectively. Computational examples are related to the arbitrary wire arrays excited by either a voltage source (antenna mode) or illuminated by an incident field (scattering mode).

6.6.1 Theoretical Background

The geometry of interest consists of an arbitrary number of parallel straight thin wires placed above a dielectric medium that is excited by a voltage pulse generator (Fig. 6.37a), operating in transmitting mode, and excited via normal incident electric field when operating in scattering mode (Fig. 6.37b).

(a)

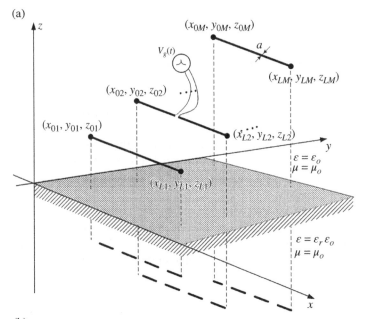

(b)

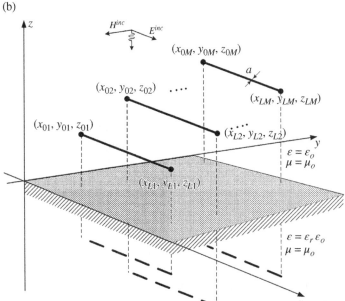

Figure 6.37 Multiple straight thin wires above a half-space. (a) Transmitting mode. (b) scattering mode.

Mathematical details regarding integral equation formulation and applied numerical method could be found elsewhere, e.g. in [15, 56].

Provided the set of space-time Hallen integral equations is numerically solved and current distribution is available, it is possible to compute the unknown charge distribution from the one-dimensional continuity equation [49]. Once the current and charge distribution are determined, it is possible to post-process them in a way to calculate desired TD measures.

6.6.1.1 Time Domain Energy Measures and Power Measure

Provided the transient current is determined by solving the Hallen equation and the charge density is computed from the one-dimensional continuity equation, the energy measures are obtained by evaluating integrals in (6.115) [49].

The original formulation of $W_q(t)$ [49] is modified, as given by (6.108), aiming to represent charge behavior at the wire ends [54]. Finally, TD power measure is calculated by using (6.117).

Mathematical details on the Hallen integral equation formulation and related numerical solution procedures could be found elsewhere, e.g. in [11].

6.6.1.2 Root Mean Square Value of Current Distribution

As it is well-known from signal theory provided the voltage, or current, respectively, generally assigned as $x(t)$ is available known, the energy W stored in the signal and dissipated at unitary resistor is given by

$$W = \int_0^{T_0} x^2(t)dt \tag{6.118}$$

where T_0 denotes the signal duration.

Thus, if one deals with time-varying current $i(t)$, the energy W dissipated on unitary resistor pertains to squared RMS, also known as effective value of current $i(t)$, is

$$I_{rms}^2 = \int_0^{T_0} i^2(t)dt \tag{6.119}$$

Applying this definition to current distribution along the straight thin wires, the distribution of RMS values of space-time dependent current $i(x, t)$, as reported in [46], is

$$I_{rms}^2(x') = \int_0^{T_0} i^2(x', t)dt \tag{6.120}$$

Note that quantities $I_{rms}^2(x')$ and $I_{rms}^2(x')$ may be used as useful energy measure on the nature of radiation and scattering from thin wires.

6.6.2 Numerical Results

The geometries analyzed in this work are geometries as shown in Fig. 6.38.

Furthermore, the four geometries depicted in Fig. 6.38 are excited by three types of excitations: the Gaussian incident plane wave

$$E^{inc}(t) = E_0 e^{-g^2(t-t_0)^2} \tag{6.121}$$

the double-exponential function (6.50), and the Gaussian pulse voltage source (6.60).

(a)

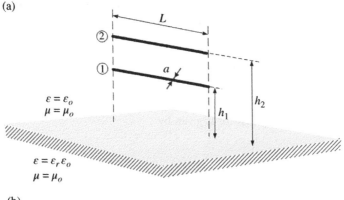

(b)

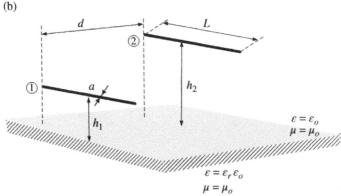

Figure 6.38 Geometries of interest. (a) Configuration 1 ($h_1 = 1$ m, $h_2 = 1.5$ m). (b) Configuration 2 ($d = 0.5$ m, $h_1 = h_2 = 1$ m). (c) Configuration 3 ($h_1 = 1$ m, $h_2 = 1.5$ m, $h_3 = 2$ m). (d) Configuration 4 ($d = 0.5$ m and $h_1 = h_2 = 1$ m, $h_3 = 1.5$ m).

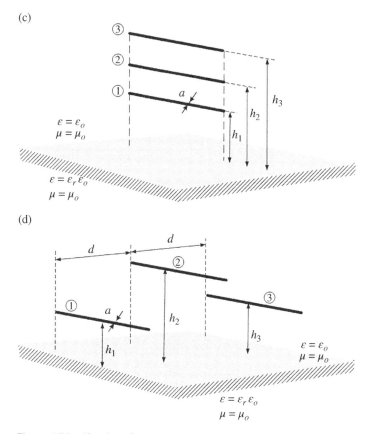

Figure 6.38 (Continued)

What follow are numerical results for the current distribution along the wires, TD energy and power measures, and distribution of RMS and squared RMS values, respectively, for different wire configurations.

6.6.2.1 Configuration 1

Configuration 1 deals with wires assigned as *geometry 1* in the scattering mode with wire length $L = 1$ m, wire radius $a = 2$ mm, heights above dielectric half-space ($\varepsilon_r = 10$) $h_1 = 1$ m, $h_2 = 1.5$ m illuminated by Gaussian plane wave (normal incidence) (6.50) with following parameters: $E_0 = 1$ V/m, $g = 2 \cdot 10^9$ s^{-1}, $t_0 = 2$ ns.

Fig. 6.39a shows the transient current induced at the center of wire 1, while TD energy measures and TD power measure are shown in Fig. 6.39b and 6.39c, respectively. Finally, Fig. 6.39d and 6.39e shows spatial distribution of the RMS and squared RMS values of the transient current along wire 1, respectively.

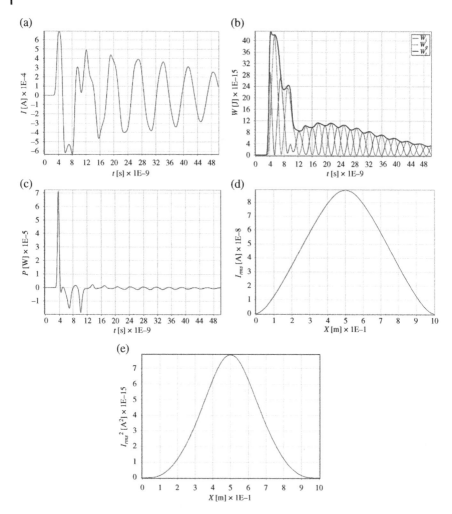

Figure 6.39 Configuration 1: (a) Transient current induced at the center of wire 1. (b) TD energy measures for wire 1. (c) TD power measure for wire 1. (d) Spatial distribution of the RMS values of the transient current along wire 1. (e) Spatial distribution of the squared RMS values of the transient current along wire 1.

Furthermore, Fig. 6.40a shows the transient current induced at the center of wire 2, while TD energy measures and TD power measure are shown in Fig. 6.40b and 6.40c, respectively. Finally, Fig 6.40d and 6.40e shows spatial distribution of the RMS and squared RMS values of the transient current along wire 2, respectively.

It is visible from Fig. 6.40b that there is increase in energy in wire 2 due to reflection from wire 1. There is no such increase in energy for wire 1 (Fig. 6.39b). Similar behavior can be seen in Figs. 6.39c and 6.40c pertaining to TD power measure.

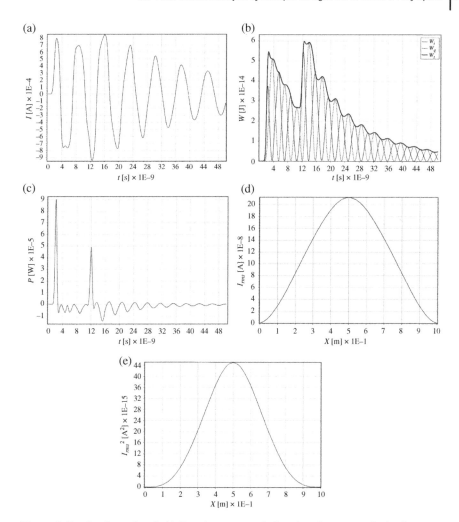

Figure 6.40 Configuration 1: (a) Transient current induced at the center of wire 2. (b) TD energy measures for wire 2. (c) TD power measure for wire 2. (d) Spatial distribution of the RMS values of the transient current along wire 2. (e) Spatial distribution of the squared RMS value of the transient current along wire 2.

Analyzing the RMS and squared RMS values of current corresponding to the energy on unitary resistance, it is obvious that the energy is dominantly stored around the wire center for both cases.

6.6.2.2 Configuration 2

Configuration 2 deals with two coupled thin wires assigned as *geometry 1* in the scattering mode with wire length $L = 10$ m, wire radius $a = 2$ cm, heights above perfectly conducting (PEC) $h_1 = 1$ m, $h_2 = 2$ m illuminated by double exponential EMP (8) with following parameters: $E_0 = 1$ V/m, $\alpha = 4 \cdot 10^7$ s^{-1}, $\beta = 6 \cdot 10^8$ s^{-1}.

Figure 6.41a shows the transient current induced at the center of wire 1, while TD energy measures and TD power measure are shown in Fig. 6.41b and 6.41c, respectively. Finally, Fig. 6.41d and 6.41e shows spatial distribution of the RMS and squared RMS values of the transient current along wire 1, respectively.

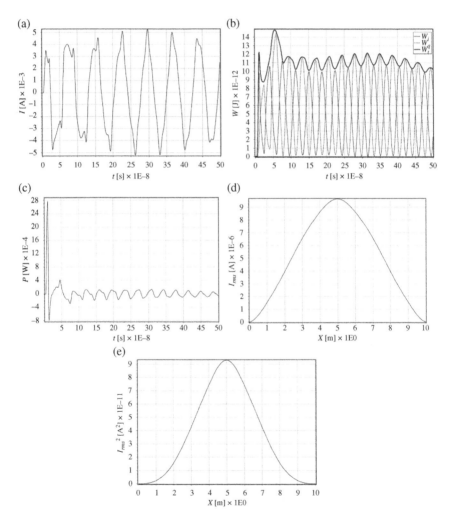

Figure 6.41 Configuration 2: (a) Transient current induced at the center of wire 1. (b) TD energy measures for wire 1. (c) TD power measure for wire 1. (d) Spatial distribution of the RMS values of the transient current along wire 1. (e) Spatial distribution of the squared RMS values of the transient current along wire 1.

Furthermore, Fig. 6.42a shows the transient current induced at the center of wire 2, while TD energy measures and TD power measure are shown in Fig. 6.42b and 6.42c, respectively. Finally, Fig. 6.42d and 6.42e shows spatial distribution of the RMS and squared RMS values of the transient current along wire 2, respectively.

It is visible from Fig. 6.41b and 6.42b that energy is confined in the antenna near field due to the PEC ground reflecting effects. This is particularly

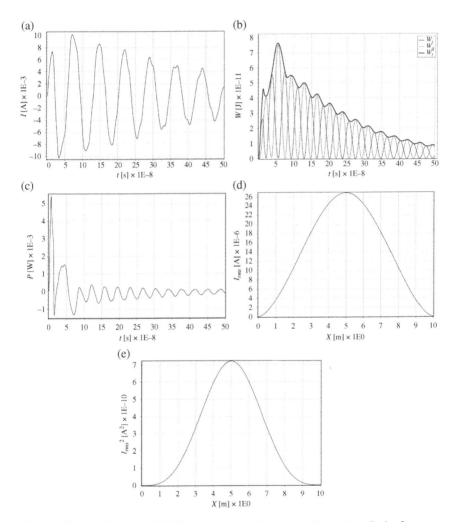

Figure 6.42 Configuration 2: (a) Transient current induced at the center of wire 2. (b) TD energy measures for wire 2. (c) TD power measure for wire 2. (d) Spatial distribution of the RMS values of the transient current along wire 2. (e) Spatial distribution of the squared RMS values of the transient current along wire 2.

pronounced for wire 1 located closer to PEC ground. Again, analyzing the RMS and squared RMS values of current corresponding to the energy on unitary resistance, it is again obvious that the energy is dominantly stored around the wire center.

6.6.2.3 Configuration 3

Configuration 3 deals with wires assigned as *geometry 2* in the scattering mode with wire length $L = 1$ m, wire radius $a = 2$ mm, heights above dielectric half-space ($\varepsilon_r = 10$) $h_1 = 1$ m, $h_2 = 1.5$ m, and distance between the wires $d = 0.5$ m illuminated by Gaussian incident plane wave (7) with following parameters: $E_0 = 1$ V/m, $g = 2 \cdot 10^9 \, \text{s}^{-1}$, $t_0 = 2$ ns. Figure 6.43a shows the transient current induced at the center of wire 1, while TD energy measures and TD power measure are shown in Fig. 6.43b and 6.43c, respectively. Finally, Fig. 6.43d and 6.43e shows spatial distribution of the RMS and squared RMS values of the transient current along wire 1, respectively.

Furthermore, Fig. 6.44a shows the transient current induced at the center of wire 2, while TD energy measures and TD power measure are shown in Fig. 6.44b and 6.44c, respectively. Finally, Fig. 6.44d and 6.44e shows spatial distribution of the RMS and squared RMS values of the transient current along wire 2, respectively.

It is visible from Fig. 6.44b that there is increase in energy due to reflection from wire 1. The same happens for TD power measure (Fig. 6.44c). Analyzing the RMS and squared RMS values of current corresponding to the energy on unitary resistance, it is obvious that the majority of energy is stored around the wire center for wire 1 and wire 2.

6.6.2.4 Configuration 4

Configuration 4 deals with wires assigned as *geometry 2* in the scattering mode with wire length $L = 10$ m, wire radius $a = 2$ cm, heights above dielectric half-space ($\varepsilon_r = 10$) $h_1 = 1$ m, $h_2 = 2$ m, and distance between the wires $d = 0.5$ m illuminated by double exponential EMP (8) with following parameters.

$$E_0 = 1 \, \text{V/m}, \quad \alpha = 4 \cdot 10^7 \, \text{s}^{-1}, \beta = 6 \cdot 10^8 \, \text{s}^{-1}$$

Figure 6.45a shows the transient current induced at the center of wire 1, while TD energy measures and TD power measure are shown in Fig. 6.45b and 6.45c, respectively. Finally, Fig. 6.45d and 6.45e shows spatial distribution of the RMS and squared RMS values of the transient current along wire 1, respectively.

Furthermore, Fig. 6.46a shows the transient current induced at the center of wire 2, while TD energy measures and TD power measure are shown in Fig. 6.46b

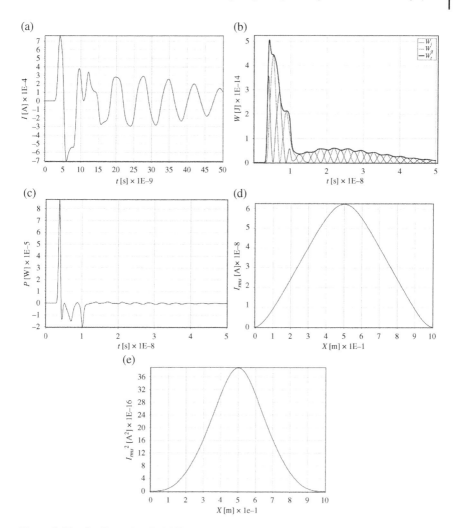

Figure 6.43 Configuration 3: (a) Transient current induced at the center of wire 1. (b) TD energy measures for wire 1. (c) TD power measure for wire 1. (d) Spatial distribution of the RMS value of the transient current along wire 1. (e) Spatial distribution of the squared RMS value of the transient current along wire 1.

and 6.46c, respectively. Finally, Fig. 6.46d and 6.46e shows spatial distribution of the RMS and squared values of the transient current along wire 2, respectively.

It is visible from Fig. 6.45b and 6.46b that there is rapid decrease in stored energy due to nature of excitation and spatial arrangements of wires. Analyzing the RMS and squared RMS values of current corresponding to the energy on unitary resistance, it is obvious that the energy is dominantly stored around the center of wires 1 and 2, respectively.

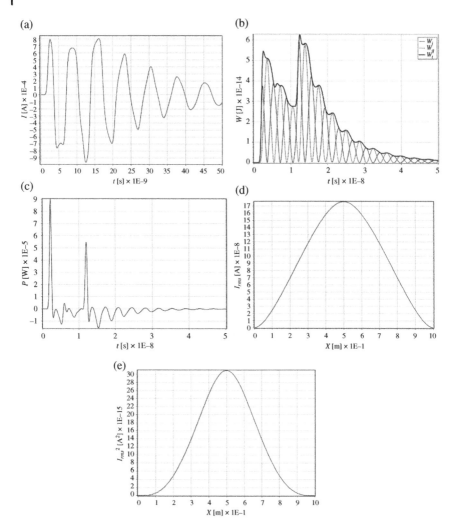

Figure 6.44 Configuration 3: (a) Transient current induced at the center of wire 2. (b) TD energy measures for wire 2. (c) TD power measure for wire 2. (d) Spatial distribution of the RMS values of the transient current along wire 2. (e) Spatial distribution of the squared RMS values of the transient current along wire 2.

6.6.2.5 Configuration 5

Configuration 5 deals with wires assigned as *geometry 3* in the antenna (radiating) mode with wire length $L = 1$ m, wire radius $a = 2$ mm, heights above dielectric half-space ($\varepsilon_r = 10$) $h_1 = 1$ m, $h_2 = 1.5$ m, $h_3 = 2$ m excited at the center of wire 3 with Gaussian pulse voltage source (9) with following parameters: $V_0 = 1$ V,

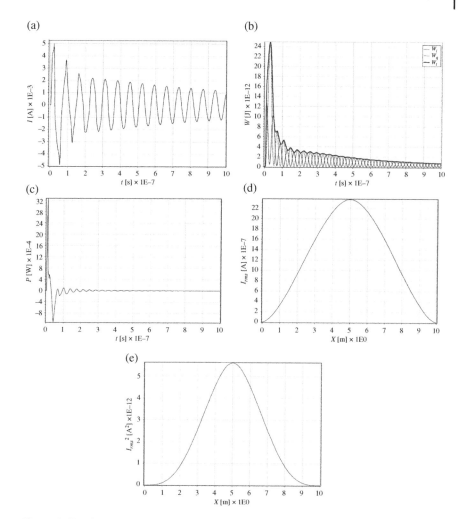

Figure 6.45 Configuration 4: (a) Transient current induced at the center of wire 1. (b) TD energy measures for wire 1. (c) TD power measure for wire 1. (d) Spatial distribution of the RMS values of the transient current along wire 1. (e) Spatial distribution of the squared RMS values of the transient current along wire 1.

$g = 2 \cdot 10^9 \, \text{s}^{-1}$, $t_0 = 2$ ns. Figure 6.47a shows the transient current induced at the center of wire 1, while TD energy measures and TD power measure are shown in Fig. 6.47b and 6.47c, respectively. Finally, Fig. 6.47d and 6.47e shows spatial distribution of the RMS and squared RMS values of the transient current along wire 1, respectively.

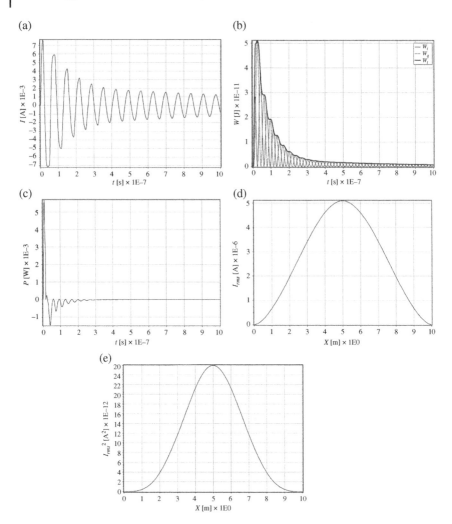

Figure 6.46 Configuration 4: (a) Transient current induced at the center of wire 2. (b) TD energy measures for wire 2. (c) TD power measure for wire 2. (d) Spatial distribution of the RMS value of the transient current along wire 2. (e) Spatial distribution of the squared RMS value of the transient current along wire 2.

Furthermore, Fig. 6.48a shows the transient current induced at the center of wire 2, while corresponding TD energy measures and TD power measure is shown in Fig. 6.48b and 6.48c, respectively. Finally, Fig. 6.48d and 6.48e shows spatial distribution of the RMS and squared RMS values of the transient current along wire 2, respectively.

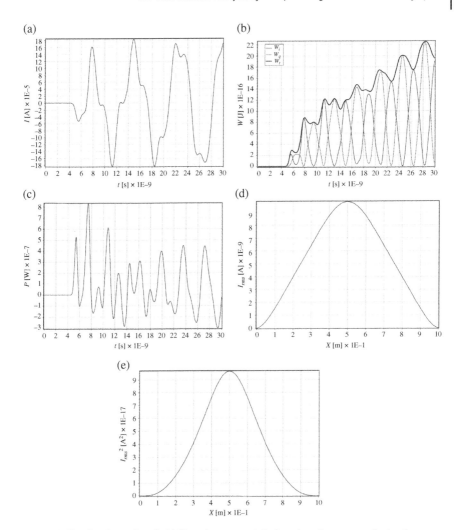

Figure 6.47 Configuration 5: (a) Transient current induced at the center of wire 1. (b) TD energy measures for wire 1. (c) TD power measure for wire 1. (d) Spatial distribution of the RMS values of the transient current along wire 1. (e) Spatial distribution of the squared RMS value of the transient current along wire 1.

Finally, Fig. 6.49a shows the transient current induced at the center of wire 3, while TD energy measures and TD power measure are shown in Fig. 6.49b and 6.49c, respectively. Finally, Fig. 6.49d and 6.49e shows spatial distribution of the RMS and squared values of the transient current along wire 3, respectively.

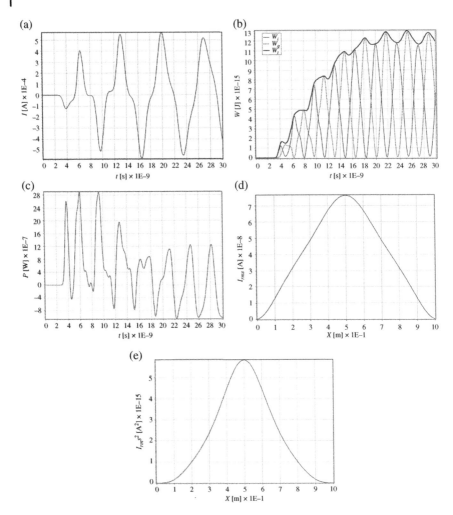

Figure 6.48 Configuration 5: (a) Transient current induced at the center of wire 2. (b) TD energy measures for wire 2. (c) TD power measure for wire 2. (d) Spatial distribution of the RMS value of the transient current along wire 2. (e) Spatial distribution of the squared RMS value of the transient current along wire 2.

It is obvious from Figs. 6.47b, 6.48b, and 6.49b that there is different behavior of the active (fed wire) from the passive wires 2 and 3. The building of the energy stored in the fields is most pronounced for wire 1 due to radiation from wire 3 and partly from reflection from the interface. The focusing of stored energy in the vicinity of feed gap area is also evident from Figs. 6.49d and 6.49e.

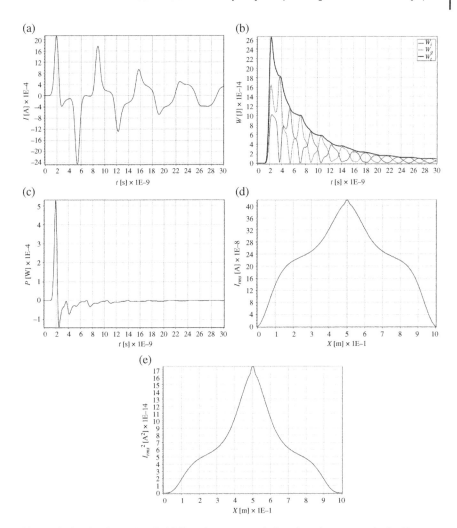

Figure 6.49 Configuration 5: (a) Transient current induced at the center of wire 3.
(b) TD energy measures for wire 3. (c) TD power measure for wire 3. (d) Spatial distribution
of the RMS value of the transient current along wire 3. (e) Spatial distribution of the
squared RMS value of the transient current along wire 3.

6.6.2.6 Configuration 6

Configuration 6 deals with thin wires assigned as *geometry 4* in the scattering mode
with wire length $L = 1$ m, wire radius $a = 2$ mm, heights above dielectric half-space
($\varepsilon_r = 10$), $h_1 = h_3 = 1$ m, $h_2 = 1.5$ m, and distance between wires $d = 0.5$ m and
illuminated with Gaussian pulse plane wave (7) with following parameters

$E_0 = 1 \, \text{V/m}$, $g = 2 \cdot 10^9 \, \text{s}^{-1}$, $t_0 = 2 \, \text{ns}$. Figure 6.50a shows the transient current induced at the center of wire 1, while TD energy measures and TD power measure are shown in Fig. 6.50b and 6.50c, respectively. Finally, Fig. 6.50d and 6.50e shows spatial distribution of the RMS and squared RMS values of the transient current along wire 1, respectively.

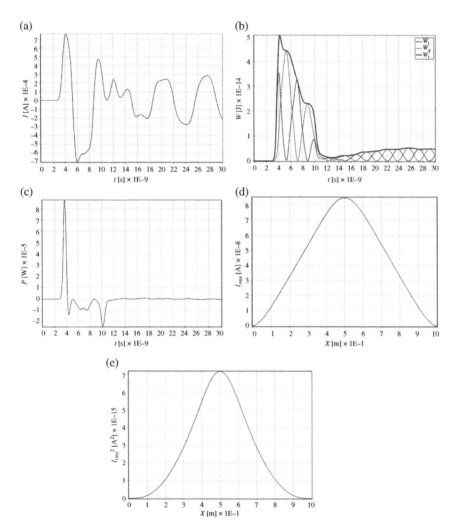

Figure 6.50 Configuration 6: (a) Transient current induced at the center of wires 1 and 3. (b) TD energy measures for wires 1 and 3. (c) TD power measure for wires 1 and 3. (d) Spatial distribution of the RMS value of the transient current along wires 1 and 3. (e) Spatial distribution of the squared RMS value of the transient current along wires 1 and 3.

Furthermore, Fig. 6.51a shows the transient current induced at the center of wire 2, while TD energy measures and TD power measure is shown in Fig. 6.51b and 6.51c, respectively. Finally, Fig. 6.51d and 6.51e shows spatial distribution of the RMS and squared RMS values of the transient current along wire 2, respectively.

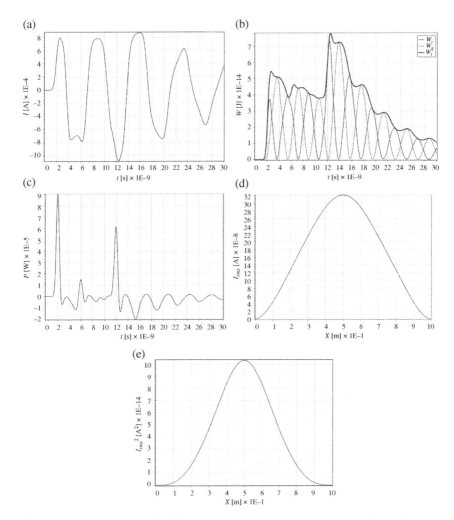

Figure 6.51 Configuration 6: (a) Transient current induced at the center of wire 2. (b) TD energy measures for wire 2. (c) TD power measure for wire 2. (d) Spatial distribution of the RMS values of the transient current along wire 2. (e) Spatial distribution of the squared RMS values of the transient current along wire 2.

It is visible from Fig. 6.50b that there is increase in energy due to reflection from wires 1 and 3, respectively. Analyzing the RMS and squared RMS values of current corresponding to the energy on unitary resistance, it is evident the energy is dominantly stored around the wire centers.

Observing all numerical results presented in Figs. 6.38 to 6.50, what can be concluded is that the energy stored in the wire near field depends on particular wire arrangement, type of half-space (PEC ground or dielectric half-space) and type and waveform of the excitation.

References

1 P. Parhami and R. Mittra, "Wire antennas over a lossy half-space," *IEEE Transactions on Antennas and Propagation*, vol. 28, no. 3, pp. 397–403, 1980.

2 S. Vitebskiy and L. Carin, "Moment method modeling of short-pulse scattering from and the resonances of a wire buried inside a lossy, dispersive half-space," *IEEE Transactions on Antennas and Propagation*, vol. 43, no. 11, pp. 1303–1312, 1995.

3 E. K. Miller, "Using adaptive sampling to minimize the number of samples needed to represent a transfer function," in *Antennas and Propagation Society International Symposium*, Baltimore, MD, USA, pp 588–591, 1996.

4 C. Y. Tham, A. McCowen, M. S. Towers and D. Poljak, "A dynamic adaptive sampling technique in frequency-domain transient analysis," *IEEE Transactions on Electromagnetic Compatibility*, vol. 44, no. 4, pp. 522–528, 2002.

5 S. Antonijević, D. Poljak, and J. Radić, "Optimized indirect time-domain analysis of the thin-wire structures," in *SoftCOM 2009, Conference Proceedings, Softcom Library*, pp. 194–198, 2009.

6 E. K. Miller and J. A. Landt, "Direct time-domain techniques for transient radiation and scattering from wires," *IEEE Transactions on Antennas and Propagation*, vol. 68, no. 11, pp. 1396–1424, 1980.

7 A. G. Tijhuis, "Toward a stable marching-on-in-time method for two-dimensional electromagnetic scattering problems," *Radio Science*, vol. 19, no. 5, pp. 1311–1317, 1984.

8 B. P. Rynne, "Stability and convergence of time marching methods in scattering problems," *IMA Journal of Applied Mathematics*, vol. 35, no. 3, pp. 297–310, 1985.

9 P. D. Smith, "Instabilities in time marching methods for scattering: cause and rectification," *Electromagnetics*, vol. 10, no. 4, pp. 439–451, 1990.

10 E. K. Miller, "A selective survey of computational electromagnetics," *IEEE Transactions on Antennas and Propagation*, vol. 36, no. 9, pp. 1281–1305, 1988.

11 D. Poljak, Advanced Modeling in Computational Electromagnetic Compatibility, New York, Nantes, France, Split: Wiley-Interscience, 2007.

12 D. Poljak and V. Roje, "Time domain calculation of the parameters of thin wire antennas and scatterers in a half-space configuration," *IEE Proceedings – Microwaves, Antennas and Propagation*, vol. 145, no. 1, pp. 57–63, 1998.

13 P. R. Barnes and F. M. Tesche, On the direct calculation of a transient plane wave reflected from a finitely conducting half-space, *IEEE Transactions on Electromagnetic Compatibility*, vol. 33, no. 2, pp. 90–6, 1991.

14 A. G. Tijhuis, Z. Q. Peng and A. R. Bretones, Transient excitation of a straight thin-wire segment: a new look at an old problem. *IEEE Transactions on Antennas and Propagation*, vol. 40, no. 10, pp. 1132–1146, 1992.

15 S. Antonijevic and D. Poljak, "Some optimizations of the Galerkin-Bubnov integral boundary element method," in *Proc. IceCOM 2013*, Dubrovnik, 2013.

16 Tesche, F. M., M. V. Ianoz and T. Karlsson, EMC Analysis Methods and Computational Methods, New York, USA: John Wiley & Sons, 1997.

17 Rothwell, E. J. and J. W. Suk, "Efficient computation of the time-domain TM plane-wave reflection coefficient", *IEEE Transactions on Antennas and Propagation*, vol. 52, no. 10, pp. 3417–3419, 2005.

18 E. J. Rothwell and M. J. Cloud, Electromagnetics, Boca Raton: CRC Press, 2001.

19 W. Yang, Z. Qinyu, Z. Naitong and X. Guangning, "Time domain calculation of UWB pulsed field reflected from a lossy half space," in *IEEE International Conference on Ultra-Wideband, 2007. ICUWB 2007*, pp. 24–26, 2007

20 W. Yang, Z. Naitong, Z. Qinyu and Z. Zhongzhao, "Simplified calculation of UWB signal transmitting through a finitely conducting slab," *Journal of Systems Engineering and Electronics*, vol. 19, no. 6, pp. 1070–1075, 2008.

21 Poljak, D. and S. Kresic, "A simplified calculation of transient plane waves in a presence of an imperfectly conducting half-space," in *Boundary Elements XXVII*, WIT Press, Southampton, pp. 541–549, 2005.

22 D. Poljak and N. Kovac, "Transient analysis of a finite length line embedded in a dielectric half-space using a simplified reflection/transmission coefficient approach," *International Conference on Electromagnetics in Advanced Applications, 2007. ICEAA 2007*, pp. 225–228, 2007.

23 H. Stehfest, "Numerical inversion of Laplace transforms algorithm, Algorithm 368", *Communication of the ACM*, vol. 13, no. 1, pp. 47–49, 1970.

24 S. Antonijevic and D. Poljak, "A novel time-domain reflection coefficient function: TM-case," *IEEE Transactions on Electromagnetic Compatibility*, vol. 55, no. 6, pp. 1147–1153, 2013.

25 H. Hassanzadeh and M. Pooladi-Darvish, "Comparison of different numerical Laplace inversion methods for engineering applications," *Applied Mathematics and Computation*, vol. 189, no. 2, pp. 1966–1981, 2007.

26 E. K. Miller, A. J. Poggio, G. J. Burke and E. S. Selden, "Analysis of wire antennas in the presence of a conducting half-space. Part I. The vertical antenna in free space," *Canadian Journal of Physics*, vol. 50, no. 9, pp. 879–888, 1972.

27 E. K. Miller, A. J. Poggio, G. J. Burke and E. S. Selden, "Analysis of wire antennas in the presence of a conducting half-space. Part 6. The horizontal antenna in free space," *Canadian Journal of Physics*, vol. 50, no. 21, pp. 2614–2627, 1972.

28 L. Pajewski, A. Benedetto, X. Derobert, A. Giannopoulos et al., "Applications of ground penetrating radar in civil engineering,", *COST Action TU1208*, 2013.

29 C. Warren, N. Chiwaridzo and A. Giannopoulos, "Radiation characteristics of a high-frequency antenna in different dielectric environments," *15th International Conference on Ground Penetrating Radar – GPR 2014*, Brussels, Belgium, pp. 796–801, 2014.

30 C. Warren, L. Pajewski, D. Poljak, A. Ventura et al., "A comparison of Finite-Difference, Finite-Integration, and Integral-Equation methods in the Time-Domain for modelling Ground Penetrating Radar antennas," in The Proceedings of 2016 16th International Conference of Ground Penetrating Radar (GPR). New York: IEEE, pp. 1–5, 2016.

31 D. Poljak, S. Sesnic, A. Susnjara, D. Paric et al., "Direct time domain evaluation of the transient field transmitted into a lossy ground due to GPR antenna radiation," *Engineering Analysis with Boundary Elements*, vol. 82, pp. 27–31, 2017.

32 L. Gürel and U. Oguz, "Three-dimensional FDTD modeling of a ground-penetrating radar," *IEEE Transactions on Geoscience and Remote Sensing*, vol. 38, no 4, pp. 1513–1521, 2000.

33 N. J. Cassidy and T. M. Millington, "The application of finite-difference time-domain modelling for the assessment of GPR in magnetically lossy materials," *Journal of Applied Geophysics*, vol. 67, no. 4, pp. 296–308, 2009.

34 P. Shangguan and I. L. Al-Qadi, "Calibration of FDTD simulation of GPR signal for asphalt pavement compaction monitoring," *Geoscience and Remote Sensing, IEEE Transactions on*, vol. 53, no. 3, pp. 1538–1548, 2015.

35 D. Poljak, M. Cvetković, O. Bottauscio, A. Hirata et al., "On the use of conformal models and methods in dosimetry for non-uniform field exposure," *IEEE Transactions on Electromagnetic Compatibility*, vol. 60, no. 2, pp. 328–337, 2018.

36 D. Poljak, S. Sesnic, D. Paric and K. El Khamlichi Drissi, "Direct time domain modeling of the transient field transmitted in a dielectric half-space for GPR applications," in *Electromagnetics in Advanced Applications (ICEAA), 2015 International Conference on*, 7–11 Sept. 2015, pp. 345–348, 2015.

37 D. Poljak, S. Antonijevic, S. Sesnic, S. Lallechere et al., "On deterministic-stochastic time domain study of dipole antenna for GPR applications," *Engineering Analysis with Boundary Elements*, vol. 73, pp. 14–20, 2016.

38 D. Poljak, S. Sesnic, A. Susnjara, D. Paric et al., "Frequency domain and time domain analysis of the transient field radiated by GPR antenna," in *Proc. SpliTECH 2018*, Split, 2018.

39 D. Poljak, S. Šesnić, A. Šušnjara, D. Parić and S. Antonijević, "Transient calculation of the electric field above a lossy ground generated by dipole antenna, direct time domain analysis," in *GPR 2018*, Rappersville, 2018.

40 D. Poljak and S. Sesnic, "Simplified assessment of specific absorption (SA) in planar human tissue model due to transient radiation from dipole antenna," in *CEM 2020*, Lyon, France, 2020.

41 J.-Y. Chen and O. P. Gandhi, "Currents induced in an anatomically based model of a human for exposure to vertically polarized electromagnetic pulses," *IEEE Transactions on Microwave Theory and Techniques*, vol. 39, no. 1, pp. 31–39, 1991.

42 J.-Y. Chen, C. M. Furse and O. P. Gandhi, "A simple convolution procedure for calculating currents induced in the human body for exposure to electromagnetic pulses," *IEEE Transactions on Microwave Theory and Techniques*, vol. 42, no. 7, pp. 1172–1175, 1994.

43 O. P. Gandhi and C. M. Furse, "Currents induced in the human body for exposure to ultrawideband electromagnetic pulses," *IEEE Transactions on Electromagnetic Compatibility*, vol. 39, no. 2, pp. 174–180, 1997.

44 D. Poljak, V. Doric, S. Antonijevic and V. Roje, "Time domain measure for power flow of thin wire electromagnetic field," *2006 IEEE Antennas and Propagation Society International Symposium*, Albuquerque, NM, USA, pp. 2947–2950, 2006.

45 O. P. Gandhi and E. E. Aslan, "Human-equivalent antenna for electromagnetic fields," U.S. Patent 5,394,164, Feb. 28, 1995.

46 D. Poljak, E. K. Miller and C. Y. Tham, "Root-mean-square measure of nonlinear effects to the transient response of thin wires," *IEEE Transactions on Antennas and Propagation*, vol. 51, no. 12, pp. 3280–3283, 2003.

47 D. Poljak, C. Y. Tham, O. Gandhi and A. Sarolic, "Human equivalent antenna model for transient electromagnetic radiation exposure," *IEEE Transactions on Electromagnetic Compatibility*, vol. 45, no. 1, pp. 141–145, 2003.

48 D. Poljak and M. Cvetkovic, Human Interaction with Electromagnetic Fields, St. Louis, USA: Elsevier, 2019.

49 E. K. Miller and J. A. Landt, "Direct time-domain techniques for transient radiation and scattering from wires," *Proceedings of the IEEE*, vol. 168, no. 11, pp 1396–1423, 1980.

50 ICNIRP Guidelines, Guidelines for limiting exposure to time-varying electric, magnetic, and electromagnetic fields (up to 300 GHz), *Health Physics*, vol. 74, no. 4, pp. 494–522, 1998.

51 ITIS Foundation, "Tissue Properties Database," [Online], Available: https://itis. swiss/virtual-population/tissue-properties/database/.

52 D. Poljak, E. K. Miller and C. Y. Tham, "Time-domain energy measures for thin-wire antennas and scatterers," *IEEE Antennas and Propagation Magazine*, vol. 44, no. 1, pp. 87–95, 2002.

53 E. K. Miller, "The proportionality between charge acceleration and radiation from a generic wire object," *Progress in Electromagnetics Research*, vol. 162, no. 15–19, pp. 15–29, 2018.

54 D. Poljak, S. Antonijevic, V. Doric, E. Miler and E. K. D. Khalil, Transient Analysis of GPR Dipole Antenna using Time Domain Energy Measures, Bordeaux, France: NSG, 2021.

55 D. Poljak, C. Y. Tham and A. McCowen, "Transient response of nonlinearly loaded wires in a two media configuration," *IEEE Transactions on Electromagnetic Compatibility*, vol. 46, no. 1, pp. 121–125, 2004.

56 D. Poljak, S. Antonijevic, K. El Khamlichi Drisi and K. Kerroum, "Transient response of straight thin wires located at different heights above a ground plane using antenna theory and transmission line approach," *IEEE Transactions on Electromagnetic Compatibility*, vol. 52, no. 1, pp. 108–116, 2010.

57 S. Antonijević, D. Poljak and C. A. Brebbia, "Time-domain analysis of multiple straight thin wires in a non-homogeneous medium," *Engineering Analysis with Boundary Elements*, vol. 33, no. 5, pp. 627–636, 2009.

58 D. Poljak, S. Antonijevic and E. K. Miller, "Transient analysis of multiple straight wires in a presence of a half-space using different time domain measures," *Journal of Electromagnetic Waves and Applications*, vol. 36, no. 11, pp. 1485–1530, 2022.

7

Bioelectromagnetics – Exposure of Humans in GHz Frequency Range

Fifth-generation (5G) mobile communication systems operating in GHz frequency range have caused significant public concern regarding potential adverse health effects. Recently released documents: IEEE-Std. C95.1, 2019 and ICNIRP-RF guidelines, 2020 have merged toward 6 GHz as transition frequency from specific absorption rate (SAR) to incident power density (IPD) – pertaining to free space, and absorbed power density (APD) – pertaining to the skin surface [1, 2]. Thus, above 6 GHz, the use of APD averaged over a specific area is suggested in the near field region instead of widely used basic restriction quantity SAR averaged over tissue volume [3–5]. In addition, related surface temperature increase in the eye and skin is of interest as a relevant biological effect. An illustrative review on the subject has been presented elsewhere, for example, in [4].

Human exposure to 5G mobile communication systems above 6 GHz may result in local surface temperature increase, as stated in [1, 2]. Note that well-established dosimetric quantity SAR averaged over tissue volume is replaced by APD averaged over a specific area above transition frequency of 6 GHz. The use of power density instead of SAR above a transition frequency has been discussed in number of papers, for example [3–8].

Note that though in many papers the quantity IPD is used, the quantity APD (S_{ab}) is used elsewhere, e.g. in [9, 10] as proposed in ICNIRP 2020 guidelines pertaining to the basic restriction, stating that IPD pertains to the external quantity, i.e. the reference level.

Furthermore, due to the appreciable skin effect in GHz frequency range, the related surface temperature increase in the eye and skin, respectively, is of interest as a relevant biological effect due to such exposures. Particularly, the correlation between the temperature elevation and S_{ab} in terms of heating factors, and choice of averaging area are of interest. The choice of averaging area 1 and $4 \, \text{cm}^2$ is discussed in ICNIRP 2020 [1]. It is worth mentioning that heating factors have been recently discussed within the framework of ICES, SC6, and WG for IPD definition featuring the dipole above a flat multilayer model of the human tissue.

Thus, the assessment of IPD for a simplified case of Hertz dipole analyzing a dependence of the IPD in free space and S_{ab} in front of a lossy medium on the incident angle, frequency, and distance from the interface has been carried out in [8, 9], respectively.

Continuation and extension of the work reported in [8, 9] have been carried out in [10], which involve planar homogeneous tissue exposed to the radiation of finite length dipole antenna. Frequency-dependent tissue parameters can be found in [11]. Assessment of TPD in single layer tissue has been reported in [12, 13], for the case of Hertz dipole excitation and in [14, 15], for the case of finite-length dipole.

Finally, the multilayer planar tissue models for the assessment of IPD, heating factors, S_{ab}, and TPD have been analyzed in number of papers, e.g. [16–18].

In next subsections, the assessment of S_{ab} and TPD in homogeneous and heterogeneous tissue models (two-layer and three-layer) is given.

7.1 Assessment of Sab in a Planar Single Layer Tissue

As discussed in detail in [18], the concept of APD is used for the estimation of human exposure to devices located in close proximity to the body. Since antenna-body distance is much smaller than body dimension, the body can be represented by planar geometry as depicted in Fig. 7.1: a dipole antenna of length L and radius a is placed at the distance h in front of the planar tissue.

Muscle tissue of relative permittivity ε_r and conductivity σ, both depending on frequency is represented by a flat single-layer medium (lossy half-space).

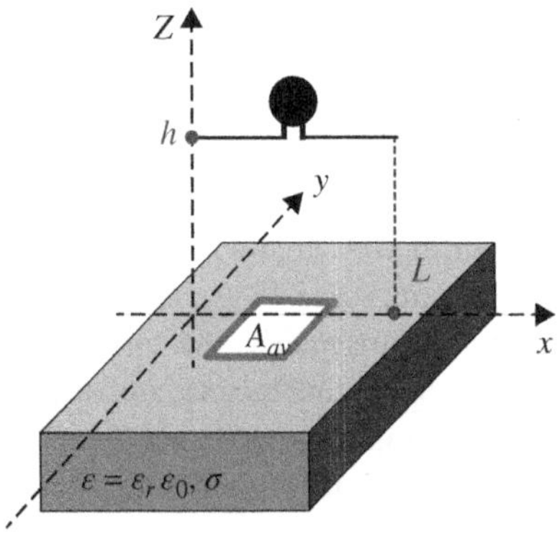

The theoretical background for the APD arises from the Poynting theorem which expresses a conservation of the energy stored in electric and magnetic fields within a volume V of interest enclosed by an area A. The time-harmonic Poynting vector S is given by:

Figure 7.1 Dipole antenna in front of planar tissue.

$$\vec{S} = \frac{1}{2} \, \mathrm{Re}\left(\vec{E} \times \vec{H}^{*}\right) \qquad (7.1)$$

where E and H are the electric and the magnetic field, respectively, due to a given radiation source.

Furthermore, absorbed power density S_{ab} adopted in ICNIRP 2020 is as follows [1, 10]:

$$S_{ab} = \frac{1}{2A_{av}} \int_{A_{av}} \text{Re}\left(\vec{E} \times \vec{H}^*\right) d\vec{A} \tag{7.2}$$

where A_{av} is the averaging area, as depicted in Fig. 7.1.

The choice of averaging area has been discussed in a number of papers. A circular area of 1 cm^2 has been used in [1, 6], while some studies suggest that the choice of averaging area of 2 or 4 cm^2 can be correlated with the average mass of 10 g for local SAR [7]. Finally, the rationale for the choice of 4 cm^2 and 1 cm^2, averaging area has been explained and adopted in ICNIRP 2020 guidelines [1].

The impact of a particular averaging area on IPD and S_{ab} due to the Hertz dipole radiation in free space and S_{ab} in the presence of a lossy half-space has been reported in [8, 9], respectively.

More realistic case of dipole antenna above an interface has been recently addressed in [10].

7.1.1 Analysis of Dipole Antenna in Front of Planar Interface

Consider dipole antenna of length L and radius a with constant current distribution I_0. The dipole is located parallel to the planar lossy half-space. According to the theory of images, one deals with a system of two dipole antennas as indicated in Fig. 7.2.

The current induced along the dipole antenna is governed by the Pocklington integro-differential equation [10]:

$$E_x^{exc} = j\omega \frac{\mu}{4\pi} \int_0^L I(x')g(x,x')dx'$$

$$- \frac{1}{j4\pi\omega\varepsilon_0} \frac{\partial}{\partial x} \int_0^L \frac{\partial I(x')}{\partial x'} g(x,x')dx' \tag{7.3}$$

where E_x^{exc} is the known excitation in the form of tangential electric field due to an equivalent voltage generator,

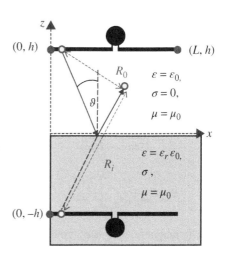

Figure 7.2 Dipole antenna and its image.

$I(x')$ is the induced current along the antenna, and $g(x, x')$ is the total Green's function:

$$g(x, x') = g_0(x, x') - R_{TM} g_i(x, x') \tag{7.4}$$

where $g_0(x, x')$ is the free space-Green's function:

$$g_0(x, x') = \frac{e^{-jk_0 R_0}}{R_0} \tag{7.5}$$

while $g_i(x, x')$ arises from the image theory:

$$g_i(x, x') = \frac{e^{-jk_0 R_i}}{R_i} \tag{7.6}$$

where R_0 and R_i denote the corresponding distance from the source to the observation point, respectively (Fig. 7.2), and k_0 is the free space wave number.

The reflection from the interface has been taken into account by the Fresnel reflection coefficient for transverse magnetic (TM) polarization [10]

$$R_{TM} = \frac{n \cos\theta - \sqrt{n - \sin^2\theta}}{n \cos\theta + \sqrt{n - \sin^2\theta}} \tag{7.7}$$

where ϑ is the incidence angle (Fig. 7.2) and refraction index n is given by:

$$n = \varepsilon_r - j \frac{\sigma}{\omega\varepsilon_0} \tag{7.8}$$

where ε_r and σ represent the relative permittivity and the electrical conductivity of the half-space, respectively, while ω is the operating angle frequency, $\omega = 2\pi f$.

The electric field components are as follows [10]:

$$E_x = \frac{1}{j4\pi\omega\varepsilon_{eff}} \left[-\int_0^L \frac{\partial I(x')}{\partial x'} \frac{\partial g(x, y, x')}{\partial x} dx' - \gamma^2 \int_{-L/2}^{L/2} I(x') g(x, x') dx' \right] \tag{7.9}$$

$$E_y = \frac{1}{j4\pi\omega\varepsilon_{eff}} \int_0^L \frac{\partial I(x')}{\partial x'} \frac{\partial g(x, y, x')}{\partial y} dx' \tag{7.10}$$

$$E_z = \frac{1}{j4\pi\omega\varepsilon_{eff}} \int_0^L \frac{\partial I(x')}{\partial x'} \frac{\partial g(x, y, x')}{\partial z} dx' \tag{7.11}$$

The magnetic field is obtained as a curl of magnetic vector potential:

$$\vec{H} = \frac{1}{\mu} \nabla \times \vec{A} \tag{7.12}$$

which for the case potential of horizontal dipole antenna has only x-component:

$$A_x = \frac{\mu}{4\pi} \int_0^L I(x')g(x,y,z,x')dx' \tag{7.13}$$

Inserting (7.13) in (7.12) and applying the curl operator, one obtains:

$$H_y = \frac{\partial A(x,y,z)}{\partial z} \tag{7.14}$$

$$H_z = -\frac{\partial A(x,y,z)}{\partial y} \tag{7.15}$$

and the magnetic field components are

$$H_y = \frac{\mu}{4\pi} \int_0^L I(x')\frac{\partial g(x,y,z,x')}{\partial z}dx' \tag{7.16}$$

$$H_z = -\frac{\mu}{4\pi} \int_0^L I(x')\frac{\partial g(x,y,z,x')}{\partial y}dx' \tag{7.17}$$

However, when computing the electric and magnetic field components, besides the reflection coefficient for TM polarization R_{TM}, the reflection coefficient for transverse electric polarization R_{TE} is also used for the computation of the corresponding field components [10].

$$R_{TE} = \frac{\cos\theta - \sqrt{n - \sin^2\theta}}{\cos\theta + \sqrt{n - \sin^2\theta}} \tag{7.18}$$

The essential step in the analysis of dipole antenna is the determination of equivalent axial current distribution. Once the current distribution is known, it is possible to calculate the field components of interest. As it is well-known, in vast antenna literature, the approach is in general twofold. First, simplified approach is based on the assumed, mostly sinusoidal current distribution, while the more rigorous approach includes the numerical solution of integro-differential equation (7.3). The approach used in this subsection previously reported in [10], similar to the analysis of ground-penetrating radar (GPR) dipole antenna in [19] is based on both analytical and numerical approaches.

Assuming the sinusoidal current distribution in the form [10]

$$I_x(x') = I_x(0)\frac{\sin\left[k_0\left(\frac{L}{2} - |x'|\right)\right]}{\sin(k_0h)} = I_m\sin\left[k_0\left(\frac{L}{2} - |x'|\right)\right] \tag{7.19}$$

and taking the derivative of current as follows:

$$\frac{\partial I(x')}{\partial x'} = -k_0 I_m \cos\left[k_0\left(\frac{L}{2} - |x'|\right)\right] \cdot \mathrm{sgn}(x') \tag{7.20}$$

The field components of electric field from Eqs. (7.9)–(7.11) become:

$$E_x = \frac{1}{j4\pi\omega\varepsilon_{eff}} \cdot \left(\begin{array}{l} \displaystyle\int_0^L k_0 I_m \cos\left[k_0\left(\frac{L}{2} - |x'|\right)\right] \mathrm{sgn}(x') \frac{\partial g(x,y,x')}{\partial x} dx' \\[2mm] + k_0^2 \displaystyle\int_0^L I_m \sin\left[k_0\left(\frac{L}{2} - |x'|\right)\right] g(x,x') dx' \end{array} \right) \tag{7.21}$$

$$E_y = \frac{-1}{j4\pi\omega\varepsilon_{eff}} \cdot \int_0^L k_0 I_m \cos\left[k_0\left(\frac{L}{2} - |x'|\right)\right] \mathrm{sgn}(x') \frac{\partial g(x,y,x')}{\partial y} dx' \tag{7.22}$$

$$E_z = \frac{-1}{j4\pi\omega\varepsilon_{eff}} \cdot \int_0^L k_0 I_m \cos\left[k_0\left(\frac{L}{2} - |x'|\right)\right] \mathrm{sgn}(x') \frac{\partial g(x,y,x')}{\partial z} dx' \tag{7.23}$$

and the magnetic field stems from (7.16)–(7.17) and is given by:

$$H_y = \frac{\mu}{4\pi} \cdot \int_{-L/2}^{L/2} I_m \sin\left[k_0\left(\frac{L}{2} - |x'|\right)\right] \frac{\partial g(x,y,z,x')}{\partial z} dx' \tag{7.24}$$

$$H_z = -\frac{\mu}{4\pi} \cdot \int_{-L/2}^{L/2} I_m \sin\left[k_0\left(\frac{L}{2} - |x'|\right)\right] \frac{\partial g(x,y,z,x')}{\partial y} dx' \tag{7.25}$$

Numerical solution of integro-differential Eq. (7.3) is carried out via the Galerkin–Bubnov variant of the Indirect Boundary Element Method (GB-IBEM) [11, 12].

Once the induced current is determined the related radiated electric and magnetic field induced at the interface is obtained via numerical/analytical evaluation of corresponding field integrals. The mathematical details on GB-IBEM procedure and numerical evaluation of field integrals (are available elsewhere, e.g. in [12].

7.1.2 Calculation of Absorbed Power Density

According to [1], S_{ab} is calculated for two averaged areas $A = 4\,\mathrm{cm}^2$ and $A = 1\,\mathrm{cm}^2$, as depicted in Fig. 7.8 Now, for the case of normal incidence considered in this work, Fig. 7.9, taking into account existing components of electromagnetic fields one obtains:

$$S_{ab} = \frac{1}{2A_{av}} \int\limits_{A_{av}} Re\left(E_x \cdot H_y^*\right) \cdot dA = \frac{1}{2\Delta x \Delta y} \int\limits_{0}^{\Delta x} \int\limits_{0}^{\Delta y} Re\left(E_x \cdot H_y^*\right) \cdot dx dy \quad (7.26)$$

where E_x and H_y are obtained from expressions (7.9), (7.16), (7.21), and (7.24), respectively.

7.1.3 Computational Examples

Computational examples deal with four frequencies of interest with dielectric half-space modeled as muscle tissue. Antenna is modeled as a thin dipole and three lengths are considered at each frequency of interest. Frequency-dependent tissue parameters taken from [11] and antenna lengths are given in Table 7.1. For each combination of frequency and antenna length, three heights above the planar tissue are observed: 5, 10, and 20 mm.

All computations are carried out for both 1 and 4 cm² control surfaces.

The tangential component of the electric field, E_x and tangential component of magnetic field, H_y used in eq. (7.26) for S_{ab} computation are depicted in Figs. 7.3 and 7.4 The frequency is 6 GHz and antenna is the half-wave dipole placed 20 mm above the tissue. One can observe that the field values obtained by using numerical and analytical current solutions differ only slightly.

The value of APD at given frequencies is depicted in Figs. 7.12a and 7.13a in case of numerical current solution. Different combinations of antenna lengths and heights are considered, i.e. $L = \lambda/2$, $\lambda/4$, and $\lambda/10$ and $h = 5$, 10, and 20 mm. S_{ab} values for analytical current solution are not given in the figure plot; instead, the absolute relative error is presented in Figs. 7.5 and 7.6.

The absolute relative error is computed as: $ERR_{ABS} = \left|(S_{ab\text{-}numerical} - S_{ab\text{-}analytical})/(S_{ab\text{-}numerical})\right|$.

Table 7.1 Frequency-dependent parameters of lossy half-space [11] and antenna lengths.

f (GHz)	6	10	30	60
ε_r	48.2	42.8	23.2	12.9
σ (S/m)	5.2	10.6	35.5	52.8
$L = \lambda/2$ (cm)	2.5	1.5	0.5	0.25
$L = \lambda/4$ (cm)	1.25	0.75	0.25	0.125
$L = \lambda/10$ (cm)	0.5	0.3	0.1	0.05

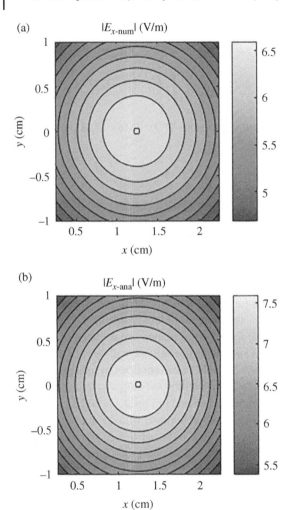

Figure 7.3 Distribution of the electric field × component on the surface $A = 4\,\text{cm}^2$. Operating frequency is 6 GHz and antenna height is 20 mm. (a) $|E_x|$ with numerical current solution, (b) $|E_x|$ with analytical current solution.

Observing the calculated results for the APD in Figs. 7.4–7.6, the following conclusions can be drawn:

- For the half-wave dipole, the difference between S_{ab} values computed using the numerical and analytical current solutions is more pronounced for frequencies of 6 GHz and 10 GHz than 30 and 60 GHz.

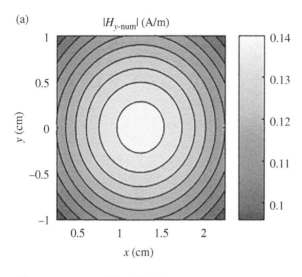

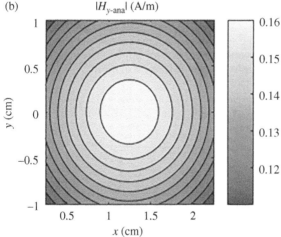

Figure 7.4 Distribution of the magnetic field y component on the surface $A = 4\,\text{cm}^2$. Operating frequency is 6 GHz and antenna height is 20 mm. (a) $|H_y|$ with numerical current solution, (b) $|H_y|$ with numerical current solution.

- This is not the case, however, for dipoles of $L = \lambda/4$ and $L = \lambda/10$ length where numerical and analytical approaches differ more at higher frequency values. Note that the Err_{abs} has the same frequency distribution for antennas $L = \lambda/4$ and $L = \lambda/10$.
- The frequency distribution of the Err_{abs} for the control surfaces of 1 and 4 cm^2 is the same.

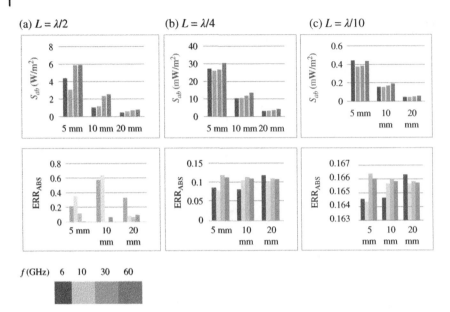

Figure 7.5 The values of absorbed power density are in the top row and absolute relative error in the bottom row for control surface of $A = 1$ cm^2 for three antenna lengths (a–c). Note that S_{ab} in (a) is in W/m^2 while in (b) and (c) S_{ab} is in mW/m^2.

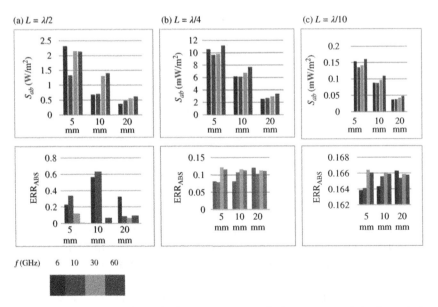

Figure 7.6 The values of absorbed power density are in the top row and absolute relative error in the bottom row for control surface of $A = 4$ cm^2 for three antenna lengths (a–c). Note that S_{ab} in (a) is in W/m^2 while in (b) and (c), S_{ab} is in mW/m^2.

- S_{ab} changes slowly with frequency for the antenna-body distance larger than 5 mm.
- Decreasing the antenna length, the S_{ab} values tend to be smaller.
- Also, Err_{abs} values decrease with antenna length.

The field integrals are solved numerically and the impact of the two-media interface is taken into account by means of the reflection coefficients arising from the plane wave assumption. The small difference in the field and APD values obtained via numerically computed and approximate current distribution can be noticed.

While increasing the operating frequency, the influence of antenna-body distance decreases for the case of half-wave dipole, the effect is quite opposite for electrically smaller antennas ($\lambda/4$, $\lambda/10$).

7.2 Assessment of Transmitted Power Density in a Single Layer Tissue

An alternative quantity to Sab in dosimetry pertaining to GHz frequency region is transmitted power density (TPD). A simplified assessment of TPD due to the Hertz dipole radiating in the presence of a lossy medium featuring Fresnel coefficient approximation and Modified Image Theory (MIT) has been carried out, e.g. in [12, 13], respectively.

An extension of the work reported in [12, 13] (pertaining to Hertz dipole case) dealing with the straight wire of finite length is available in [14, 15]. The work presented in [14, 15] includes the description of the mathematical model for the assessment of parameters of interest and some illustrative results for current distribution along the dipole antenna. Human tissue is modeled as a flat conducting half-space with muscle properties. As in [12, 13], the antenna current assessment is undertaken twofold: first the approximate sinusoidal current distribution is used; an alternative pertains to a calculated current distribution which is governed by the Pocklington integro-differential equation for the wires above a lossy half-space.

Once the current distribution is either assumed or calculated, the field transmitted into the lossy medium representing the human tissue and volume power density (VPD) can be evaluated. Finally, TPD versus tissue depth can be computed by integrating the squared value of the electric field. This task has been carried out in [15] in which the results for the transmitted field, VPD, and TPD obtained analytically and numerically, respectively, depending on the determination of the current distribution are reported.

7.2.1 Formulation

Dipole of radius a and length L horizontally located in front of a lossy half-space (planar tissue) (Fig. 7.7) is considered.

The prerequisite to evaluate the fields, VPD and TPD is the knowledge of current distribution.

The first step in the analysis is to determine the current distribution along the wire.

As it is well-known from the wire antenna theory, the current distribution $I(x')$ along the dipole antenna is governed by the Pocklington integro-differential equation [14]:

$$E_x^{exc} = \frac{1}{j4\pi\omega\varepsilon_0} \left[\int_{-L/2}^{L/2} \frac{\partial I(x')}{\partial x'} \frac{\partial g_a\left(\vec{r},x'\right)}{\partial x} dx' - k_0^2 \int_{-L/2}^{L/2} I(x') g_a\left(\vec{r},x'\right) dx' \right] \tag{7.27}$$

where k_0 denotes the wave number of free space and the total Green's function is:

$$g_a(x,x') = g_{a0}(x,x') - \Gamma_{TM}^{ref}(\theta') g_{ai}(x,x') \tag{7.28}$$

with the Green function of an unbounded lossless medium given by:

$$g_a\left(\vec{r},x'\right) = \frac{e^{-jkR_a}}{R_a} \tag{7.29}$$

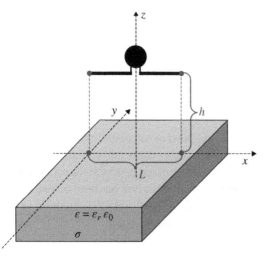

Figure 7.7 Horizontal dipole antenna in front of a lossy half space (planar tissue).

The influence of the two-media interface is taken into account via the Fresnel plane wave reflection coefficient [14]:

$$\Gamma_{TM}^{ref}(\theta) = \frac{n\cos\theta - \sqrt{n - \sin^2\theta}}{n\cos\theta + \sqrt{n - \sin^2\theta}} \tag{7.30}$$

where the refraction index n is:

$$n = \varepsilon_r - j\frac{\sigma}{\omega\varepsilon_0} \tag{7.31}$$

and ε_r is relative permittivity of a tissue.

The assumed sinusoidal current distribution:

$$I(x') = I_m \sin k_0 \left(\frac{L}{2} - |x'|\right) \tag{7.32}$$

stems from the analytical solution of integro-differential equation (7.27) in free space under set of approximations [14]. Furthermore, so-called exact current distribution can be obtained by numerically solving Pocklington equation (7.27).

The solution of (7.27) via Galerkin–Bubnov Indirect Boundary Element Method (GB-IBEM) has been reported elsewhere, e.g. in [20, 21].

Next step is the electric field computation in the lower medium. Provided the current distribution is known, either assumed or calculated, the transmitted field components are as follows:

$$E_x\left(\vec{r}\right) = \frac{1}{j4\pi\omega\varepsilon_{eff}} \left[\begin{array}{c} \displaystyle\int_{-L/2}^{L/2} \frac{\partial I(x')}{\partial x'} \frac{\partial g\left(\vec{r},x'\right)}{\partial x} dx' - \\[4mm] \displaystyle -\gamma^2 \int_{-L/2}^{L/2} I(x')g\left(\vec{r},x'\right) dx' \end{array} \right] \tag{7.33}$$

$$E_y\left(\vec{r}\right) = \frac{1}{j4\pi\omega\varepsilon_{eff}} \int_{-L/2}^{L/2} \frac{\partial I(x')}{\partial x'} \frac{\partial g\left(\vec{r},x'\right)}{\partial y} dx' \tag{7.34}$$

$$E_z\left(\vec{r}\right) = \frac{1}{j4\pi\omega\varepsilon_{eff}} \int_{-L/2}^{L/2} \frac{\partial I(x')}{\partial x'} \frac{\partial g\left(\vec{r},x'\right)}{\partial z} dx' \tag{7.35}$$

where ε_{eff} is the complex permittivity of the ground while γ is the complex propagation constant of a lossy medium (human tissue):

$$\gamma = \sqrt{(j\omega\mu\sigma - \omega^2\mu\varepsilon)} \tag{7.36}$$

The total Green's function can be written in the form:

$$g\left(\vec{r},x'\right) = \Gamma_{TM}(\theta)g_0\left(\vec{r},x'\right) \tag{7.37}$$

where Green's function of unbounded lossy medium is:

$$g_0\left(\vec{r},x'\right) = \frac{e^{-\gamma R}}{R} \tag{7.38}$$

and plane wave transmission coefficient is given by:

$$\Gamma_{TM}^{tr}(\theta) = \frac{2n\cos\theta}{n\cos\theta + \sqrt{n-\sin^2\theta}} \tag{7.39}$$

where θ is an angle of incidence.

To calculate the field by assuming the sinusoidal current distribution, one has to determine the current derivative, i.e.

$$\frac{\partial I(x')}{\partial x'} = -k_0 I_m \cos\left(k_0\left(\frac{L}{2}-|x'|\right)\right) \cdot \frac{x}{|x'|} \tag{7.40}$$

And the expressions for the field components are:

$$E_x\left(\vec{r}\right) = -\frac{1}{j4\pi\omega\varepsilon_{eff}} \left[\begin{array}{l} -\displaystyle\int_{-L/2}^{L/2} k_0 I_m \cos\left(k_0\left(\frac{L}{2}-|x'|\right)\right) \cdot \frac{x}{|x'|} \cdot \frac{\partial g\left(\vec{r},x'\right)}{\partial x} dx' - \\ \\ -\gamma^2 \displaystyle\int_{-L/2}^{L/2} I_m \sin k_0\left(\frac{L}{2}-|x'|\right) g\left(\vec{r},x'\right) dx' \end{array} \right] \tag{7.41}$$

$$E_y\left(\vec{r}\right) = -\frac{1}{j4\pi\omega\varepsilon_{eff}} \int_{-L/2}^{L/2} k_0 I_m \cos\left(k_0\left(\frac{L}{2}-|x'|\right)\right) \cdot \frac{x}{|x'|} \cdot \frac{\partial g\left(\vec{r},x'\right)}{\partial y} dx' \tag{7.42}$$

$$E_z\left(\vec{r}\right) = -\frac{1}{j4\pi\omega\varepsilon_{eff}} \int_{-L/2}^{L/2} k_0 I_m \cos\left(k_0\left(\frac{L}{2}-|x'|\right)\right) \cdot \frac{x}{|x'|} \cdot \frac{\partial g\left(\vec{r},x'\right)}{\partial z} dx' \tag{7.43}$$

Evaluation of integrals (7.41)–(7.43) can be performed numerically combined with finite difference approximation of derivatives of the total Green's function to avoid quasi-singularity problems [21].

Furthermore, numerical computation of field integrals (7.33)–(7.35) using the numerically obtained current distribution via GB-IBEM is available elsewhere, e.g. in [20, 21].

Finally, one deals with the evaluation of the VPD and TPD.

The VPD in lossy media is given by:

$$VPD = \frac{1}{2}\sigma \left| E\left(\vec{r}\right) \right|^2 \tag{7.44}$$

The broadside field component is of interest [12, 13] and it can be written as:

$$VPD = \frac{1}{2}\sigma |E_z(x,y,z)|^2 \tag{7.45}$$

Now, inserting z-component of the electric field (7.43) into (7.44) yields:

$$VPD = \frac{1}{2}\sigma \left| \frac{1}{j4\pi\omega\varepsilon_{eff}} \int_{-L/2}^{L/2} \frac{\partial I(x')}{\partial x'} \frac{\partial g\left(\vec{r},x'\right)}{\partial z} dx' \right|^2 \tag{7.46}$$

Furthermore, the TPD is defined as follows [7]:

$$TPD = \frac{1}{2} \int_0^r \sigma \left| E\left(\vec{r}\right) \right|^2 dr \tag{7.47}$$

where r is the variable perpendicular to the surface of the human body and point $r = 0$ refers to the air–body interface, while σ stands for tissue conductivity.

For the geometry considered in this paper, Fig. 7.7, TPD is given by:

$$TPD(x,y) = \frac{1}{2} \int_0^r \sigma |E(x,y,z)|^2 dz \tag{7.48}$$

And inserting (7.32) into (7.40), one obtains

$$TPD(x,y) = \frac{1}{2} \int_0^z \sigma \left| \frac{1}{j4\pi\omega\varepsilon_{eff}} \int_{-L/2}^{L/2} \frac{\partial I(x')}{\partial x'} \frac{\partial g(x,y,z,x')}{\partial z} dx' \right|^2 dz \tag{7.49}$$

If the sinusoidal current distribution of (7.32) and (7.40) is assumed, expressions (7.46) and (7.49) become:

$$VPD(x,y) = \frac{1}{2}\sigma \left| -\frac{1}{j4\pi\omega\varepsilon_{eff}} \int_{-L/2}^{L/2} k_0 I_m \cos\left(k_0\left(\frac{L}{2} - |x'|\right)\right) \cdot \frac{x}{|x'|} \cdot \frac{\partial g\left(\vec{r},x'\right)}{\partial z} dx' \right|^2 \tag{7.50}$$

$$\text{TPD}(x,y) = \frac{1}{2}\int_0^z \sigma \left| -\frac{1}{j4\pi\omega\varepsilon_{\text{eff}}} \int_{-L/2}^{L/2} k_0 I_m \cos\left(k_0\left(\frac{L}{2} - |x'|\right)\right) \cdot \frac{x}{|x'|} \cdot \frac{\partial g\left(\vec{r},x'\right)}{\partial z} dx' \right|^2 dz$$

$$(7.51)$$

Otherwise, if the numerical approach is used, the numerical integration is applied to integrals in (7.46) and (7.49) directly.

7.2.2   Results for Current Distribution

Illustrative results pertain to current distribution along $L = \lambda/4$ and $L = \lambda/2$ dipoles at $f = 30$ GHz ($\lambda = 1$ cm), with wire radius $a = L/(10N)$, where N is a number of elements. The dipoles are assumed to be excited with unitary voltage source $V_0 = 1$ V, horizontally located in front of the planar lossy half-space at a certain distance $h = 0.5$ cm, which corresponds to $h = \lambda/2$. At operating frequency $f = 30$ GHz, the tissue properties are as follows: $\sigma = 35.49$ S/m and $\varepsilon_r = 23.16$.

Fig. 7.8 shows the absolute value of the calculated current distribution along the $L = \lambda/4$ antenna compared with the sinusoidal assumed current distribution at $f = 30$ GHz.

In this case, the maximal discrepancy between calculated and approximate sinusoidal current distribution (taking the waveform of triangular current) is 4%.

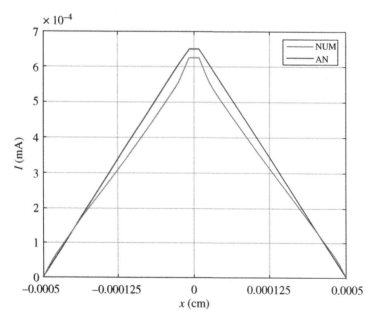

Figure 7.8   Current distribution along the straight wire at $f = 30$ GHz. NUM and AN in the legend stand for numerical and analytical results, respectively.

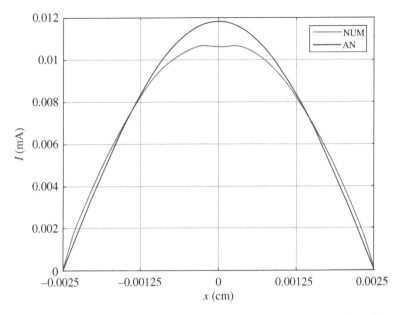

Figure 7.9 Current distribution along the half-wave dipole at f = 30 GHz. NUM and AN in the legend stand for numerical and analytical results, respectively.

Fig. 7.9 shows absolute value of the calculated current distribution along the half-wave dipole compared with the sinusoidal assumed current distribution at $f = 30$ GHz.

The maximal discrepancy between approximate and calculated current distribution is, as expected in this case somewhat higher, around 12%.

7.2.2.1 Results for Transmitted Field, VPD, and TPD

This subsection deals with results for transmitted field VPD and TPD based on the analytical/numerical determination of the current distribution.

Namely, once the current distribution is either assumed or calculated, the field transmitted into the lossy medium representing the human tissue and VPD are evaluated. Finally, TPD versus tissue depth is computed as integral over VPD. Some illustrative results for the transmitted field, VPD, and TPD for different parameters of interest are presented in the paper.

All computational examples deal with dipole antenna length L, with wire radius $a = L/(10N)$, where N is a number of elements, excited with unitary voltage source $V_0 = 1$ V, horizontally located in front of the planar tissue at a certain distance h. The tissue properties at several frequencies of interest are given in Table 7.2.

Table 7.2 Tissue properties at different frequencies.

Frequency	Wavelength	Conductivity	Relative permittivity
$f = 6\,\text{GHz}$	$\lambda = 3\,\text{cm}$	$\sigma = 5.2\,\text{S/m}$	$\varepsilon_r = 48.2$
$f = 10\,\text{GHz}$	$\lambda = 3\,\text{cm}$	$\sigma = 10.63\,\text{S/m}$	$\varepsilon_r = 42.76$
$f = 30\,\text{GHz}$	$\lambda = 1\,\text{cm}$	$\sigma = 35.49\,\text{S/m}$	$\varepsilon_r = 23.16$
$f = 60\,\text{GHz}$	$\lambda = 0.5\,\text{cm}$	$\sigma = 52.83\,\text{S/m}$	$\varepsilon_r = 12.86$

7.2.2.2 Different Distance from the Interface

Figs. 7.10–7.13 show the transmitted field, VPD, and TPD, respectively, versus tissue depth oriented in z-direction for several distances of dipole from the interface for assumed analytical (AN) and numerically calculated (NUM) current distribution, respectively.

Fig. 7.10a–c deal with $f = 6$ GHz case.

(a)

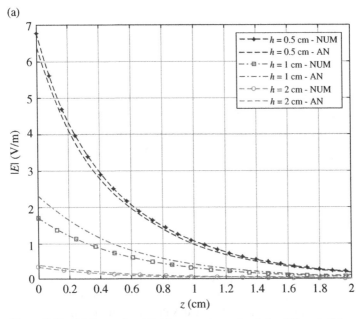

Figure 7.10 (a) Transmitted field versus tissue depth for different distances of the half-wave dipole antenna from the interface at f = 6 GHz, (b) VPD versus tissue depth for different distances of the half-wave dipole antenna from the interface at f = 6 GHz, (c) TPD versus tissue depth for different distances of the half-wave dipole antenna from the interface at f = 6 GHz.

(b)

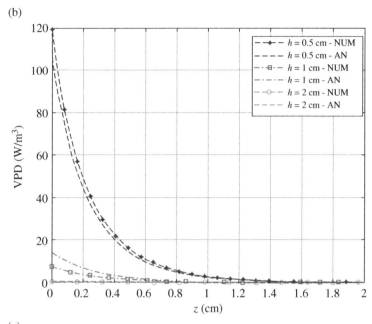

(c)

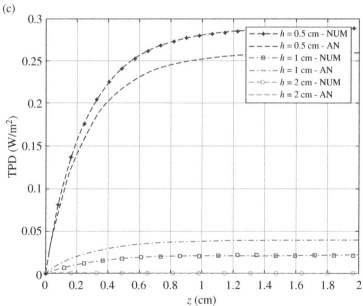

Figure 7.10 (Continued)

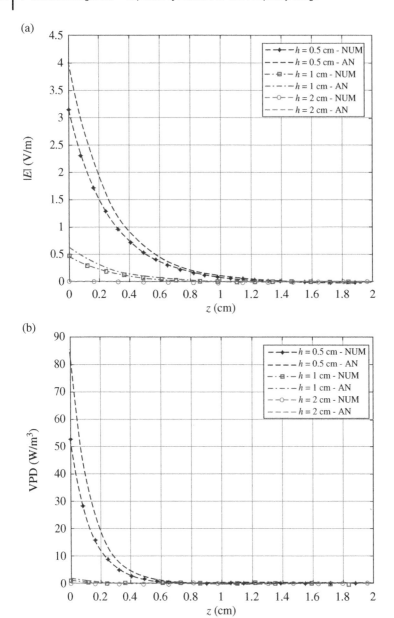

Figure 7.11 (a) Transmitted field versus tissue depth for different distances of the half-wave dipole antenna from the interface at f = 10 GHz, (b) VPD versus tissue depth for different distances of the half-wave dipole antenna from the interface at f = 10 GHz, (c) TPD versus tissue depth for different distances of the half-wave dipole antenna from the interface at f = 10 GHz.

(c)

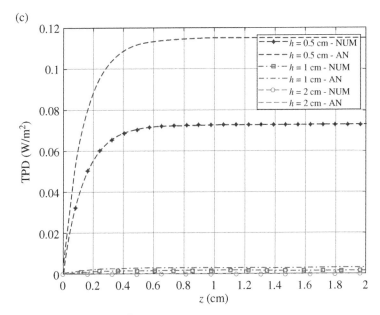

Figure 7.11 (Continued)

(a)

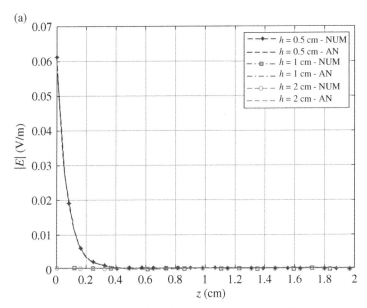

Figure 7.12 (a) Transmitted field versus tissue depth for different distances of the half-wave dipole antenna from the interface at f = 30 GHz, (b) VPD versus tissue depth for different distances of the half-wave dipole antenna from the interface at f = 30 GHz, (c) TPD versus tissue depth for different distances of the half-wave dipole antenna from the interface at f = 30 GHz.

(b)

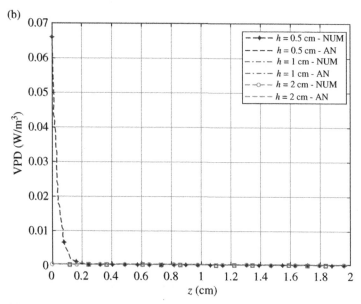

(c)

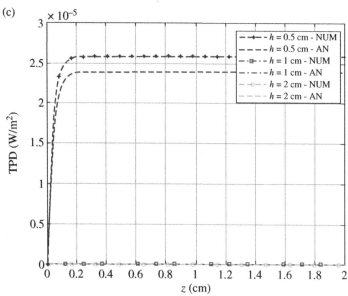

Figure 7.12 (Continued)

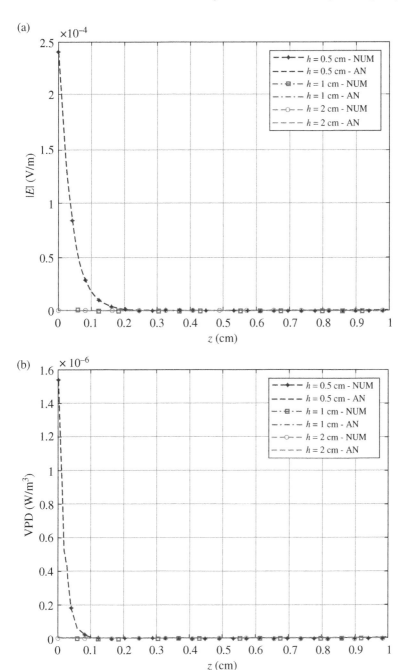

Figure 7.13 (a) Transmitted field versus tissue depth for different distances of the half-wave dipole antenna from the interface at $f = 60$ GHz, (b) VPD versus tissue depth for different distances of the half-wave dipole antenna from the interface at $f = 60$ GHz, (c) TPD versus tissue depth for different distances of the half-wave dipole antenna from the interface at $f = 60$ GHz.

(c)

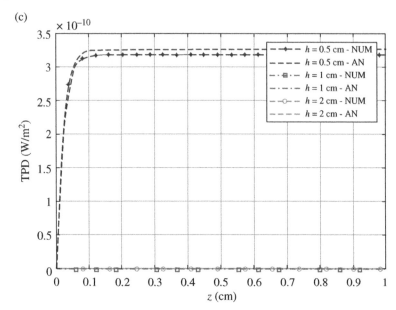

Figure 7.13 (Continued)

From Fig. 7.10a,b, it is obvious that the values of transmitted field and VPD rapidly decrease as the dipole moves away from the interface at $f = 6$ GHz. Furthermore, TPD increases up to depth $z = 1.2$ cm and then goes to saturation (Fig. 7.10c).

Some discrepancies in the results for transmitted field at $f = 10$ GHz obtained by using assumed sinusoidal and calculated current distribution can be noticed for lower tissue depths in Fig. 7.11. The field practically vanishes for $z = 1.5$ cm and $h = 2$ cm. Fig. 7.11b shows VPD to attenuate till $z = 0.8$ cm and then falls to zero. For TPD, as it is visible from Fig. 7.11c, there are some discrepancies between different approaches for $h = 0.5$ cm.

Fig. 7.12a–c deal with $f = 30$ GHz.

Comparing the filed values in Figs. 7.10 and 7.12, it is visible that as the frequency increases, in accordance with the surface effect, there is a decrease in the field strength for 2 orders of magnitude. Discrepancies in the results obtained via different approaches decrease. Furthermore, VPD falls to zero (Fig. 7.12b) while TPD increases up to depth of $z = 0.2$ cm and then goes to saturation (Fig. 7.12c).

The last frequency of interest is $f = 60$ GHz. The corresponding results are shown in Fig. 7.13a–c.

Comparing the results for the transmitted field for $f = 6$ GHz, $f = 10$ GHz, and $f = 30$ GHz with the curve for $f = 60$ GHz, it is obvious that the field strength decreases with frequency increase. From Fig. 7.12b, it is visible that VPD falls to zero for $z = 0.1$ cm. Finally, TPD increases (Fig. 7.12c) till depth $z = 0.1$ cm and then goes to saturation.

7.2.2.3 Different Antenna Length

Figures 7.14–7.17 show the transmitted field, VPD, and TPD for different antenna lengths for fixed antenna distance from the interface $h = 0.5$ cm at frequency of interest.

It is visible from Fig. 7.14 that the field increases for greater antenna lengths. Furthermore, VPD attenuates for shorter antenna length (Fig. 7.14b).

From Fig. 7.14c, it is visible that TPD increases up to $z = 1.5$ cm and then goes to saturation. Discrepancies in the results obtained via different approaches are the most pronounced for $L = \lambda/2$.

Fig. 7.15a–c deal with $f = 10$ GHz.

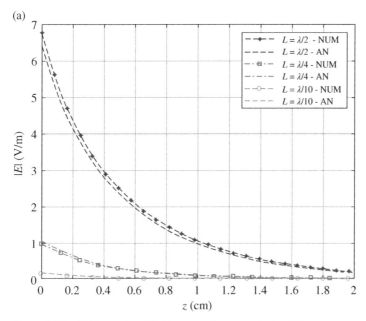

(a)

Figure 7.14 (a) Transmitted field versus tissue depth for different antenna lengths L at distance $h = 0.5$ cm from the interface at $f = 6$ GHz, (b) VPD versus tissue depth for different antenna lengths L at distance $h = 0.5$ cm from the interface at $f = 6$ GHz, (c) TPD versus tissue depth for different antenna lengths L at distance $h = 0.5$ cm from the interface at $f = 6$ GHz.

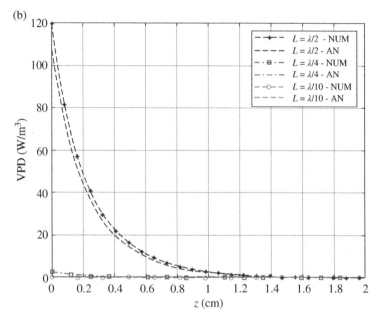

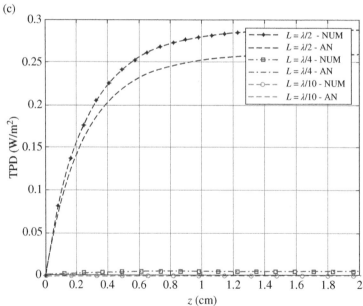

Figure 7.14 (Continued)

(a)

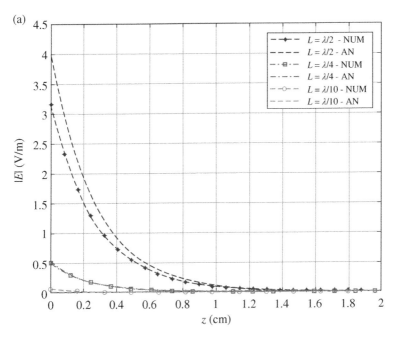

(b)

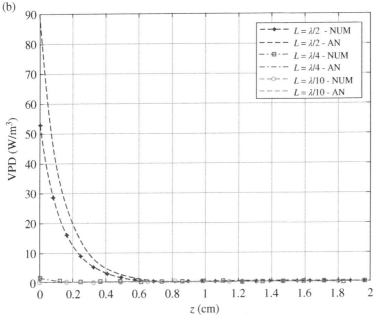

Figure 7.15 (a) Transmitted field vs. tissue depth for different antenna lengths at distance $h = 0.5$ cm from the interface at $f = 10$ GHz, (b) VPD versus tissue depth for different antenna lengths L at distance $h = 0.5$ cm from the interface at $f = 10$ GHz, (c) TPD versus tissue depth for different antenna lengths L at distance $h = 0.5$ cm at $f = 10$ GHz.

(c)

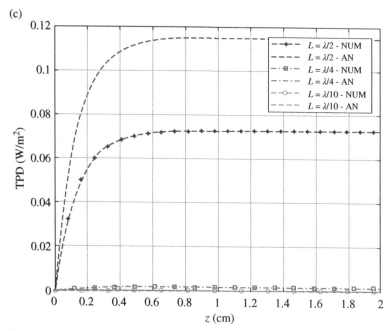

Figure 7.15 (Continued)

(a)

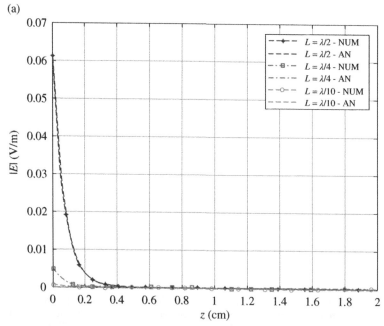

Figure 7.16 (a) Transmitted field versus tissue depth for different antenna lengths at distance $h = 0.5$ cm from the interface at $f = 30$ GHz, (b) VPD versus tissue depth for different antenna lengths L at distance $h = 0.5$ cm from the interface at $f = 30$ GHz, (c) TPD versus tissue depth for different antenna lengths L at distance $h = 0.5$ cm from the interface at $f = 30$ GHz.

(b)

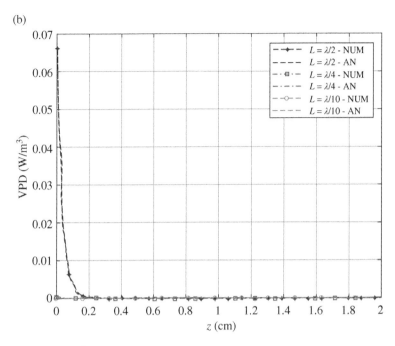

(c)

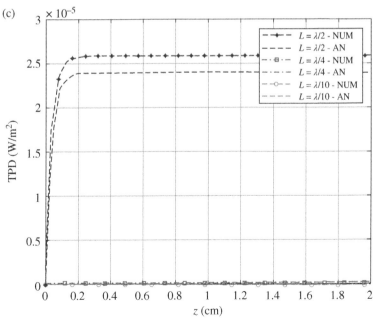

Figure 7.16 (Continued)

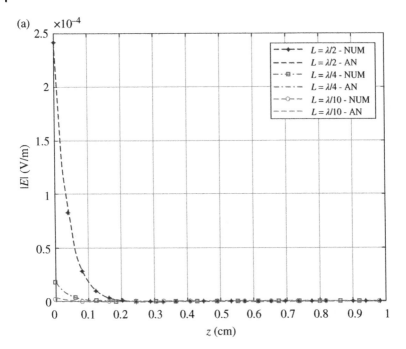

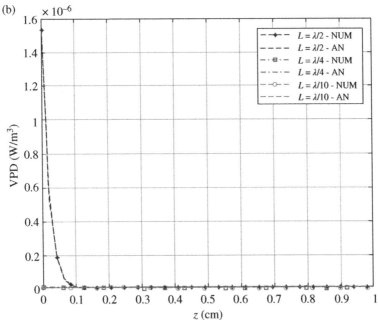

Figure 7.17 (a) Transmitted field versus tissue depth for different antenna lengths L at distance h = 0.5 cm from the interface at f = 60 GHz, (b) VPD versus tissue depth for different antenna lengths L at distance h = 0.5 cm from the interface at f = 60 GHz, (c) TPD versus tissue depth for different antenna lengths L at distance h = 0.5 cm from the interface at f = 60 GHz.

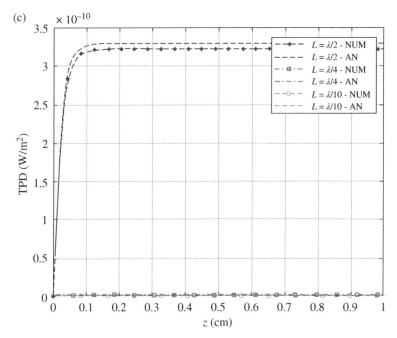

Figure 7.17 (Continued)

Discrepancies in the results obtained via different approaches are the most pronounced for $L = \lambda/2$, as it is visible in Fig. 7.15a. Field falls to zero at $z = 1.5$ cm. VPD falls to zero at $z = 0.8$ cm (Fig. 7.15b) while at the same depth, TPD goes to saturation (Fig. 7.15c).

Next frequency of interest (Fig. 7.16a–c) is $f = 30$ GHz.

It is visible from Fig. 7.23a that field falls to zero at $z = 0.4$ cm for half-wave dipole and at $z = 0.2$ cm for quarter-wave dipole.

It is visible from Fig. 7.16b that VPD falls to zero at $z = 0.2$ cm for the half-wave dipole, while TPD value is negligible (Fig. 7.16c).

The last frequency of interest in this set of results is $f = 60$ GHz (Fig. 7.17a–c).

Comparing the results for transmitted field at $f = 60$ GHz (Fig. 7.17) with the waveforms obtained for lower frequencies, it could be concluded that by frequency increase, there is a decrease in the discrepancies of the results obtained via different approaches. Also, by increasing the frequency, transmitted field attenuates more rapidly.

Observing Fig. 7.17b,c and comparing them for the cases of lower frequency, one concludes that by frequency increase, VPD attenuates faster. VPD practically vanishes for shorter wires. TPD penetrates faster through the tissue for higher frequencies and rapidly goes to saturation as well.

7.2.2.4 Different Frequencies

Fig. 7.18 shows the transmitted field, VPD, and TPD for the fixed distance $h = 0.5$ cm of the quarter wave dipole with radius $a = 0.0051$ mm for different frequencies.

It is visible that there are smaller discrepancies for different approaches. At frequencies greater than 30 GHz, VPD and TPD are negligible.

The results depicted in Figs. 7.10–7.18 show similar behavior as the results reported in [12, 13] pertaining to the case of planar tissue exposed to Hertz dipole radiation.

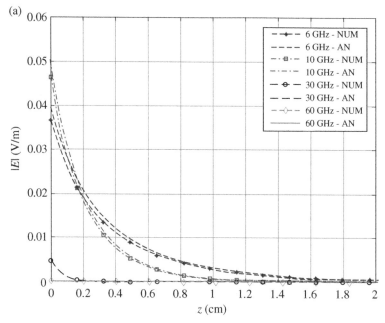

Figure 7.18 (a) Transmitted field versus tissue depth for dipole antenna of length $L = \lambda/4$ at distance $h = 0.5$ cm from the interface for different frequencies, (b) VPD versus tissue depth for dipole antenna of length $L = \lambda/4$ at distance $h = 0.5$ cm from the interface for different frequencies, (c) TPD versus tissue depth for dipole antenna of length $L = \lambda/4$ at distance $h = 0.5$ cm from the interface for different frequencies.

(b)

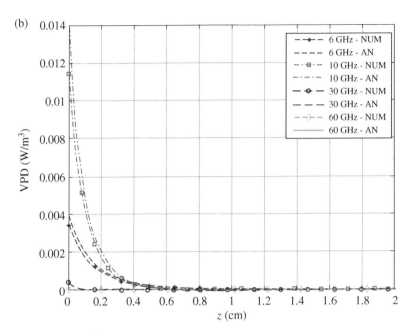

(c)

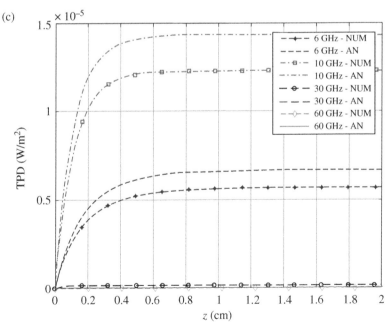

Figure 7.18 (Continued)

7.3   Assessment of S_{ab} in a Multilayer Tissue Model

Analytical/numerical assessment of S_{ab} in the multilayer planar model of the human tissue due to radiation of horizontal dipole antenna in GHz frequency range, thus extending a simplified homogeneous tissue model reported in [10] has been carried out elsewhere, e.g. in [16, 17]. The procedure is outlined in this section.

The analysis in [17] is performed for the case of assumed sinusoidal current distribution, and for the current obtained by numerically solving the Pocklington integro-differential equation using the GB-IBEM. The influence of the multilayer medium (properties of the tissue) is taken into account via the corresponding Fresnel reflection/transmission coefficient. Once the current distribution is known, the irradiated electric and magnetic fields are determined by evaluating the corresponding field integrals via boundary element formalism. The last step is to evaluate S_{ab}.

According to ICNIRP 2020 guidelines, S_{ab} is averaged over area of 4 and 1 cm^2, respectively, depending on the frequency range of interest. The case of the two-layer and three-layer model, respectively, of human tissue is considered and the results S_{ab} are obtained for frequencies: 6, 10, 30, 60, and 90 GHz. The results obtained via assumed and calculated current distribution are compared to the results available from the relevant literature.

7.3.1   Theoretical Background

APD (S_{ab}) is basic restriction quantity for the assessment of human exposure to electromagnetic fields operating above 6 GHz and located in close proximity to the human body [2].

Given the electric and magnetic fields at the interface are determined the APD S_{ab} averaged over area proposed in ICNIRP 2020 is defined [1]:

$$S_{ab} = \frac{1}{2A_{av}} \int_{A_{av}} Re\left(\vec{E} \times \vec{H}^{*}\right) d\vec{A} \tag{7.52}$$

where E and H are the electric and the magnetic fields, respectively, due to a given radiation source, while A_{av} is the averaging area.

The rationale for the choice of 4 and 1 cm^2, averaging area has been explained in ICNIRP 2020 guidelines [1].

Electric and magnetic fields are obtained from the corresponding integral expressions as given in [17], provided the current distribution along the dipole is assumed or calculated.

Provided the antenna-body distance is appreciably smaller than the body dimensions, the tissue can be represented by simplified planar multilayered configuration as depicted in Fig. 7.19, which shows a dipole antenna of length L and radius a placed at the distance d in front of (a) two-layer tissue model consisting of skin and muscle denoted as SM and (b) three-layer tissue model consisting of skin, fat, and muscle denoted as SFM. The skin thickness is $\Delta s_1 = 0.15$ mm in both SM and SFM models while fat thickness in SFM model is $\Delta s_2 = 4$ mm. The last

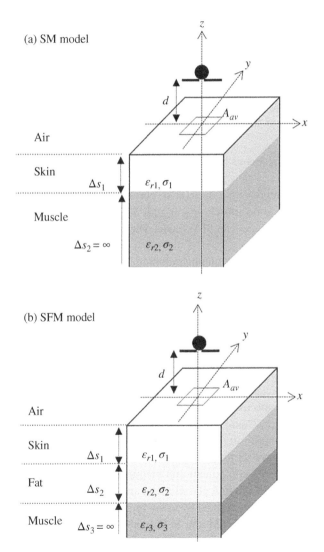

Figure 7.19 Dipole antenna in front of planar multilayered tissue. (a) Two-layered model consisting of skin and muscle layers, (b) three-layered model consisting of skin, fat, and muscle.

Table 7.3 Relative permittivity (ε_r) and electric conductivity (σ) of skin, fat, and muscle tissues at different operating frequencies (f).

	Skin		Fat		Muscle	
f (GHz)	ε_r	σ (S/m)	ε_r	σ (S/m)	ε_r	σ (S/m)
6	34.9	3.89	9.8	0.872	48.2	5.2
10	31.3	8.01	8.8	1.71	42.8	10.6
30	15.5	27.1	5.91	5.33	23.2	35.5
60	7.98	36.4	4.4	8.39	12.9	52.8
90	5.94	39	3.79	10.2	9.3	60.7

Relative permittivity and conductivity of skin, fat, and muscle correspond to data reported by Gabriel in [4] gathered in online tissue properties database [5].

layer in both models is a homogeneous half-space, i.e. the thickness of muscle is (a) $\Delta s_2 = \infty$ and (b) $\Delta s_3 = \infty$.

Furthermore, each layer is represented by relative permittivity ε_r and electric conductivity σ: (a) skin (ε_{r1}, σ_1) and muscle (ε_{r2}, σ_2) for SM model and (b) skin (ε_{r1}, σ_1), fat (ε_{r2}, σ_2), and muscle (ε_{r3}, σ_3) for SFM model. Dielectric property values are given in Table 7.3.

The basic step in the analysis of dipole antenna radiation in the presence of non-homogeneous media is the assessment of equivalent axial current distribution. Provided the current distribution is evaluated, or assumed, the radiated electric and magnetic field, respectively, can be calculated above, or within (which is of interest in the paper), a multilayer.

7.3.2 Results

Fig. 7.20 depicts electric field, magnetic field, and APD obtained by using analytical and numerical solutions, respectively, for current distribution along the wire, for various antenna lengths, ($L = \lambda/2$, $\lambda/4$, and $\lambda/10$), operating frequencies ($f = 6$, 10, 30, 60, and 90 GHz) and antenna-interface distances ($d = 2, 5, 10, 15, 20, 30, 50$, and 150 mm). Control surfaces are set to $A = 1$ and 4 cm^2.

Absolute values of electric field x component and magnetic field y component are depicted in Fig. 7.20 for dipole antenna placed 5 mm above the tissue surface operating at 6 GHz. The field values obtained by using numerical and analytical approaches agree satisfactorily for both SM and SFM models. Numerical results are somewhat higher when compared to the analytical ones. Analyzing the field values obtained by means of the SM and SFM models, it can be concluded that SFM results in lower field values for the given exposure scenario.

Figs. 7.21 and 7.22 show the results for APD for control areas 1 and 4 cm^2, respectively. APD decreases as frequency increases. It is worth noting that differences in

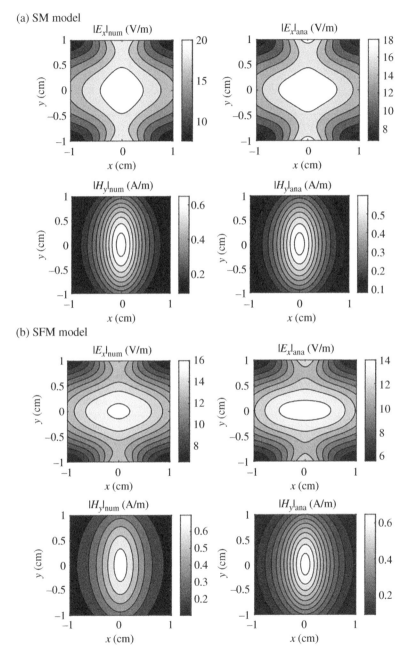

Figure 7.20 Absolute value of electric field x component and magnetic field y component at the tissue surface for $f = 6$ GHz, $d = 5$ mm, $L = \lambda/2$, and $A = 4$ cm^2 obtained by using numerical (left) and analytical current solution (right) for two-layered model (a) and three-layered tissue model (b).

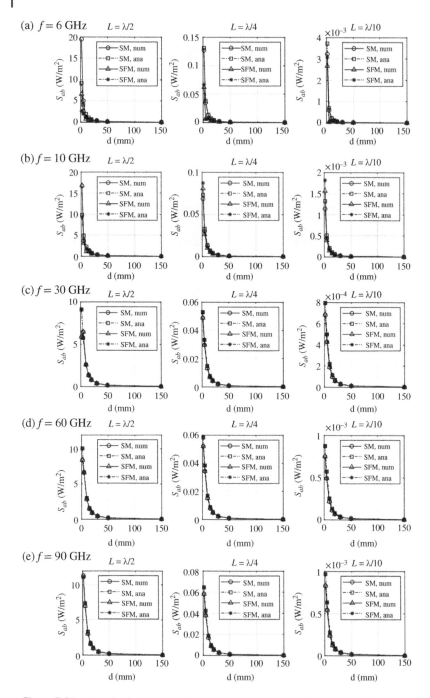

Figure 7.21 Absorbed power density versus antenna-body distance at different frequencies and for different antenna lengths. Control surface is $A = 1\ cm^2$.

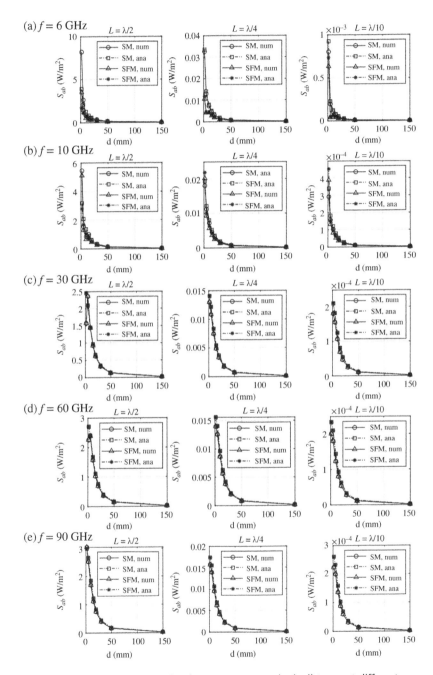

Figure 7.22 Absorbed power density versus antenna-body distance at different frequencies and for different antenna lengths. Control surface is $A = 4$ cm^2.

S_{ab} values at frequencies of 60 and 90 GHz are small. Decreasing the antenna length, the APD decreases as well. Expectedly, as antenna-body distance increases, the APD decreases. The antenna-body distance at which S_{ab} falls to zero is larger when frequency is increased. However, as previously stated, the overall S_{ab} value is smaller as frequency gets higher.

Finally, by changing the control surface from 1 to 4 cm^2, the S_{ab} decreases. Also, the distance at which S_{ab} falls to zero is larger when control surface is 4 cm^2. Namely, in case of $A = 1$ cm^2, at $f = 6$ and 10 GHz, S_{ab} falls to zero after 20 mm, while for $f = 30$, 60, and 90 GHz, S_{ab} is zero at 50 mm. In case of $A = 4$ cm^2, these distances are 50 and 150 mm, for $f = 6$ and 10 GHz, and $f = 30$, 60, and 90 GHz, respectively.

The metrics for comparison of APD values by using (i) analytical versus numerical approach and (ii) three-layered (SFM) versus two-layered (SM) tissue models are computed in the form of relative absolute errors.

Discrepancy in computed S_{ab} values with numerical and analytical current solutions depends mostly on antenna length and frequency. Relative absolute error for numerical and analytical results computed as $Err_{num-ana} = 100 (S_{ab-num} - S_{ab-ana})/S_{ab-num}$ is around 10% and 20% for antenna length $L = \lambda/4$ and $\lambda/10$, respectively. These values do not vary greatly with frequency or antenna-body distance.

On the other hand, for half-wave dipole, $Err_{num-ana}$ is up to 80% for $f = 6$ GHz and antenna-body distances below 30 mm. As frequency is increased, both the upper limit and maximal value of $Err_{num-ana}$ reduce from 30 mm and 80%, respectively to 2 mm and 20% at 60 GHz. However, for larger distances, regardless of the frequency, $Err_{num-ana}$ is around 10% for halfwave dipole, too.

Furthermore, comparing the S_{ab} values with respect to different tissue models, it can be concluded that the discrepancies between SM and SFM models decrease as frequency and antenna-body distance are increased, while the impact of antenna length is almost negligible. Relative absolute error in S_{ab} results for SM and SFM models defined as $Err_{SFM-SM} = 100 (S_{ab-SFM} - S_{ab-SM})/S_{ab-SFM}$ is around 55%, 25%, 0.12%, 0.08%, and 0.02% for 6, 10, 30, 60, and 90 GHz, respectively. However, at 6 GHz and antenna-body distance of 2 and 5 mm, Err_{SFM-SM} is higher than 100% for all antenna lengths, being the highest for $L = \lambda/10$.

Therefore, some general conclusions regarding the S_{ab} behavior with respect to operating frequency, antenna length, and antenna-body distance are as follows:

- The APD decreases rapidly as the antenna-body distance and operating frequency increase.
- Increasing the antenna length, the S_{ab} also increases.
- The distance at which S_{ab} falls to zeros increases with the frequency.

Furthermore, changing the control surface, using the two planar models and two different solutions of antenna current lead to the following conclusions:

- Discrepancies in computed S_{ab} values with numerical and analytical current solutions depend mostly on antenna length and frequency for antennas with $L = \lambda/4$ and $\lambda/10$. The difference is quite large for halfwave dipole at lower frequencies and smaller antenna-body distances.
- Differences in S_{ab} values obtained by two-layered (SM), and three-layered (SFM) models are almost entirely negligible at higher frequencies. At 6 GHz, the difference is still very large.
- Decreasing the control surface, the S_{ab} increases. Also, the distance at which S_{ab} falls to zeros increases for the bigger control surface.

The future work is likely to deal with nonplanar multilayered tissue representations dealing with cylindrical, spherical, and more complex realistic geometries.

7.4 Assessment of Transmitted Power Density in the Planar Multilayer Tissue Model

Analytical/numerical approach to determine TPD in planar multilayered model of the human tissue exposed to the dipole antenna radiation has been addressed in [18]. Analytical approach deals with assumed sinusoidal current distribution, while numerical approach pertains to the determination of current by solving the corresponding Pocklington integro-differential equation via GB-IBEM. The effects of multilayer geometry are considered via the corresponding Fresnel plane wave reflection/transmission approximation. The two-layered model consisting of skin and muscle tissues and three-layered model with fat tissue inserted between skin and muscle are considered. Some illustrative results for current distribution, transmitted field, VPD, and TPD at various frequencies and distances of the antenna from the interface are given.

Thus, the work presented in [18] further extends the approach used in [14, 15] by representing a tissue in terms of planar multilayer. It is noteworthy that the effect of multilayered half-space is taken into account via the corresponding reflection/transmission coefficient already used in the analysis of GPR [19].

7.4.1 Formulation

Dipole antenna of radius a and length L horizontally located in front of a multi-layered planar tissue, as shown in Fig. 7.23, driven by a voltage source is considered.

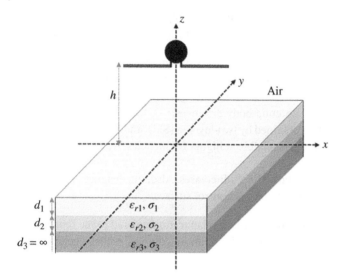

Figure 7.23 Horizontal dipole antenna in front of a planar multilayered tissue.

Key parameter in the analysis of antenna properties is the current distribution along the wire. Provided the current distribution is known, the electric field transmitted into the tissue, VPD and TPD can be determined. As it is well-known from the wire antenna theory, the current distribution along the dipole antenna is governed by the corresponding Pocklington integro-differential equation.

The TPD is defined [6] as follows:

$$\text{TPD} = \frac{1}{2} \int_0^r \sigma(\vec{r}) \left| E(\vec{r}) \right|^2 dr \tag{7.53}$$

where r is the variable perpendicular to the surface of the human body and point $r = 0$ refers to the air–body interface, while σ stands for tissue conductivity.

For the geometry considered in this paper, depicted in Fig. 7.23, TPD is given by

$$\text{TPD}(z) = \frac{1}{2} \int_0^{z_{end}} \sigma(z) \left(|E_x(z)|^2 + |E_z(z)|^2 \right) dz \tag{7.54}$$

where z_{end} stands for the tissue depth taken into account within the TPD calculation.

Full mathematical details are available in [21].

7.4.2 Results

Computational examples are obtained using the two-layered and three-layered models, respectively. Illustrative results pertain to the transmitted field, VPD, and TPD for various distances of the wire from the interface, antenna lengths, and frequencies. In all cases, dipole antenna is excited by the unitary voltage source $V_0 = 1$ V and wire radius is chosen as $a = L/(10N)$, where N is a number of wire segments within GB-IBEM solution procedure.

Frequency-dependent parameters of skin, fat, and muscle tissues in ITIS database [11] are given in Table 7.4.

7.4.2.1 Two-Layer Model

Two-layer tissue model is characterized by layers of skin ($d_1 = 1.5$ mm) and muscle ($d_2 + d_3 = \infty$) (Fig. 7.23). Dipole antenna is placed at a distance h horizontally to the interface.

Figure 7.24 shows the current distribution along the antenna at distance $h = 1$ cm from the interface, compared with the assumed sinusoidal current distribution at $f = 6$ GHz. The discrepancy between calculated and approximate sinusoidal current distribution appears in the feed-gap area. Considering all combinations of antenna heights, antenna lengths, and frequencies, the maximal value of the discrepancy is approximately 4.5%.

The next set of figures (Figs. 7.25–7.27) represents results for the transmitted field, VPD, and TPD for different values of antenna height. The set of figures to follow deals with different antenna lengths and various frequencies, respectively.

It is obvious that the field values rapidly decrease with tissue depth ($|E|$ versus z). Observing the transmitted field at fixed antenna length and frequency, it can be seen that field values decrease as antenna-body distance increases ($|E|$ versus h). However, the so-called skin depth does not depend on antenna-body distance. Namely, the skin depth is the distance from the air–body interface at which the

Table 7.4 Frequency-dependent parameters of body tissues.

f (GHz)	Skin		Fat		Muscle	
	ε_r	σ (S/m)	ε_r	σ (S/m)	ε_r	σ (S/m)
6	34.9	3.89	9.80	0.872	48.2	5.20
10	31.3	8.01	8.80	1.71	42.8	10.6
30	15.5	27.1	5.91	5.33	23.2	35.5
60	7.98	36.4	4.40	8.39	12.9	52.8

Source: Adapted from Li et al. [16].

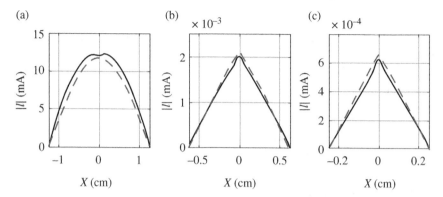

Figure 7.24 The comparison of numerical and analytical current distribution along the antenna of length $L = \lambda/2$ (a), $L = \lambda/4$ (b), and $L = \lambda/10$ (c). The operating frequency is $f = 6$ GHz and antenna is at distance $h = 1$ cm above the three-layered tissue model.

transmitted field falls to appx. 36% of its maximal value. Figs. 7.25–7.27 show that, no matter the antenna length, the skin depth at 6 and 10 GHz is in the muscle part of the model, while at frequencies 30 and 60 GHz, the skin depth is in the skin part of the body model.

It can be concluded that although the field increases with the frequency ($|E|$ versus f), the higher the frequency, the more rapidly the field falls to zero. Also, electric field decreases with antenna length ($|E|$ versus L).

As for analytical and numerical approaches in the current solution, the difference in the electric field computation is small and tends to decrease as frequency and antenna length decrease.

Furthermore, VPD depends on antenna length, antenna-body distance, and frequency, in a similar way as electric field but with faster decay with respect to the tissue depth. However, the discrepancy between the analytical and numerical approaches in VPD computation is more pronounced, although they still exhibit a good agreement.

Finally, it can be observed that the TPD grows until saturation. The depth inside the tissue at which TPD enters the saturation mode depends on frequency and it tends to be closer to the body surface as frequency increases, which is equivalent to the dependence of the skin depth on frequency.

Naturally, the discrepancy between analytical and numerical results is higher for TPD when compared with electric field and VPD computation. In particular, TPD results show the greatest discrepancies for $L = \lambda/2$ at 6 GHz and 10 GHz with maximum difference of approximately 33%. This difference decreases with antenna length. On the contrary, the smaller the antenna-body distance, the larger is the discrepancy between the analytical and numerical solutions.

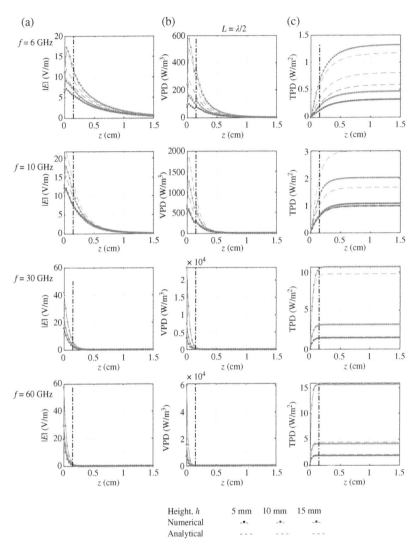

Figure 7.25 Absolute value of electric field transmitted into the tissue ($|E|$, (a)), volume power density (VPD, (b)), and transmitted power density (TPD, (c)) versus tissue depth for halfwave dipole ($L = \lambda/2$) above a two-layered tissue.

7.4.2.2 Three-Layer Model

Three-layer tissue model consists of layers of skin ($d_1 = 1.5$ mm), fat ($d_2 = 4$ mm), and muscle ($d_3 = \infty$) (Fig. 7.23). Dipole antenna is placed at a distance h horizontally to the interface.

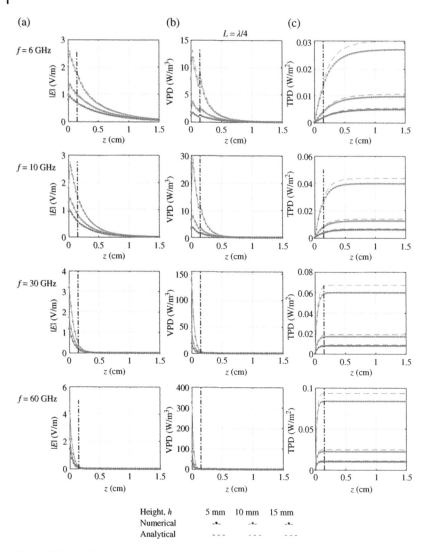

Figure 7.26 Absolute value of electric field transmitted into the tissue ($|E|$, (a)), volume power density (VPD, (b)), and transmitted power density (TPD, (c)) versus tissue depth for quarter-wave dipole ($L = \lambda/4$) above a two-layered tissue.

The current distribution along the antenna at horizontal distance $h = 5\,\text{mm}$ from the interface, compared with the assumed sinusoidal current distribution at $f = 60\,\text{GHz}$ is depicted in Fig. 7.28 The maximal discrepancy, considering all combinations, appears in the feed gap area and it does not exceed 8%.

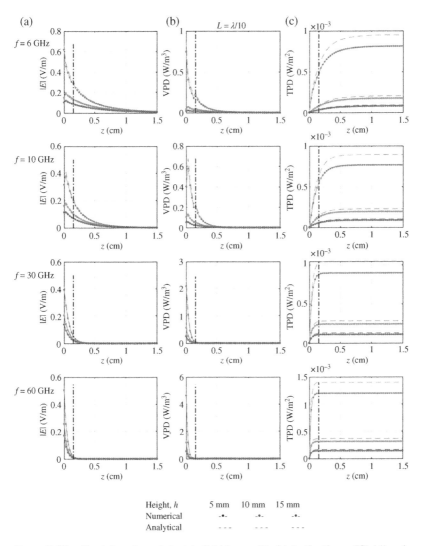

Figure 7.27 Absolute value of electric field transmitted into the tissue (|E|, (a)), volume power density (VPD, (b)), and transmitted power density (TPD, (c)) versus tissue depth for tenth-wave dipole (L = λ/10) above a two-layered tissue.

Figures 7.29–7.31 show the results for the transmitted field, VPD, and TPD for different values of antenna height, antenna length, and various frequencies.

Observing the electric field distribution, it can be concluded that the maximal magnitude occurs farther from the air–body interface when compared to position

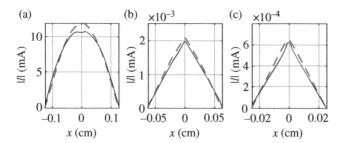

Figure 7.28 The comparison of numerical and analytical current distribution along the antenna of length $L = \lambda/2$ (a), $L = \lambda/4$ (b), and $L = \lambda/10$ (c). The operating frequency is $f = 6$ GHz and antenna is at distance $h = 1$ cm above the three-layered tissue model.

of maximal magnitude of electric field in the two-layer model. Furthermore, the field decreases more slowly with the tissue depth, i.e. the point at which it reaches zero value is located deeper in the tissue. Due to reflections from skin–fat and fat–muscle interfaces, the distribution is characterized by three different regions. Namely, in the first region pertaining to skin tissue, the electric field increases to its maximal value, which occurs in the proximity of the skin–fat interface. Away from the interface, the field starts to decrease with slope narrower in the fat (second region) than in the muscle tissue (third region). However, the difference in the distribution character between the regions tends to fade out as the frequency is increased. Also, as frequency rises, the maximal magnitude of electric field moves closer to the air–skin interface and the slope of the field descent levels in all tissues.

Also, it can be observed that field values are somewhat lower at frequencies of 6 and 10 GHz for the same antenna lengths when compared to a two-layer model. Still, by increasing the frequency and antenna-body distance, the difference in electric field magnitudes between the two models decreases.

VPD distribution, naturally, follows the electric field distribution with three characteristic regions. VPD values are much greater in the skin than in fat tissue while their magnitude in the muscle tissue rapidly approaches to zero in all cases.

As for TPD, the saturation depth is different when compared to a two-layer model at frequencies of 6 and 10 GHz, respectively. Increasing the frequency, the difference between the two-layer and three-layer models becomes insignificant.

Analytical and numerical approaches do not differ greatly as is the case for the two-layer model. Generally, it can be concluded that this difference decreases as frequency and antenna-body distance are increased. Antenna length and number of tissue layers have no impact on the difference between the analytical and numerical approaches. Analyzing the results presented in Figs. 7.29–7.31, it can be concluded that the appreciable discrepancies between analytical/numerical results occur for $L = \lambda/2$. In particular, there are highest discrepancies in TPD values for $L = \lambda/2$ at 6 and 10 GHz, respectively (50% difference).

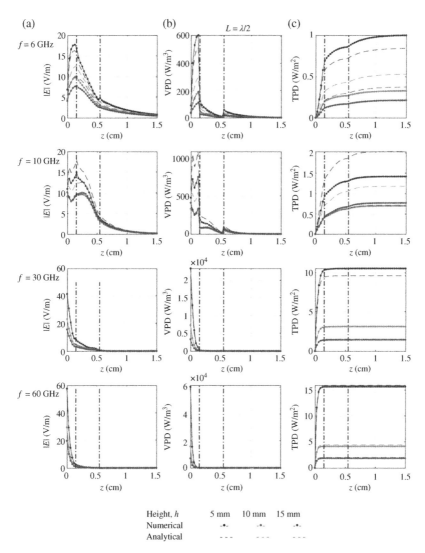

Figure 7.29 Absolute value of electric field transmitted into the tissue ($|E|$, (a)), volume power density (VPD, (b)), and transmitted power density (TPD, (c)) versus tissue depth for half-wave dipole ($L = \lambda/2$) above a three-layered tissue.

7.4.2.3 Skin Depth and Saturation Depth

Aiming to compare the results obtained for different tissue models for the same input parameter set, i.e. the same combination of different antenna lengths, antenna heights, and frequencies, the results for skin depth and saturation depth are presented in Fig. 7.32. For this analysis, the saturation depth ($D_{saturation}$) is

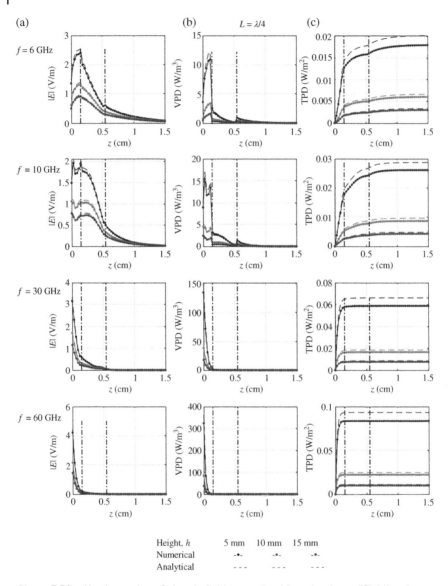

Figure 7.30 Absolute value of electric field transmitted into the tissue (|E|, (a)), volume power density (VPD, (b)), and transmitted power density (TPD, (c)) versus tissue depth for quarter-wave dipole (L = λ/4) above a three-layered tissue.

defined as the distance in the tissue from the air–body interface at which TPD reaches 98% of its maximal value. The skin depth (δ), as previously defined, is the distance in the tissue from the air–body interface at which electric field magnitude drops to 36% of its maximal value.

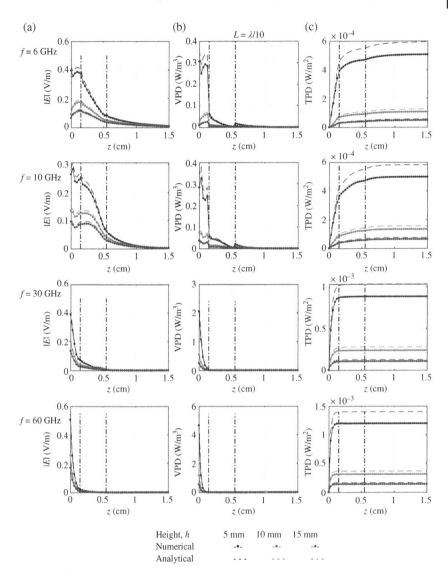

Figure 7.31 Absolute value of electric field transmitted into the tissue (|E|, (a)), volume power density (VPD, (b)), and transmitted power density (TPD, (c)) versus tissue depth for tenth-wave dipole ($L = \lambda/10$) above a three-layered tissue.

It can be observed that in both models, the skin depth at 6 and 10 GHz occurs in the second tissue, which is muscle for a two-layer model and fat for a three-layer model. The skin depth at 30 and 60 GHz occurs in skin tissue, no matter the number of layers, antenna height, and antenna length.

The saturation depth for the two-layer model is in muscle tissue at 6 and 10 GHz and practically at skin–muscle interface at 30 GHz. As for the three-layer model, the saturation depth is in the muscle tissue at 6 and 10 GHz and at fat–skin interface at 30 GHz. The saturation depth is in skin tissue at 60 GHz for both models. This does not depend on antenna length.

The actual skin depth and saturation depth values can be easily extracted from Fig. 7.32.

Conclusions drawn from the presented results are:

- Higher antenna-body distances cause small increase in both skin depth and saturation depth. Therefore, the antenna height above the planar tissue has small impact on both quantities.
- Reducing the antenna length, the influence of the antenna height decreases even more.
- The frequency variation has the biggest impact on both skin depth and saturation depth. Namely, as frequency increases, the impact of the antenna height, the antenna length, and number of layers, respectively, become negligible as

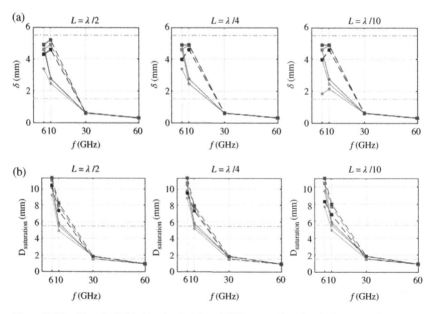

Figure 7.32 Electric field skin depth (a) and TPD saturation depth (b) versus frequency at three antenna heights: $h = 5$ mm, $h = 10$ mm and $h = 15$ mm. The yellow horizontal dotted line represents skin–muscle interface for a two-layered model and skin–fat interface for a three-layered model. The brown horizontal dotted line represents the fat–muscle interface for a three-layered model.

all curves converge to same values, both in case of skin depth and saturation depth.

- Although for some combinations of input parameters (antenna-body distance, antenna length, and frequency) there are some discrepancies between the analytical and numerical approaches, they are not found to be significant. This is an important conclusion as the analytical approach is appreciably less demanding in terms of computational cost.

References

1 ICNIRP, "Guidelines for limiting exposure to electromagnetic fields (100kHz to 300 GHz)," *Health Physics*, vol. 118, no. 5, pp. 483–524, 2020.

2 IEEE-C95.1, IEEE standard for safety levels with respect to human exposure to radio frequency electromagnetic fields, 3 kHz to 300 GHz, NY, USA: IEEE, 2019.

3 A. Hirata, D. Funahashi and S. Kodera, "Setting exposure gidelines and product safety standards for radio-frequency exposure at frequencies above 6GHz: brief review," *Annals of Telecommunications*, vol. 74, pp. 17–24, 2019.

4 S. Pfeifer, E. Carrasco, P. Crespo-Valero, E. Neufeld, S. Kühn, T. Samaras, ... N. Kuster, "Total field reconstruction in the near field using pseudo-vector E-field measurements," *IEEE Transactions on Electromagnetic Compatibility*, vol. 61, no. 2, pp. 476–486, 2019.

5 B. Thors, D. Colombi, Z. Ying and C. Törnevik, "Exposure to RF EMF from array antennas in 5G mobile communication equipment," *IEEE Access*, vol. 4, pp. 7469–7478, 2016.

6 D. Funahashi, A. Hirata, S. Kodera and K. R. Foster et al., "Area-averaged transmitted power density at skin surface as metric to estimate surface temperature elevation," *IEEE Access*, vol. 6, pp. 77665–77674, 2018.

7 D. Funahashi, T. Ito, A. Hirata, T. Iyama, T. Onishi, "Averaging area of incident power density for human exposure from patch antenna arrays," *IEICE Transactions on Electronics*, vol. E101-C, no. 8, pp. 644–646, 2018.

8 D. Poljak and M. Cvetković, "On the Incident Power Density Calculation in GHz Frequency Range - A case of Hertz dipole," in *SpliTech2019 - 4th International Conference on Smart and Sustainable Technologies*, Split, Croatia, 2019.

9 D. Poljak and M. Cvetković, "Assessment of Absorbed Power Density (Sab) at the Surface of Flat Lossy Medium in GHz Frequency Range: A case of Hertz dipole," in *5th International Conference on Smart and Sustainable Technologies (SpliTech)*, Split, Croatia, 2020.

10 D. Poljak, V. Dorić and A. Šušnjara, Absorbed Power Density at the Surface of Planar Tissue due to Radiation of Dipole Antenna, Splitech, Split, Croatia, 2021.

11 ITIS Foundation, "Copyright © 2010–2021 IT'IS Foundation," 2019. [Online]. Available: https://itis.swiss/virtual-population/tissue-properties/database/

12 D. Poljak and V. Doric, On the Concept of the Transmitted Field and Transmitted Power Density for Simplifed Case of Hertz Dipole, Rome, Italy: EMC Europe, 2020.

13 D. Poljak and V. Doric, Assessment of Transmitted Power Density due to Hertz Dipole Radiation using the Modified Image Theory Approach, Proc. SoftCOM, 2020.

14 D. Poljak, A. Susnjara and A. Dzolic, Assessment of Transmitted Power Density due to Radiation from Dipole Antenna of Finite Length, Part I: Theoretical Background and Current Distribution, Proc. SoftCOM, 2021.

15 D. Poljak, A. Susnjara and A. Dzolic, Assessment of Transmitted Power Density due to Radiation from Dipole Antenna of Finite Length, Part II: Transmitted Field, Volume Power Density and Transmitted Power Density, Proc. SoftCOM, 2021.

16 K. Li, Y. Diao, K. Sasaki, A. Prokop, D. Poljak, V. Doric, J. Xi, S. Kodera, A. Hirata, W. El Hajj. Intercomparison of Calculated Incident Power Density and Temperature Rise for Exposure from Different Antennas at 10-90 GHz. IEEE Access, 2021.

17 D. Poljak, A. Šušnjara, L. Kraljević. Absorbed Power Density in a Multilayer Tissue Model due to Radiation of Dipole Antenna at GHz Frequency Range Part I: Theoretical Background, Splitech, 2022

18 D. Poljak, A. Susnjara and A. Fisic, Assessment of Transmitted Power Density in the Planar Multilayer Tissue Model due to Radiation from Dipole Antenna, JCOMSS, 2022.

19 A. Šušnjara, V. Dorić and D. Poljak, "Electric Field Radiated By a Dipole Antenna and Transmitted Into a Two-Layered Lossy Half Space: Comparison of Plane Wave Approximation with the Modified Image Theory Approach," in *2018 3rd International Conference on Smart and Sustainable Technologies (SpliTech)*, Split, Croatia, 2018.

20 D. Poljak, Advanced Modelling in Computational Electromagnetic Compatibility, New York: John Wiley and Sons, 2007.

21 D. Poljak, K. Drissi, Computational Methods in Electromagnetic Compatibility: Antenna Theory Approach versus Transmission Line Models, Hoboken, New Jersey: John Wiley & Sons, 2018.

8

Multiphysics Phenomena

Multiphysics simulation, grasping different physical phenomena, plays an important role in a number of science and engineering areas, e.g. [1, 2]. Multiphysics phenomena involve coupling of different physical phenomena, i.e. different types of physical fields in a macroscopic sense. Namely, one physical field influences the other to a certain extent. If one physical field affects the other field, it provided the impact on this other field, which is negligible. The phenomenon being considered is then modeled as two separate problems, one at a time, in a sequential manner [1]. Examples of such a problem are radiofrequency (RF) hyperthermia modeling, exposure of humans to microwave radiation (e.g. base station antennas, mobile phones), etc.

On the other hand, if the behavior of one physical field affects the other in the same temporal scale with a strong nonlinear coupling, then two-way simultaneous analysis is necessary, i.e. both the fields should be calculated at the same time, or through iterative procedures. Typical problem is plasma confinement modeling within magnetohydrodynamics (MHD) framework for fusion-related applications [3, 4].

Multiphysics computational models addressed in this chapter involve electromagnetic-thermal dosimetry, MHD models for plasma confinement, transport equations, and the Schrodinger equations. First, numerical solution of Stratton-Chu integral expressions/Helmholtz equation for the assessment of specific absorption rate (SAR) induced in a realistic model of the human head featuring the use of finite element method (FEM) and hybrid finite element method/boundary element method (FEM/BEM) is outlined. This is followed by thermal analysis based on the FEM solution of bio-heat transfer equation. Furthermore, MHD model of plasma confinement is based on the solution of the Grad-Shafranov equation (GSE), while the transport equations for plasma dynamics in tokamaks are addressed. Finally, the chapter deals with analytical and

Deterministic and Stochastic Modeling in Computational Electromagnetics: Integral and Differential Equation Approaches, First Edition. Dragan Poljak and Anna Šušnjara.
© 2024 The Institute of Electrical and Electronics Engineers, Inc.
Published 2024 by John Wiley & Sons, Inc.

numerical solutions of the Schrodinger equation as many phenomena in quantum transport, condensed matter physics, optics, nanodevices, etc., are governed by the Schrodinger equation. The main difficulty in realistic scenarios is to prescribe appropriate boundary conditions, as the Schrodinger equation itself is posed in an unbounded domain.

Of particular interest in both physics and engineering is modeling of semiconductor structures. In some applications, such as analysis of nanowires, a combined classical/quantum physics approach is used featuring a hybrid Poisson/Schrodinger equation approach. Namely, as the sizes of nanowires approach to nanoscales, quantum effects become crucial to understanding and designing such structures. Thus, charge or scalar potential distribution could be determined by solving the Poisson/Schrodinger equation.

8.1 Electromagnetic-Thermal Modeling of Human Exposure to HF Radiation

This section deals with electromagnetic-thermal analysis of the human head exposed to a high-frequency (HF) range below 6 GHz.

8.1.1 Electromagnetic Dosimetry

The principal task of HF dosimetry is to quantify heating effects which are expressed by *SAR*. A well-known definition of SAR is the rate of electromagnetic energy W absorbed by the unit body mass m:

$$SAR = \frac{dP}{dm} = \frac{\sigma}{2\rho}|E|^2 \tag{8.1}$$

where P is the dissipated power, E is the peak value of the internal electric field, ρ is the tissue density, and σ is the tissue conductivity.

Plane-wave illuminating the human head is an unbounded scattering problem formulated by the Stratton-Chu integral relation and Helmholtz equation. Thus, the domain exterior to the head is expressed via the boundary integral equation [5–7]

$$\alpha \vec{E}'_{ext} = \vec{E}'_{inc} - j\omega\mu \oint_{\partial V} \left(\vec{n} \times \vec{H}_{ext}\right) G\left(\vec{r}, \vec{r}'\right) dS$$

$$+ \oint_{\partial V} \left[\left(\vec{n} \times \vec{E}'_{ext}\right) \times \nabla G\left(\vec{r}, \vec{r}'\right) - \frac{1}{\sigma + j\omega\mu} \nabla_s \cdot \left(\vec{n} \times \vec{H}_{ext}\right) \nabla G\left(\vec{r}, \vec{r}'\right)\right] dS \tag{8.2}$$

where $\vec{n}$ is an outer normal to surface ∂V bounding the volume V and α is the solid angle subtended at the observation point, $\vec{E}'_{ext}$ and $\vec{E}'_{inc}$ are the total and the incident electric field, respectively, while $G(\vec{r}, \vec{r}')$ denotes the Green's function for the free space given by

$$G(\vec{r}, \vec{r}') = \frac{e^{-jk|\vec{r}-\vec{r}'|}}{4\pi|\vec{r}-\vec{r}'|} \tag{8.3}$$

where $|\vec{r} - \vec{r}'|$ is the distance from the observation point to the source point, and k denotes the wave number.

Furthermore, the behavior of the field in an interior region is governed by a partial differential equation

$$\nabla \times \left(\frac{j}{\omega\mu}\nabla \times \vec{E}_{int}\right) - (\sigma + j\omega\varepsilon)\vec{E}_{int} = 0 \tag{8.4}$$

which is required to be coupled with (8.2).

The hybrid FEM/BEM solution procedure presented in [7] is outlined below.

Unknown electric and magnetic fields are approximated in terms of edge elements preserving the tangential continuity of the fields on the boundary

$$\vec{E} = \sum_{i=1}^{n} \delta_i \vec{w}_i e_i \tag{8.5}$$

$$\vec{H} = \sum_{i=1}^{n} \delta_i \vec{w}_i h_i \tag{8.6}$$

The unknown coefficients e_i and h_i, respectively, associated with each edge of the model, are determined from the global system of equations with $\delta_i = 1$ if the local edge direction coincides with chosen global edge direction, otherwise one has $\delta_i = -1$.

By taking the dot product of (8.4) with test function w_i, according to the weighted residual approach, it follows

$$\int_V \left[\nabla \times \left(\frac{j}{\omega\mu}\nabla \times \vec{E}_{int}\right) - (\sigma + j\omega\varepsilon)\vec{E}_{int}\right]\delta_i\vec{w}_i dV = 0 \tag{8.7}$$

Having applied some standard vector identities, and the Gauss divergence theorem, the weak formulation is obtained

$$\int_V \left[\nabla \times \delta_i\vec{w}_i \cdot \vec{E}_{int} - (\sigma + j\omega\varepsilon)\delta_i\vec{w}_i \cdot \vec{E}_{int}\right] dV = \oint_{\partial V} \delta_i\vec{w}_i \cdot d\vec{S} \times \vec{H}_{int} \tag{8.8}$$

Now FEM/BEM are coupled by forcing the tangential components of electric and magnetic fields to be continuous across the surface ∂V

$$\oint_{\partial V} \delta_i \vec{w}_i \cdot d\vec{S} \times \vec{E}'_{int} = \oint_{\partial V} \delta_i \vec{w}_i \cdot d\vec{S} \times \vec{E}'_{ext} \tag{8.9}$$

Now substituting E_{ext} and H_{ext} in (8.4) with E_{int} and H_{int}, respectively, and after inserting (8.2) into (8.9), yields

$$\oint_{\partial V} \delta_i \vec{w}_i \cdot d\vec{S} \times \alpha_i \vec{E}'_{ext} = \oint_{\partial V} \delta_i \vec{w}_i \cdot d\vec{S} \times \vec{E}'_{inc} - j\omega\mu \oint_{\partial V} \delta_i \vec{w}_i \cdot d\vec{S} \times \oint_{\partial V} \vec{n}$$

$$\times \vec{H}_{int} G\left(\vec{r}, \vec{r}'\right) d\vec{S} + \oint_{\partial V} \delta_i \vec{w}_i \cdot d\vec{S} \times \oint_{\partial V} \left(\vec{n} \times \vec{E}_{int}\right)$$

$$\times \nabla G\left(\vec{r}, \vec{r}'\right) d\vec{S} - \frac{1}{\sigma + j\omega\mu} \oint_{\partial V} \delta_i \vec{w}_i \cdot d\vec{S}$$

$$\times \oint_{\partial V} \nabla_s \cdot \left(\vec{n} \times \vec{H}_{ext}\right) \times \nabla G\left(\vec{r}, \vec{r}'\right) d\vec{S}$$

$$\tag{8.10}$$

Finally, having performed the FEM/BEM procedure, the following system of equations is obtained:

$$[E_{BEM}]\{e_{BEM}\} = \{e_{inc}\} + [H_{BEM}]\{h_{BEM}\} \tag{8.11}$$

$$[E_{FEM}]\{e_{FEM}\} = [H_{FEM}]\{h_{FEM}\} \tag{8.12}$$

where $\{e_{BEM}\}$ and $\{h_{BEM}\}$ are unknown coefficients associated with boundary surface of the scattering problem, $\{e_{inc}\}$ are known coefficients calculated from the incident field, while matrices $[E_{BEM}]$, $\{h_{BEM}\}$, $[E_{FEM}]$, and $[H_{FEM}]$ arise from corresponding integral and differential equations, respectively.

8.1.2 Thermal Dosimetry

Once the SAR distribution is determined, the related temperature increase can be obtained by solving the steady-state bio-heat equation [5–7]. Derivation of the bio-heat transfer equation can be found elsewhere, e.g. [6, 7] and it is outlined in this section.

Heat conduction through media is described by Fourier law:

$$\vec{q} = -\lambda \nabla T \tag{8.13}$$

where q is the heat flow density, T denotes a temperature, and λ [J/kg/°C] is the heat conductivity of the medium.

The heat through surface element dS in differential dt is

$$\overline{Q} = \int_t \int_S \vec{q} d\vec{S} dt - \int_t \int_S \lambda \nabla T d\vec{S} dt \tag{8.14}$$

Now, the Gauss divergence theorem yields

$$\overline{Q} = - \int_t \int_S \lambda \nabla T d\vec{S} dt = - \int_t \int_V \nabla(\lambda \nabla T) dV dt \tag{8.15}$$

Furthermore, differential of total heat delivered into a volume V, due to internal and external heat sources in time differential dt, is

$$d\overline{Q}_{tot} = d\overline{Q}_{int} + d\overline{Q}_{ext} \tag{8.16}$$

On the other hand, differential of internal heat sources can be written as:

$$d\overline{Q}_{int} = Q_i dV dt \tag{8.17}$$

while a differential of external heat sources is

$$d\overline{Q}_v = \nabla(\lambda \nabla T) dV dt \tag{8.18}$$

Finally, the differential of total heat is:

$$d\overline{Q}_{tot} = C_V \rho \frac{\partial T}{\partial t} dV dt \tag{8.19}$$

Combining previous equations, one obtains

$$C_V \rho \frac{\partial T}{\partial t} dV dt = Q_i dV dt + \nabla(\lambda \nabla T) dV dt \tag{8.20}$$

Taking spatial and temporal integration yields

$$\int_t \int_V \left[C_V \rho \frac{\partial T}{\partial t} - Q_i - \nabla(\lambda \nabla T) \right] dV dt = 0 \tag{8.21}$$

while the function under spatial-temporal integral can be referred to as *Fourier heat conduction equation*:

$$C_V \rho \frac{\partial T}{\partial t} = Q_i + \nabla(\lambda \nabla T) \tag{8.22}$$

For stationary phenomena, (8.22) simplifies to *Poisson equation*:

$$\nabla(\lambda \nabla T) = Q_i \tag{8.23}$$

For source-free areas, (8.23) becomes homogeneous (*Laplace equation*)

$$\nabla(\lambda\nabla T) = 0 \tag{8.24}$$

Expressions (8.23) and (8.24) are widely used to study the heat transfer from power cables.

Next, convective heat transfer has to be addressed.

Convection, as a phenomenon, pertains to heat transfer between a surface and a fluid. Figure 8.1 shows convective heat transfer between a fluid at temperature T_e flowing at velocity v and a surface at temperature T_s.

In an external forced flow, the rate of the heat transfer is approximately proportional to the difference between the surface temperature T_s, and the temperature of the free stream fluid T_e.

Therefore, the heat flux density q can be expressed as:

$$q_s = -\lambda\frac{\partial T}{\partial n} = H(T_s - T_a) \tag{8.25}$$

which represents the Neumann boundary condition for the air–body interface, where h_c is the convective heat transfer coefficient [W/m^{2}K].

The final step is to obtain the Pennes' equation.

Volume blood perfusion of tissue Q_b can be expressed as:

$$Q_b = W_b C_{pb}(T_a - T) \tag{8.26}$$

where W_b is the volumetric blood perfusion rate and T_a is the arterial blood temperature.

Finally, the space-time bio-heat transfer equation can be written in the form:

$$\nabla(\lambda\nabla T) + Q_b + Q_m + Q_{SAR} = C_V\rho\frac{\partial T}{\partial t} \tag{8.27}$$

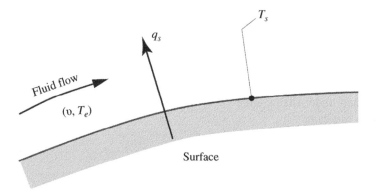

Figure 8.1 Convective heat transfer.

where Q_m is the metabolic heat generation per unit volume of tissue while the electromagnetic power deposition due to external radiation sources is given by:

$$Q_{EM} = \frac{\sigma}{2}|E|^2 = \rho \cdot SAR \tag{8.28}$$

and represents the volumetric heat source due to external field.

For stationary phenomena, the Pennes' equation (8.27) simplifies into

$$\nabla(\lambda\nabla T) + W_b C_{pb}(T_a - T) + Q_m + Q_{EM} = 0 \tag{8.29}$$

The bioheat transfer Eqs. (8.27) and (8.29), respectively, express the energy balance between conductive heat transfer in a volume control of tissue, heat loss due to perfusion effect, metabolism, and energy absorption due to radiation.

Parameter $W_b C_{pb}$ can be also written as $\rho_b c_b w$ where ρ_b is the blood mass density, c_b is the specific heat capacity of blood, and w is the perfusion rate.

The integral formulation of (8.29) convenient for FEM solution can be written as follows [5]

$$\int_{V'}\left[\lambda\nabla f_j \cdot \nabla T + \rho_b c_b w \cdot T \cdot f_j\right]dV' = \int_{V'}(\rho_b c_b w \cdot T_a + Q_m + Q_{EM}) \cdot f_j dV'$$
$$+ \oint_{\partial V'}\lambda\frac{\partial T}{\partial n}f_j\vec{n} \cdot d\vec{S'} \tag{8.30}$$

The boundary condition at the interface between skin and air is given in terms of the heat flux density

$$\lambda\frac{\partial T}{\partial n} = -H(T_s - T_a) \tag{8.31}$$

where H, T_s, and T_a are, respectively, the convection coefficient, the temperature of the skin, and the temperature of the air.

Standard finite element discretization of Helmholtz equation yields the following matrix equation

$$[K]\{T\} = \{M\} + \{P\} \tag{8.32}$$

where [K] is the finite element matrix of the form

$$K_{ji} = \int_{\Omega_e}\nabla f_j(\lambda\nabla f_i)d\Omega_e + \int_{\Omega_e}W_b C_{pb}f_j f_i d\Omega_e \tag{8.33}$$

while {M} denotes the flux vector

$$M_j = \int_{\Gamma_e}\lambda\frac{\partial T}{\partial n}f_j d\Omega_e \tag{8.34}$$

and {P} stands for the source vector

$$P_j = \int_{\Omega_e} \left(W_b C_{pb} T_a + Q_m + Q_{EM} \right) \cdot f_j d\Omega_e \tag{8.35}$$

More details can be found elsewhere, e.g. in [5–7].

8.1.3 Computational Examples

Figure 8.2 shows the computational model of the human head including various head and eye tissues constructed from the magnetic resonance imaging (MRI) of a 24-year-old male [8]. The interior domain of the model is discretized to 10^6 tetrahedral elements, while the boundary surface is discretized to 6838 triangles.

The computational example pertains to the head exposed to vertically polarized plane wave at $f = 1.8$ GHz (1 V/m amplitude) incident on the anterior side.

Figures 8.3–8.5 show the obtained numerical results for the induced electric field, SAR and the temperature increase, respectively at sagittal and transverse cross-sections with the brain and eye tissues included.

It is visible from Fig. 8.3 that the attenuation of the electric field inside the head model, while the highest values of SAR and the respective thermal increase, as shown in Figs. 8.4 and 8.5, are found at the superficial parts of the head and eye.

It is important to note that according to [9], the temperature elevation in the human brain as well as in the eye is listed as one of the most important concerns [10].

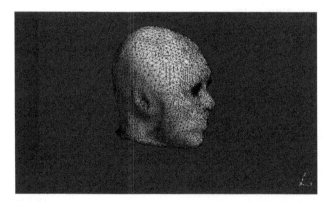

Figure 8.2 Geometry of the human head for electromagnetic-thermal analysis.

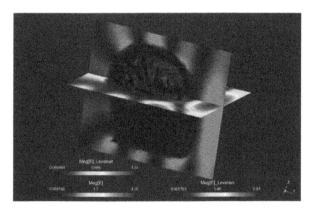

Figure 8.3 Magnitude of the internal electric field.

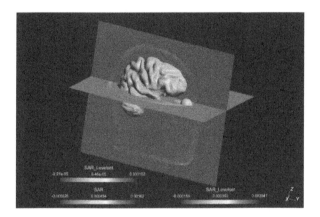

Figure 8.4 SAR induced in the head.

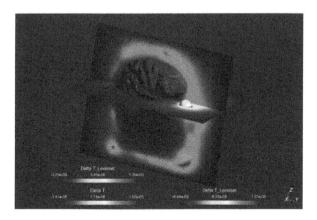

Figure 8.5 Temperature increase in the head.

8.2 Magnetohydrodynamics (MHD) Models for Plasma Confinement

MHD models represent simplest theoretical description of ionized gases, thus providing analysis tool for plasma shaping and confinement in tokamaks by high-intensity magnetic fields.

The use of MHD models is the simplest theory for the description of ionized gases, i.e. the simplest theoretical approach for the analysis of electrically conducting fluids, i.e. plasma. Though MHD approach is far from rigorous description of any realistic plasma configuration of interest, it still provides the theoretical basis for understanding the global configuration for magnetized plasma. As plasma represents a good conductor at high temperatures, plasma shaping and confinement can be carried out by high-intensity magnetic fields. Particular configurations of plasma are governed by GSE whose general form cannot be solved analytically [11–13]. In last few decades, there were a number of papers dealing with the numerical treatment of tokamak plasma, e.g. [14, 15].

Numerical modeling of main plasma region is a less difficult task compared to the modeling of plasma edge in tokamak [15], as properly prescribing appropriate boundary conditions to be imposed at the plasma edge is still a demanding task [15].

A review of the use of the standard FDM and FEM procedures featuring isoparametric elements for the solution of rectangular plasma assuming the source term (given in the form of current density) to be monomial [13] has been reported in [4].

Specific plasma configurations are governed by GSE.

8.2.1 The Grad-Shafranov Equation

MHD equilibrium in axisymmetric plasma shape, shown in Fig. 8.6, representing dynamic phenomena in tokamaks is of interest. The behavior of dynamics phenomena in tokamaks can be analyzed by solving combined equations of electromagnetics and fluid dynamics.

GSE can be derived in a few steps.

The force balance equation is given by [13]

$$\vec{J} \times \vec{B} = \nabla p \tag{8.36}$$

where J is the current density, B is magnetic flux density, and p stands for the kinetic pressure.

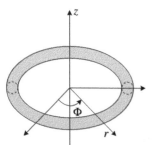

Figure 8.6 Toroidal tokamak geometry.

In cylindrical coordinates (r, ϕ, z), (8.36) is expressed as follows

$$\vec{J}_{pol} \times B_\phi \vec{e}_\phi + J_\phi \vec{e}_\phi \times \vec{B}_{pol} = \nabla p \tag{8.37}$$

where $\vec{J}_{pol}$ is the poloidal current density, $\vec{B}_{pol}$ is the poloidal magnetic flux density, and $\vec{e}_\phi$ is the unit vector in toroidal direction.

Now, the poloidal quantities can be written as follows:

$$\vec{B}_{pol} = \frac{1}{r}\left(\nabla \Psi \times \vec{e}_\phi\right) \tag{8.38}$$

$$\vec{J}_{pol} = \frac{1}{r}\left(\nabla f \times \vec{e}_\phi\right) \tag{8.39}$$

where Ψ is the poloidal magnetic flux, while f is defined as

$$f = \frac{rB_\phi}{\mu_0} \tag{8.40}$$

Now, taking into account

$$\vec{e}_\phi \cdot \nabla \Psi = \vec{e}_\phi \cdot \nabla f = 0 \tag{8.41}$$

relation (8.37) becomes

$$-\frac{1}{r}B_\phi \nabla f + \frac{1}{r}J_\phi \nabla \psi = \nabla p \tag{8.42}$$

Utilizing relations:

$$\nabla p(\Psi) = \frac{\partial p}{\partial \Psi}\nabla \Psi \tag{8.43}$$

$$\nabla f(\Psi) = \frac{\partial f}{\partial \Psi}\nabla \Psi \tag{8.44}$$

expression (8.42) can be written as

$$J_\phi = r\frac{\partial p}{\partial \psi} + B_\phi \frac{\partial f}{\partial \psi} = r\frac{\partial p}{\partial \psi} + \mu_0 \frac{1}{r}f\frac{\partial f}{\partial \psi} \tag{8.45}$$

As the magnetic flux density and current density are related by Ampere law

$$\nabla \times \vec{B} = \mu_0 \vec{J} \tag{8.46}$$

by taking into account (8.38), one obtains

$$-\mu_0 rJ_\phi = r\frac{\partial}{\partial r}\frac{1}{r}\frac{\partial \psi}{\partial r} + \frac{\partial^2 \psi}{\partial z^2} \tag{8.47}$$

Finally, combining (8.45) and (8.47) yields

$$r\frac{\partial}{\partial r}\frac{1}{r}\frac{\partial \psi}{\partial r} + \frac{\partial^2 \psi}{\partial z^2} = -\mu_0 r^2 \frac{\partial p}{\partial \psi} - \mu_0^2 f \frac{\partial f}{\partial \psi} \tag{8.48}$$

which is one of the commonly used forms of GSE.

An alternative useful form of GSE is given by using elliptic operator

$$\left[r\frac{\partial}{\partial r}\left(\frac{1}{r}\frac{\partial \psi}{\partial r}\right) + \frac{\partial^2 \psi}{\partial z^2} \right] = -\mu_0 r J_\phi \tag{8.49}$$

where J is the toroidal component of the plasma current.

Equation (8.49) can be solved analytically only for a few special cases and for additional problems with a higher degree of complexity, numerical solution is necessary.

8.2.1.1 Analytical Solution

Various analytical solutions to GSE have been derived by researchers in recent years [16, 17]. The analytical solutions are essential in describing various parameters involved in real tokamak scenarios and they are useful for benchmarking various numerical codes. In this subsection, four different analytical solutions will be outlined, as well as the emphasis on their applications.

Starting from a slightly modified version of (8.48), i.e. from equation

$$\frac{\partial^2 \psi}{\partial r^2} - \frac{1}{r}\frac{\partial \psi}{\partial r} + \frac{\partial^2 \psi}{\partial z^2} = -\mu_0^2 f \frac{df}{d\psi} - \mu_0 r^2 \frac{dP}{d\psi} \tag{8.50}$$

in order to obtain any solution corresponding to the realistic source functions that appear on the right-hand side of GSE, it is necessary to determine possible solutions to the homogeneous equation

$$\frac{\partial^2 \psi}{\partial r^2} - \frac{1}{r}\frac{\partial \psi}{\partial r} + \frac{\partial^2 \psi}{\partial z^2} = 0 \tag{8.51}$$

Solution of (8.51) can be obtained by separation of variables and is given by

$$\psi_0(r,z) = (c_1 r J_1(kr) + c_2 r Y_1(kr))(c_3 e^{kz} + c_4 e^{-kz}) \tag{8.52}$$

On the other hand, the solutions can also be based on the series expansion [18]

$$\psi_0 = \sum_{n=0,2,\dots} f_n(r) z^n \tag{8.53}$$

provided that each expansion term satisfies the following equation [18]

$$r\frac{d}{dr}\left(\frac{1}{r}\frac{df_n(r)}{dr}\right) = -(n+1)(n+2)f_{n+2}, n = 0, 2, \ldots \tag{8.54}$$

One of the possible solutions satisfying these conditions and suitable for further implementation is given by [18]

$$\psi_0(r,z) = c_1 + c_2 r^2 + c_3\left(r^4 - 4r^2 z^2\right) + c_4\left(r^2 \ln r - z^2\right) \tag{8.55}$$

The Solov'ev equilibrium is the simplest solution of the inhomogeneous GSE [17], being widely used in studies of plasma equilibrium studies, transport, and MHD stability analysis, respectively.

The source functions in Solov'ev equilibrium are linear in ψ and are given by [19]

$$P(\psi) = \frac{A}{\mu_0}\psi, f^2(\psi) = 2B\psi + F_0^2 \tag{8.56}$$

Using source functions (8.55) yields the following variant of GSE

$$\frac{\partial^2 \psi}{\partial r^2} - \frac{1}{r}\frac{\partial \psi}{\partial r} + \frac{\partial^2 \psi}{\partial z^2} = Ar^2 + B \tag{8.57}$$

with the corresponding solution

$$\psi(r,z) = \psi_0(r,z) - \frac{A}{8}r^4 - \frac{B}{2}z^2 \tag{8.58}$$

It is worth noting that a number of plasma shapes can be generated using (8.58). However, the current profile of this solution is restricted, as two free parameters A and B allow one to choose only the plasma current and the ratio of the volume-averaged particle pressure to the average poloidal magnetic field pressure along the plasma boundary.

The Herrnegger–Maschke solutions of GSE for a parabolic source function are originally reported in [17] and can be written as follows

$$P(\psi) = \frac{C}{2\mu_0}\psi^2, \quad f^2(\psi) = D\psi^2 + F_0^2 \tag{8.59}$$

In this case, GSE simplifies into

$$\frac{\partial^2 \psi}{\partial r^2} - \frac{1}{r}\frac{\partial \psi}{\partial r} + \frac{\partial^2 \psi}{\partial z^2} = Cr^2\psi + D\psi \tag{8.60}$$

The solution of (8.60) can be given in the form of Coulomb wave functions

$$\psi = \alpha(F_0(\eta, x) + \gamma G_0(\eta, x))\cos(kz) \tag{8.61}$$

As in the case of the Solov'ev equilibrium, the Herrnegger-Maschke solutions have only two free parameters (C and D), which provide one to independently specify the plasma current and pressure ratio, respectively.

Innovative source functions were introduced by McCarthy [17]. These source functions involve a linear dependence of pressure and a quadratic dependence of the current profile

$$P(\psi) = \frac{S}{\mu_0}\psi, f^2(\psi) = T\psi^2 + 2U\psi + F_0^2 \tag{8.62}$$

Inserting (8.62) in (8.50) yields

$$\frac{\partial^2 \psi}{\partial r^2} - \frac{1}{r}\frac{\partial \psi}{\partial r} + \frac{\partial^2 \psi}{\partial z^2} = -Sr^2 - T\psi - U \tag{8.63}$$

Note that the term $Sr^2 + U$ satisfies the equation $\Delta^*(Sr^2 + U) = 0$; thus, the following homogeneous equation has to be solved

$$\frac{\partial^2 \psi_h}{\partial r^2} - \frac{1}{r}\frac{\partial \psi_h}{\partial r} + \frac{\partial^2 \psi_h}{\partial z^2} + T\psi_h = 0 \tag{8.64}$$

Once ψ_h is found, the solution can be written in the form

$$\psi = \psi_h - \frac{S}{T}r^2 - \frac{U}{T} \tag{8.65}$$

and the following differential equation is obtained

$$\frac{\partial^2 \psi}{\partial r^2} - \frac{1}{r}\frac{\partial \psi}{\partial r} + \frac{\partial^2 \psi}{\partial z^2} = -T\psi - Sr^2 - U \tag{8.66}$$

Equation (8.66) can be solved by the separation of variables which yields the following two ordinary differential equations:

$$\frac{\partial^2 H(z)}{\partial z^2} + k^2 H(z) = 0 \tag{8.67}$$

$$\frac{\partial^2 G(r)}{\partial r^2} - \frac{1}{r}\frac{\partial G(r)}{\partial r} - (k^2 - T)G(r) = 0 \tag{8.68}$$

The corresponding solutions of (8.67) and (8.68) are given by:

$$H(z) = c_1 e^{jkz} + c_2 e^{-jkz} \tag{8.69}$$

$$G(r) = rB_1(ar) \tag{8.70}$$

where B_1 represents the family of Bessel functions and parameter a satisfies the equation [17]

$$a^2 = \pm(T - k^2) \tag{8.71}$$

More mathematical details on this analytical solution procedure can be found in [17].

This family of solutions has a current profile with three independent parameters, thus providing one to independently specify the plasma current density, pressure ratio, and one shape moment such as the internal inductance [17]. Consequently, it is possible to fit experimental configurations in a manner consistent with external magnetic measurements.

To obtain exact solution out of Eqs. (8.69) and (8.70) for various real scenarios, the numerical solution of the free boundary problem (with a conventional equilibrium solver) and subsequent projection of the numerically calculated solution onto the exact solutions via a least squares fitting procedure is implemented [17]. The obtained solution can be written as

$$
\begin{aligned}
\psi = {} & c_1 + c_2 r^2 + r J_1(pr)(c_3 + c_4 z) + c_5 \cos pz + c_6 \sin pz \\
& + r^2 (c_7 \cos pz + c_8 \sin pz) + c_9 \cos p\sqrt{r^2 + z^2} \\
& + c_{10} \sin p\sqrt{r^2 + z^2} + r J_1(vr)(c_{11}\cos qz + c_{12}\sin qz) \\
& + r J_1(qr)(c_{13}\cos vz + c_{14}\sin vz) \\
& + r Y_1(vr)(c_{15}\cos qz + c_{16}\sin qz) \\
& + r Y_1(qr)(c_{17}\cos vz + c_{18}\sin vz)
\end{aligned}
\tag{8.72}
$$

where the vector of coefficients c_i can be found in [17].

An example of a predictive reversed shear equilibrium using ASDEX upgrade field coils and vessel can be described in a similar fashion with the equation [17]

$$
\begin{aligned}
\psi = {} & c_1 + c_2 r^2 + r I_1(pr)(c_3 + c_4 z) + c_5 \cosh pz + c_6 \sinh pz \\
& + r^2 (c_7 \cosh pz + c_8 \sinh pz) + c_9 \cosh p\sqrt{r^2 + z^2} \\
& + c_{10} \sinh p\sqrt{r^2 + z^2} + r I_1(vr)(c_{11}\cosh qz + c_{12}\sinh qz) \\
& + r I_1(qr)(c_{13}\cosh vz + c_{14}\sinh vz) \\
& + r K_1(vr)(c_{15}\cosh qz + c_{16}\sinh qz) \\
& + r K_1(qr)(c_{17}\cosh vz + c_{18}\sinh vz)
\end{aligned}
\tag{8.73}
$$

More mathematical details are available in [17].

8.2.1.2 Analytical Results

The results presented in this subsection correspond to the results for tokamak equilibrium obtained using analytical solutions (8.72) and (8.73) derived in [17]. In Fig. 8.7, the results for poloidal flux contours for ASDEX Upgrade discharge # 10 966, $t = 1.242$ s, calculated using (8.72) are presented.

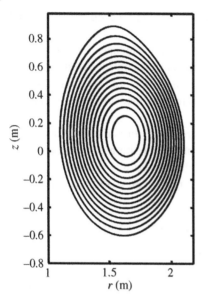

Figure 8.7 Exact GSE solution for ASDEX Upgrade discharge # 10 966, $t = 1.24$ s.

The highest value for the poloidal magnetic flux $\psi_{\text{max}} = 1.3\,\text{Tm}^2$ is observed at the center of tokamak plasma, as it is expected, while the final contour (called separatrix) defines area where the value of the magnetic flux is equal to zero.

For the solution depicted in Fig. 8.8, similar behavior of plasma flux can be observed, with a somewhat different shape of plasma due to the defined discharge pulse.

The maximum value of the magnetic flux is somewhat higher, i.e. $\psi_{\text{max}} = 1.4\,\text{Tm}^2$.

Figure 8.9 shows a predictive reversed shear equilibrium using ASDEX upgrade field coils and vessel geometry obtained by using (8.73).

The value of the maximum flux is significantly lower than in previous examples, i.e. $\psi_{\text{max}} = 0.6\,\text{Tm}^2$.

8.2.1.3 Solution by the Finite Difference Method (FDM)

For application of FDM to solve GSE (8.48), it could be written in the following form

$$-\left(\frac{\partial^2 \Psi}{\partial r^2} + \frac{\partial^2 \Psi}{\partial z^2} - \frac{1}{r}\frac{\partial \Psi}{\partial r}\right) = \mu_0 r J_\phi \tag{8.74}$$

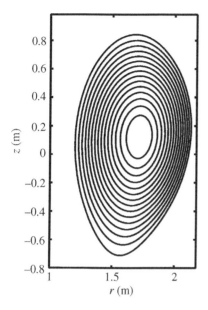

Figure 8.8 Exact GSE solution for ASDEX Upgrade discharge # 10 958, t = 5.20 s.

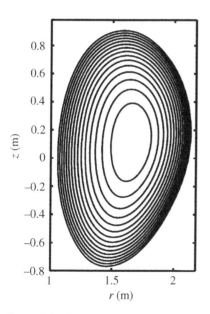

Figure 8.9 Exact GSE solution for a predictive reversed ASDEX Upgrade equilibrium.

Now FDM discretization yields

$$
\begin{aligned}
&\frac{\psi(r + \Delta r, z) - 2\psi(r, z) + \psi(r - \Delta r, z)}{\Delta r^2} \\
&+ \frac{\psi(r, z + \Delta z) - 2\psi(r, z) + \psi(r, z - \Delta z)}{\Delta z^2} \\
&+ \frac{1}{r} \frac{\psi(r + \Delta r, z) - \psi(r - \Delta r, z)}{2\Delta r} = -\mu_0 r J_\phi
\end{aligned}
\tag{8.75}
$$

Finite difference Eq. (8.75) is applied to each node and can be solved by prescribing certain boundary conditions [3, 4].

8.2.1.4 Solution by the Finite Element Method (FEM)

To implement FEM solution, GSE form (8.74) is convenient as well. Taking the scalar product over the calculation domain yields:

$$
-\int_\Omega \left(\frac{\partial^2 \Psi}{\partial r^2} + \frac{\partial^2 \Psi}{\partial z^2} - \frac{1}{r} \frac{\partial \Psi}{\partial r} \right) W_j d\Omega = \int_\Omega \mu_0 r J_\phi W_j d\Omega
\tag{8.76}
$$

and performing some mathematical manipulations leads to the weak formulation of GSE:

$$
\begin{aligned}
&-\int_\Gamma \frac{\partial \Psi}{\partial n} W_j d\Gamma + \int_\Omega \left(\frac{\partial \Psi}{\partial r} \frac{\partial W_j}{\partial r} + \frac{\partial \Psi}{\partial z} \frac{\partial W_j}{\partial z} \right) d\Omega + \int_\Omega \frac{1}{r} \frac{\partial \Psi}{\partial r} W_j d\Omega \\
&= \int_\Omega \mu_0 r J_\phi W_j d\Omega
\end{aligned}
\tag{8.77}
$$

Using the triangular elements and linear shape functions, the solution over an element is given by:

$$
\psi(r, z) = \sum_{i=1}^{3} \alpha_i f_i(r, z)
\tag{8.78}
$$

Choosing the same shape and test functions (Galerkin-Bubnov scheme)

$$
W_j(r, z) = f_j(r, z)
\tag{8.79}
$$

the local matrix system on the element is obtained:

$$
[A]\{\alpha\} = \{B\}
\tag{8.80}
$$

where FEM matrix and excitation vector coefficients are:

$$A_{ji} = \int_{\Omega_e} \left[\frac{\partial f_j(r,z)}{\partial r} \frac{\partial f_i(r,z)}{\partial r} + \frac{\partial f_j(r,z)}{\partial z} \frac{\partial f_i(r,z)}{\partial z} \right] d\Omega_e$$
$$+ \int_{\Omega_e} \frac{1}{r} f_j(r,z) \frac{\partial f_i(r,z)}{\partial r} d\Omega_e \tag{8.81}$$

$$B_{ji} = \mu_0 \int_{x_1}^{x_2} r J_\phi f_j(r,z) d\Omega_e \tag{8.82}$$

Integral (8.81) is solved analytically, while integral (8.82) can generally be calculated numerically using the Gaussian quadrature formulas.

In this case, the Gaussian three-point quadrature rule for the integration over a triangle has been used. Thus, each triangular element has been transformed into a unitary triangle, as indicated in Fig. 8.10.

The coordinate transformation is given by:

$$r = \sum_{i=1}^{3} r_i \phi_i(\xi', \xi''); \quad z = \sum_{i=1}^{3} z_i \phi_i(\xi', \xi'') \tag{8.83}$$

where

$$\phi_1(\xi', \xi'') = 1 - \xi' - \xi''$$
$$\phi_2(\xi', \xi'') = \xi' \tag{8.84}$$
$$\phi_3(\xi', \xi'') = \xi''$$

More details can be found elsewhere, e.g. in [3].

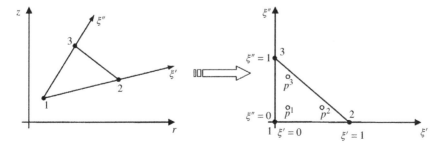

Figure 8.10 Geometry of triangular element and numerical integration points.

8.2.1.5 Computational Examples

The computational examples pertain to rectangular plasma. Figs. 8.11–8.13 show the distribution of flux $\Psi(Wb)$ for rectangular plasma obtained by FDM and FEM $\mu_0 r J_\varphi = 1$, $\mu_0 r J_\varphi = r^3 z^2$, and $\mu_0 r J_\varphi = r^2 z^3$, respectively.

The numerical results obtained by FEM are in quite satisfactory agreement with the results obtained via FDM and with the results published in [13].

(a)
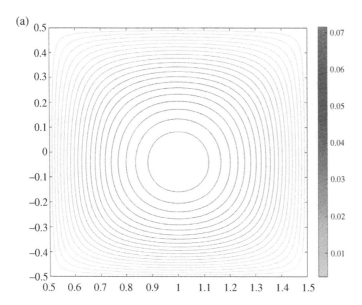

(b)
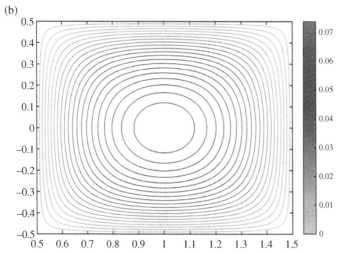

Figure 8.11 $\Psi(Wb)$ for rectangular plasma $-\mu_0 r J_\varphi = 1$. (a) FDM solution; (b) FEM solution.

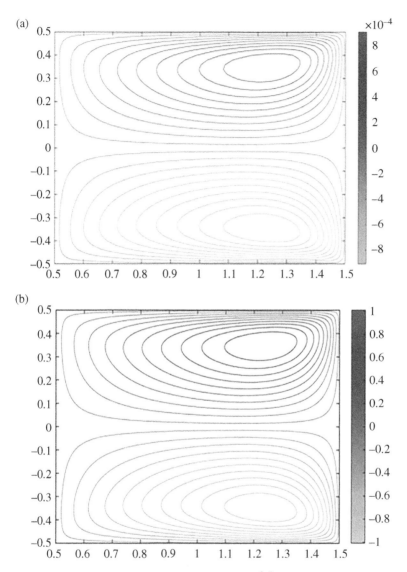

Figure 8.12 Ψ(*Wb*) for rectangular plasma $-\mu_0 r J_\varphi = r^2 z^3$. (a) FDM solution. (b) FEM solution.

8.2.2 Transport Phenomena Modeling

Solution of transport equations is crucial for analyzing the plasma behavior in tokamaks. Complexity of phenomena occurring in tokamak plasma, an integrated approach for the simulation of the global behavior of a tokamak discharge is required [20]. In particular, taking into account the axisymmetry of a tokamak

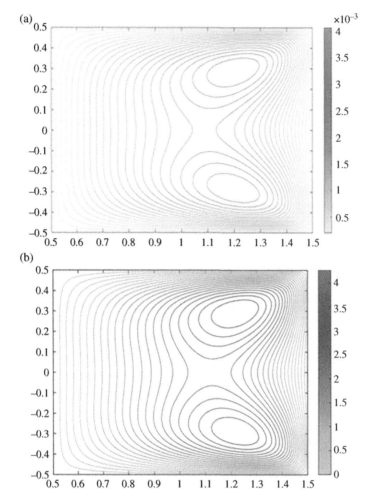

Figure 8.13 $\Psi(Wb)$ for rectangular plasma $-\mu_0 r J_\varphi = r^3 z^2$. (a) FDM solution. (b) FEM solution.

and relevant temporal and spatial scales orderings, time evolution equations for the transport of macroscopic plasma quantities, such as current density, pressure, and density, can be cast in time and a single space coordinate. The space coordinate arises from flux surface averaging, which is possible due to the much faster transport along the magnetic field lines compared to the perpendicular transport. Typically, this coordinate is the (normalized) toroidal magnetic flux. The flux surface topology is determined by the MHD equilibrium, described by GSE [21]. Rigorous derivation of the transport equations, starting from the kinetic description and involving gyromotion averaging, fluid and MHD equations, and flux surface averaging, is reviewed in [22].

CDE describing the current (or, equivalently, the magnetic field) diffusion through the plasma as a conductive medium is derived by flux-averaging Ohm's law. CDE is coupled with the equilibrium governed in terms of GSE. Implementation of higher-order numerical schemes, featuring FEM using Hermite basis functions and an analytical solution of CDE has been reported in [23]. The results obtained via different approaches agree satisfactorily.

8.2.2.1 Transport Equations

Generally, there are six transport equations [24] for the analysis of the transport phenomena dealing with the distribution of the following quantities and their flux densities (derivatives): current, ion density, electron density, ion temperatures, electron energy, and rotation transport.

A generic equation representing all six transport equations can be written in the following form [24]

$$
\frac{a(x)Y(x,t)b(x)Y(x,t-1)}{\Delta t} + \frac{1}{c(x)}\frac{\partial}{\partial x}\left[-d(x)\frac{\partial Y(x,t)}{\partial x} + e(x)Y(x,t)\right]
$$
$$
= f(x) - g(x)Y(x,t)
$$

(8.85)

where $a(x)$, $b(x)$, $c(x)$, $d(x)$, $e(x)$, $f(x)$, and $g(x)$ are transport coefficients, while $Y(x,t)$ and $Y(x,t-1)$ are the values of interest at the present instant and the previous time step Δt, respectively.

Differential Eq. (8.85) is accompanied by generic boundary conditions [24].

An example of transport equation is the current diffusion equation (CDE) [23].

8.2.2.2 Current Diffusion Equation and Equilibrium in Tokamaks

The evolution of the magnetic field configuration in axisymmetric tokamak devices is represented by a coupled system of CDE and GSE [25]. The coupling considered here is the adiabatic evolution of the equilibrium, which is regarded as a generalized (queer) differential equation problem, as reported in [26, 27].

The CDE derived by flux-averaging the parallel Ohm's law can be written in the following form [25]:

$$
\frac{\partial \psi}{\partial t} - \frac{\langle|\nabla\rho|^2/R^2\rangle}{\mu_0\sigma_\parallel\rho_1^2\langle 1/R^2\rangle}\frac{\partial^2\psi}{\partial x^2} - \left[\frac{\langle|\nabla\rho|^2/R^2\rangle}{\mu_0\sigma_\parallel\rho_1^2\langle 1/R^2\rangle}\frac{\partial}{\partial x}\ln\left(\frac{V'\langle|\nabla\rho|^2/R^2\rangle}{F}\right) + \frac{x}{\rho_1}\frac{d\rho_1}{dt} + \frac{x}{2B_0}\frac{dB_0}{dt}\right]
$$
$$
\times\frac{\partial\psi}{\partial x} = \frac{B_0}{\sigma_\parallel F\langle 1/R^2\rangle}j_{ni}
$$

(8.86)

where B_0 the vacuum magnetic field at a given R_0, $\sigma_\parallel$ the parallel conductivity, V the plasma volume enclosed by a flux surface, $V' = dV/d\rho$, and j_{ni} is the non-inductive current density. The flux coordinate $x = \rho/\rho_1$, $\rho = \sqrt{\Phi/\pi B_0}$ with ρ_1 its value at the plasma boundary (the last closed flux surface) and Φ the toroidal magnetic field flux in a given flux surface. The flux surface averaging operator is defined as [12]

$$\langle A \rangle \equiv \frac{\partial}{\partial V} \int_V A \, dV \tag{8.87}$$

Introducing the following substitutions

$$\langle 1/R^2 \rangle = g_1 \tag{8.88}$$

$$\langle |\nabla \rho|^2 / R^2 \rangle = g_2 \tag{8.89}$$

and expression (8.86) becomes

$$\frac{\partial \psi}{\partial t} - \frac{g_2}{\mu_0 \sigma_\parallel \rho_1^2 g_1} \frac{\partial^2 \psi}{\partial x^2} - \left[\frac{g_2}{\mu_0 \sigma_\parallel \rho_1^2 g_1} \frac{\partial}{\partial x} \ln\left(\frac{V' g_2}{F}\right) + \frac{x}{\rho_1} \frac{d\rho_1}{dt} + \frac{x}{2B_0} \frac{dB_0}{dt} \right] \frac{\partial \psi}{\partial x}$$
$$= \frac{B_0}{\sigma_\parallel F g_1} j_{ni} \tag{8.90}$$

Now coupling of the current diffusion and equilibrium has to be carried out.

In the Grad-Hogan coupling scheme [28], considered in this paper as well as in CRONOS [25] and ASTRA [26] transport suites or in the European Transport Simulator (ETS) [24], the current diffusion (as well as other transport equations) is advanced in time using the geometry from the last GSE solution. GSE is then solved using the resulting $p'(\overline{\psi})$ and $FF'(\overline{\psi})$ profiles. The two equations are iterated until they yield consistent results in terms of the magnetic flux and the plasma current profiles. Here, $\overline{\psi} = (\psi - \psi_0)/(\psi_1 - \psi_0)$ is the normalized poloidal flux with $\overline{\psi} = 0$ on the magnetic axis and $\overline{\psi} = 1$ on the plasma boundary. While p' can be calculated directly from the transport equations results, FF' must be solved using the averaged GSE:

$$FF' = \frac{\mu_0}{g_1} \left(\left\langle \frac{j_\phi}{R} \right\rangle - p' \right) \tag{8.91}$$

The average current term can be calculated as

$$\left\langle j_\phi / R \right\rangle = - \frac{\frac{\partial}{\partial x}\left(V' g_2 \frac{\partial \psi}{\partial x}\right)}{\rho_1^2 \mu_0 V'} \tag{8.92}$$

As the current density is continuous, it follows that

$$\psi \in C^2, \quad g_2 \in C^1, \quad V' \in C^1 \tag{8.93}$$

The iteration between CDE and GSE via AGSE is prone to numerical instabilities, as discussed recently in [26] and the references therein. Consequently, of great importance is to implement a stable, high-accuracy numerical solver for the diffusion equation(s) that inherently yields smooth solutions. A plausible choice is FEM which provides higher-order accuracy. Namely, finite domain methods discretize the domain and determine the results in the set of nodes, while FEM through sophisticated interpolation scheme provides the results at any point of the domain.

Final step in the formulation is to prescribe proper boundary conditions. On the magnetic axis, the geometry of the problem yields

$$\left.\frac{\partial \psi}{\partial x}\right|_{x=0} = 0 \tag{8.94}$$

The most common boundary condition posed on the plasma boundary is the total plasma current (see e.g. [29]) given by

$$I_p = -\frac{1}{2\pi\mu_0} V' g_2 \left.\frac{\partial \psi}{\partial \rho}\right|_{x=1} \tag{8.95}$$

However, if the plasma current is not prescribed, a different boundary condition has to be considered. This is often the case with free boundary equilibrium (FBE) simulations. As the magnetic flux is required to be consistent in the transport and equilibrium equations, the natural boundary condition at the last closed flux surface ($x = 1$) would be defined by the equality of the boundary magnetic flux, i.e.

$$\psi^{diff} = \psi^{equi} \tag{8.96}$$

where ψ^{diff} and ψ^{equi} come, respectively, from the solutions of the CDE and the equilibrium equation.

This is similar to prescribing the loop voltage in fixed boundary simulation, which is known to be prone to numerical errors. Therefore, one can use a plasma current predictor I_p^*, derived from (8.96) using $L_i I_p = \psi_0 - \psi_1$. Here, the subscripts 0 and 1 denote the values on the magnetic axis ($x = 0$) and the plasma boundary ($x = 1$) and L_i is the internal inductance of the plasma. Finally, one obtains

$$I_p^* = I_p \left(1 + \frac{\psi_1^{diff} - \psi_1^{equi}}{\psi_0^{equi} - \psi_1^{equi}}\right). \tag{8.97}$$

What follow are numerical and analytical solutions, respectively.

8.2.2.3 FEM Solution of CDE

Generally, FEM is considered to be a more sophisticated numerical technique than finite difference method (FDM) schemes [30] and is highly automatized and convenient for computer implementation based on step-by-step procedures. Furthermore, FEM provides efficient modeling of complex shape geometries and inhomogeneous domains. FEM implementation results in high band and symmetric matrices and also provides accuracy refinement in terms of higher-order approximation and automatic inclusion of Neumann (natural) boundary conditions within the framework of mathematical formulation.

If it can be assumed that $dB_0/dt = 0$, as it is defined in [11], the current diffusion Eq. (8.90) can be written in the following form

$$\mu_0 \sigma_{\|} \rho_1^2 \frac{g_1}{g_2} \frac{\partial \psi}{\partial t} = \frac{\partial^2 \psi}{\partial x^2} + \left[\frac{\partial}{\partial x} \ln\left(\frac{V' g_2}{F}\right) + \mu_0 \sigma_{\|} \rho_1 x \frac{d\rho_1}{dt} \frac{g_1}{g_2} \right] \frac{\partial \psi}{\partial x} + \frac{\mu_0 B_0 \rho_1^2}{F g_2} j_{ni}$$

(8.98)

The ln term is differentiated analytically

$$\frac{\partial}{\partial x} \ln\left(\frac{V' g_2}{F}\right) = \frac{V'' g_2 F + V' g'_2 F - V' g_2 F'}{V' g_2 F} \equiv LN(x)$$

(8.99)

Taking the scalar product of CDE with test functions, W_j over the calculation domain yields

$$\mu_0 \rho_1^2 \int_0^1 \sigma \frac{g_1}{g_2} \frac{\partial \psi}{\partial t} W_j dx = \int_0^1 \frac{\partial^2 \psi}{\partial x^2} W_j dx + \int_0^1 \left[LN(x) + \mu_0 \sigma \rho_1 x \frac{d\rho_1}{dt} \frac{g_1}{g_2} \right] \frac{\partial \psi}{\partial x} W_j dx$$
$$+ \int_0^1 \frac{\mu_0 B_0 \rho_1^2}{F g_2} j_{ni} W_j dx$$

(8.100)

Note that the second-order differentiation of the unknown function is avoided by taking the integration by parts of the first term on the right-hand side

$$\int_0^1 \frac{\partial^2 \psi}{\partial x^2} W_j dx = \frac{\partial \psi}{\partial x} W_j \Big|_{x=0}^{x=1} - \int_0^1 \frac{\partial \psi}{\partial x} \frac{\partial W_j}{\partial x} dx$$

(8.101)

Term $\dfrac{\partial \psi}{\partial x} W_j \Big|_{x=0}^{x=1}$ represents the Neumann boundary conditions given with (8.94) and (8.95) and it can be directly included in matrix equation.

Now the weak form of the CDE is given by

$$\mu_0 \rho_1^2 \int \sigma \frac{g_1}{g_2} \frac{\partial \psi}{\partial t} W_j dx = - \int \frac{\partial \psi}{\partial x} \frac{\partial W_j}{\partial x} dx + \int \left[LN(x) + \mu_0 \sigma \rho_1 x \frac{d\rho_1}{dt} \frac{g_1}{g_2} \right] \frac{\partial \Psi}{\partial x} W_j dx$$
$$+ \mu_0 B_0 \rho_1^2 \int \frac{j_{ni}}{Fg_2} W_j dx$$

$$(8.102)$$

According to the standard finite element procedure, the unknown magnetic flux ψ is expressed in terms of linearly independent basis functions $\{N_i\}$ with unknown complex coefficients ψ_i, i.e.

$$\psi = \sum_i \psi_i(t) N_i(x) \tag{8.103}$$

Choosing the same basis functions as test functions $W_j = N_j$, the Gallerkin-Bubnov procedure is introduced to the weak formulation of the CDE

$$\mu_0 \rho_1^2 \sum_i \int_0^1 \sigma \frac{g_1}{g_2} N_i N_j dx \cdot \frac{\partial \psi_i}{\partial t} = \sum_i \left\{ \int_0^1 \left[LN(x) + \mu_0 \sigma \rho_1 x \frac{d\rho_1}{dt} \frac{g_1}{g_2} \right] \frac{\partial N_i}{\partial x} N_j dx \right.$$
$$\left. - \int_0^1 \frac{\partial N_i}{\partial x} \frac{\partial N_j}{\partial x} dx \right\} \psi_i + \mu_0 B_0 \rho_1^2 \int_0^1 \frac{j_{ni}}{Fg_2} N_j dx, j = 1, 2, ..., n$$

$$(8.104)$$

which can, for the sake of simplicity, be written in the form of matrix equation

$$[M] \frac{\partial}{\partial t} \{\psi\} = [D]\{\psi\} + \{K\} \tag{8.105}$$

where

$$M = \mu_0 \rho_1^2 \int_0^1 \sigma \frac{g_1}{g_2} N_i N_j dx \tag{8.106}$$

$$D = - \int_0^1 \frac{\partial N_i}{\partial x} \frac{\partial N_j}{\partial x} dx + \int_0^1 \left[LN(x) + \mu_0 \sigma \rho_1 x \frac{d\rho_1}{dt} \frac{g_1}{g_2} \right] \frac{\partial N_i}{\partial x} N_j dx \tag{8.107}$$

$$K = \mu_0 B_0 \rho_1^2 \int_0^1 \frac{j_{ni}}{Fg_2} N_j dx \tag{8.108}$$

According to the time domain discretization scheme [26], the solution for the magnetic flux at $n + 1$ time instant is calculated using the solution for the n-th time instant given with

$$([M] - \Delta t[D])\{\psi\}^{n+1} = [M]\{\psi\}^n + \Delta t[K] \tag{8.109}$$

FEM solution of CDE featuring the use of linear and Hermite shape functions has been reported in [23], and it is outlined below.

Linear shape functions chosen for the basis functions are given by

$$N_1^e = \frac{x_2^e - x}{x_2^e - x_1^e}; \quad N_2^e = \frac{x - x_1^e}{x_2^e - x_1^e} \tag{8.110}$$

and matrices M, D, and vector K, according to finite elements procedure, are assembled from the local matrices as follows

$$[M]^{e_k} = \mu_0 \rho_1^2 \int_{x_k}^{x_{k+1}} \sigma \frac{g_1}{g_2} N_i^e N_j^e dx \tag{8.111}$$

$$[D]^{e_k} = -\int_{x_k}^{x_{k+1}} \frac{\partial N_i^e}{\partial x} \frac{\partial N_j^e}{\partial x} dx + \int_{x_k}^{x_{k+1}} \left[LN(x) + \mu_0 \sigma \rho_1 x \frac{d\rho_1}{dt} \frac{g_1}{g_2} \right] \frac{\partial N_i^e}{\partial x} N_j^e dx \tag{8.112}$$

$$[K]^{e_k} = \mu_0 B_0 \rho_1^2 \int_{x_k}^{x_{k+1}} \frac{j_{ni}}{F g_2} N_j^e dx \tag{8.113}$$

Integrals in (8.111)–(8.113) are evaluated via the Gaussian four-point quadrature rule. The values of the coefficients g_1, g_2, F, and V' at the Gaussian points over the element are obtained using linear interpolation. First and second derivatives of the resulting poloidal flux are obtained using the smooth noise-robust differentiators described in [31].

Hermite interpolation on $[-1;1]$ involves choosing a set of ordered nodes $x_1, \ldots, x_n \in [-1; 1]$ and approximating a smooth function $f(x)$ using its values and derivatives on the nodal set

$$p(x) = \sum_{k=1}^{n} f(x_k) l_k^0(x) + f'(x_k) l_k^1(x) \tag{8.114}$$

where l_k^0 and l_k^1 are the Hermite interpolating polynomials and they satisfy conditions

$$l_k^0(x_j) = \delta_{jk}, \quad \frac{dl_k^0}{dx}(x_j) = \delta_{jk}, \quad l_k^1(x_j) = 0 \quad \frac{dl_k^1}{dx}(x_j) = 0 \tag{8.115}$$

Having solved Eq. (8.104) with Hermite shape functions, the results comprise both poloidal flux and its first derivative. The second derivative is obtained using smooth robust differentiators [31].

8.2.2.4 Analytical Solution Procedure

In order to obtain analytical solution of (8.98), time derivative of ρ_1 is neglected and current density driven by non-inductive sources is equal to zero. Taking this into account, (8.98) becomes

$$\frac{\partial \psi}{\partial t} - \frac{g_2}{\mu_0 \sigma_\| \rho_1^2 g_1} \frac{\partial^2 \psi}{\partial x^2} - \left[\frac{g_2}{\mu_0 \sigma_\| \rho_1^2 g_1} \frac{\partial}{\partial x} \ln\left(\frac{V' g_2}{F}\right) \right] \frac{\partial \psi}{\partial x} = 0 \tag{8.116}$$

Introducing the following expressions

$$f(x) = \frac{g_2}{\mu_0 \sigma_\| \rho_1^2 g_1} \tag{8.117}$$

$$g(x) = \frac{g_2}{\mu_0 \sigma_\| \rho_1^2 g_1} \frac{\partial}{\partial x} \ln\left(\frac{V' g_2}{F}\right) \tag{8.118}$$

(8.116) can be written as

$$\frac{\partial \psi}{\partial t} - f(x) \frac{\partial^2 \psi}{\partial x^2} - g(x) \frac{\partial \psi}{\partial x} = 0 \tag{8.119}$$

The particular solution of (8.119) can be obtained in the form [29]

$$\psi(x, t) = At\Phi(x) + A \int F(x) \int \frac{\Phi(x) dx}{f(x) F(x)} dx \tag{8.120}$$

where A is an arbitrary constant and auxiliary functions $F(x)$ and $\Phi(x)$ are given by

$$F(x) = \exp\left(- \int \frac{g(x)}{f(x)} dx \right)$$

$$\Phi(x) = \int F(x) dx \tag{8.121}$$

Undertaking additional mathematical manipulation, the expression for the space-time dependent poloidal flux is obtained

$$\psi(x, t) = At \int \frac{F}{V' g_2} dx + A\mu_0 \rho_1^2 \int \int \frac{\sigma_\| g_1 V'}{F} \int \frac{F}{V' g_2} dx dx dx \tag{8.122}$$

where arbitrary constant A is calculated from the boundary condition

$$\left.\frac{\partial \psi}{\partial x}\right|_{x=1} = -\left(\frac{2\pi \mu_0 \rho_1}{V' g_2} I_P\right)_{x=1} \tag{8.123}$$

and is given by

$$A = -\frac{2\pi}{\rho_1 \sqrt{\dfrac{F}{V' g_2}} \displaystyle\int \dfrac{\sigma_\| g_1 V'}{F} dx dx} I_p \tag{8.124}$$

The functions appearing under the integral (F, V', g_1, g_2) are obtained as an input in the form of discrete data sets which are transformed into polynomials of arbitrary order in a least squares sense.

The first derivative is readily obtained from (8.122) and is given with

$$\frac{\partial \psi(x,t)}{\partial x} = A \frac{F}{V' g_2} \left(t + \mu_0 \rho_1^2 \int \frac{\sigma_\| g_1 V'}{F} \int \frac{F}{V' g_2} dx dx\right) \tag{8.125}$$

The second derivative of the poloidal flux is obtained in a straightforward manner as

$$\frac{\partial^2 \psi(x,t)}{\partial x^2} = A \frac{\partial}{\partial x}\left(\frac{F}{V' g_2}\right) t + A\mu_0 \rho_1^2 \left(\frac{\partial}{\partial x}\left(\frac{F}{V' g_2}\right)\int \frac{\sigma_\| g_1 V'}{F} \int \frac{F}{V' g_2} dx dx \right.$$
$$\left. + \sigma_\| \frac{g_1}{g_2} \int \frac{F}{V' g_2} dx\right) \tag{8.126}$$

It is worth emphasizing that the second derivative does not suffer from numerical instabilities.

8.2.2.5 Numerical Results

Numerical and analytical models presented in [6] and [23] are applied to an ITER hybrid scenario CRONOS simulation from [32] and the obtained results are subsequently compared. Two types of shape functions are used within the proposed FEM approach: Hermite and linear shape functions. In CRONOS, the average current term is computed using cubic spline interpolation which is used for all derivatives with the default tension of the function.

The poloidal flux and its derivatives at a time slice $t = 1095$ s from the late flat-top phase, where the current is almost fully diffused and thus all profiles tend to be stationary, obtained by FEM and by means of analytical approach are shown in Figs. 8.14–8.16, respectively.

The results show a satisfactory agreement between the results obtained by FEM featuring two types of shape functions. Analytical results have shown an expected

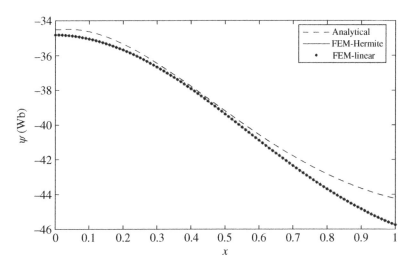

Figure 8.14 Poloidal magnetic flux for *t* = 1095s.

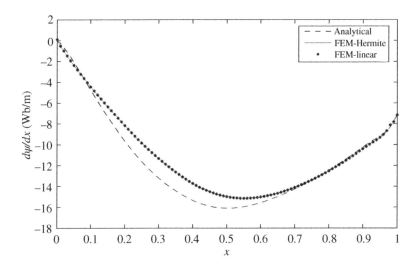

Figure 8.15 1st derivative of the poloidal magnetic flux for *t* = 1095s.

discrepancy as the noninductive current sources are not taken into account [23]. It is obvious that Hermite shape functions are better suited for the computation of the second derivative at the beginning of the domain, as they do not generate non-physical oscillations. This is due to the fact that Hermite basis functions are polynomial functions of higher order than linear basis functions which enables

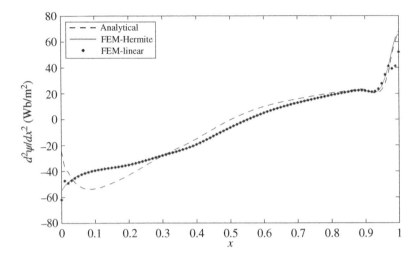

Figure 8.16 2nd derivative of the poloidal magnetic flux for t = 1095 s.

the use of known first derivative as expansion coefficients resulting in more accurate results. Furthermore, it can be observed that analytical solution shows no oscillations and is generally in satisfactory agreement with the numerical solution.

All in all, the results show a good agreement at the edge of plasma, as well as overall agreement in the waveform, with some discrepancy shown in the middle.

8.3 Modeling of the Schrodinger Equation

In classical physics, one deals with particles and fields, while in quantum physics, wave function is considered. Thus, in classical physics, there are point-like particles interacting with different forces carried by fields (e.g. point charges and electromagnetic forces and fields in classical electromagnetism) while in quantum physics, there is only wave function which when observed shows particle-like behavior (the collapse of wave function).

Therefore, the task in classical physics is to prescribe a specific location and velocity of a particle and to study how the particle evolves in accordance with the Newton's mechanics. In quantum physics, the particle is assigned a specific wave function and the evolution of such a system, on the other hand, is governed by the solution of the Schrodinger equation.

Modeling of the Schrodinger equation has been discussed elsewhere, e.g. in [33–36].

8.3.1 Derivation of the Schrodinger Equation

Derivation of the Schrodinger equation is based on the fact that a particle captured in a small volume can have only discrete energy levels. Discretization of energy levels is obvious if the considered volume of space is comparable to the de Broglie wavelength.

It is possible to derive spatially dependent Schrodinger equation starting from classical wave equation (Helmholtz equation) [37, 38]

$$\nabla^2 \psi + k^2 \psi = 0 \tag{8.127}$$

where ψ is the wave function and k is the wave number

$$k = \frac{2\pi}{\lambda} \tag{8.128}$$

According to de Broglie, particles with energy E and momentum p also have wave properties, i.e. wavelength

$$\lambda = \frac{h}{p} \tag{8.129}$$

where p is the momentum of a wave-particle defined as

$$p = mv \tag{8.130}$$

Now, it simply follows

$$k = \frac{2\pi}{\lambda} = \frac{2\pi p}{m} = \frac{2\pi mv}{h} \tag{8.131}$$

Inserting (8.131) into (8.127) yields

$$\nabla^2 \psi + \frac{4\pi^2 m^2 v^2}{h^2} \psi = 0 \tag{8.132}$$

The total energy of a quantum particle can be expressed as the sum of its kinetic and potential energies, respectively

$$E = \frac{1}{2}mv^2 + V \tag{8.133}$$

Now combining (8.132) and (8.133), it follows

$$-\frac{\hbar^2}{2m}\nabla^2 \psi + V\psi = E\psi \tag{8.134}$$

where

$$\hbar = \frac{h}{2\pi} \tag{8.135}$$

Expression (8.134) represents the three-dimensional Schrodinger equation. One-dimensional version is then given by

$$-\frac{\hbar^2}{2m}\frac{\partial^2\psi(x)}{\partial x^2} + V(x)\psi(x) = E\psi(x) \tag{8.136}$$

and for some scenarios can be solved analytically.

8.3.2 Analytical Solution of the Schrodinger Equation

The classical form of the time independent Schrodinger equation can be written as follows [39]

$$H\psi = E\psi \tag{8.137}$$

where ψ is the wave function, E is the energy of the quantum particle of mass m moving within the interval $(0, L)$, and H is the Hamiltonian

$$H = -\frac{\hbar^2}{2m}\frac{\partial^2}{\partial x^2} + V(x) \tag{8.138}$$

where $V(x)$ is the corresponding potential function, while the reduced Planck constant is given by

$$\hbar = \frac{h}{2\pi} \tag{8.139}$$

The energy spectrum of the quantum particle can be determined by solving Schrodinger equation (8.139) provided that the boundary conditions for the wave function ψ are prescribed.

Assuming the particle to be captured inside the potential well with $V = 0$ and cannot be located outside the interval $(0, L)$ with the boundary conditions:

$$\psi(0) = \psi(L) = 0 \tag{8.140}$$

One-dimensional Schrodinger equation simplifies into

$$-\frac{\partial^2\psi}{\partial x^2} = k^2\psi \tag{8.141}$$

The analytical solution of (8.141) is given by

$$\psi(x) = A\sin kx + B\cos kx \tag{8.142}$$

Inserting the boundary conditions (8.140) yields

$$B = 0 \tag{8.143}$$

and from

$$\sin kL = 0 \tag{8.144}$$

The wave number k is obtained

$$k = \frac{n\pi}{L} k = \frac{n\pi}{L} \tag{8.145}$$

and the wave function is

$$\psi(x) = A \sin\left(\frac{n\pi}{L} k\right) \tag{8.146}$$

Probability of the existence of the quantum particle within the well, i.e. within the observed interval $(0, L)$, is equal to one, i.e.

$$\int_0^L |\psi(x)|^2 dx = 1 \tag{8.147}$$

Thus, inserting (8.146) into (8.147), one obtains

$$A^2 \int_0^L \sin^2\left(\frac{n\pi}{L} k\right) dx = 1 \tag{8.148}$$

and it simply follows

$$A = \sqrt{\frac{2}{L}} \tag{8.149}$$

Finally, the solution of Schrodinger equation (8.137) is:

$$\psi(x) = \sqrt{\frac{2}{L}} \sin\left(\frac{n\pi}{L} x\right) \tag{8.150}$$

The Schrodinger equation for the case of three-dimensional potential well can be handled by using a similar procedure.

8.3.3 FDM Solution of the Schrodinger Equation

Application of FDM to one-dimensional Schrodinger equation

$$-\frac{\hbar}{2m} \frac{\partial^2 \psi(x)}{\partial x^2} + V(x)\psi = E\psi(x) \tag{8.151}$$

for the case of particle inside potential well with $V = 0$ yields

$$-\frac{\hbar}{2m}\frac{\psi(x+\Delta x) - 2\psi(x) + \psi(x - \Delta x)}{\Delta x^2} + V(x)\psi(x) = E\psi(x) \qquad (8.152)$$

and results in the system of N equations with N unknowns.

8.3.4 FEM Solution of the Schrodinger Equation

Applying the weighted residual approach to Schrodinger equation (8.151), one obtains

$$-\int_0^L \frac{\hbar}{2m}\frac{\partial^2 \psi}{\partial x^2} W_j dx + \int_0^L V(x)\psi W_j dx = \int_0^L E\psi W_j dx \qquad (8.153)$$

and utilizing the weak formulation, it follows

$$\int_0^L \frac{\partial \psi}{\partial x}\frac{\partial W_j}{\partial x}dx + \frac{2m}{\hbar^2}\int_0^L V(x)\psi W_j dx = \frac{2m}{\hbar^2}E\int_0^L \psi W_j dx + \frac{2m}{\hbar^2}\frac{\partial \psi}{\partial x}W_j\Big|_0^L$$

$$(8.154)$$

The approximate solution is given in terms of linear combination of coefficients α_i and shape functions f_i:

$$\psi(x) = \{f\}^T\{\alpha\} \qquad (8.155)$$

where linear shape functions are given by:

$$f_1(x) = \frac{x_2 - x}{\Delta x}, f_2(x) = \frac{x - x_1}{\Delta x}, \Delta x = x_2 - x_1 \qquad (8.156)$$

Discretizing the calculation domain and applying the Galerkin-Bubnov procedure $W_j = f_j$, the following matrix equation is obtained

$$[A]\{\alpha\} = [B]\{\alpha\}[E] \qquad (8.157)$$

where $[E]$ is a diagonal matrix representing the particle energy levels in different states.

FEM matrix and excitation vector coefficients are:

$$A_{ji} = \int_{x_1}^{x_2} \frac{\partial f_j(x)}{\partial x}\frac{\partial f_i(x)}{\partial x}dx + \frac{2m}{\hbar^2}\int_{x_1}^{x_2} V(x)f_j(x)f_i(x)dx \qquad (8.158)$$

$$B_{ji} = \frac{2m}{\hbar^2}\int_{x_1}^{x_2} f_j(x)f_i(x)dx \qquad (8.159)$$

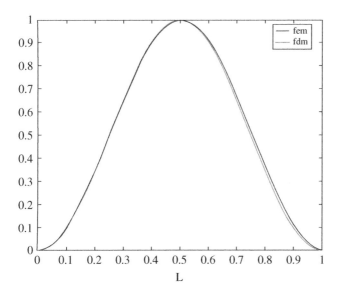

Figure 8.17 $|\psi|^2$ Probability density function for N = 100 and n = 1.

Expression (8.157) is an eigenvalue equation. Each element of $[E]$ matrix is then an eigenvalue (certain energy level). Number of solutions corresponds to the number of quantized energy levels that a particle can occupy inside an infinite potential well.

Figures 8.17–8.19 show the probability density function (PDF) $|\psi|^2$ for different values of n calculated via FDM and FEM.

The numerical results obtained by FDM and FEM agree satisfactorily. The analysis presented is a useful starting point for the analysis of practical scenarios whose solution cannot be obtained in close form.

8.3.5 Neural Network Approach to the Solution of the Schrodinger Equation

An efficient approach to handle the Schrodinger equation is the use of neural networks [40]. An overview of the solution of the Schrodinger equation for the case of one-dimensional infinite potential well with a neural network approach is presented in [39]. It has been shown in [39] that using a single hidden layer neural network, which is proved to be a universal function approximator, and by exploiting the automatic differentiation capabilities, it is possible to achieve very accurate values of the wave function and eigenvalues of the ground state. The loss function with integrated physical knowledge is set up as an unconstrained nonlinear problem and parameters of a neural network are being learnt in a completely

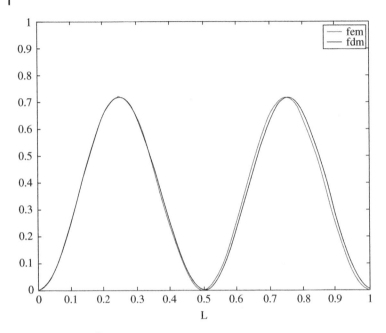

Figure 8.18 $|\psi|^2$ Probability density function for $N = 100$ and $n = 2$.

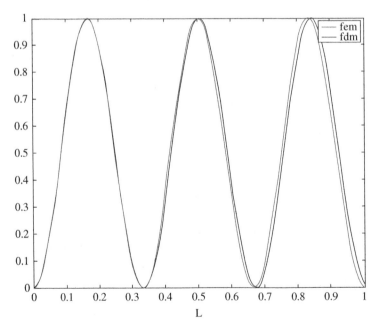

Figure 8.19 $|\psi|^2$ Probability density function for $N = 100$ and $n = 3$.

unsupervised manner. Such a technique could potentially serve as a door opener for solving high-dimensional quantum mechanics problems, otherwise tedious to set up for standard mesh-based numerical methods. The approach is outlined in this subsection.

When dealing with non-relativistic velocities from classical mechanics, it follows that the total energy E of a particle of mass m is simply given as the sum of its kinetic and potential energy U, respectively, i.e. it can be written as

$$\frac{p^2}{2m} + U = E \tag{8.160}$$

where the linear momentum (quantity of movement) of a particle is

$$p = mv \tag{8.161}$$

Replacing linear momentum with the corresponding quantum mechanical operator and introducing the wave function ψ, one obtains the 3-D non-relativistic Schrodinger equation

$$-\frac{\hbar^2}{2m} \nabla^2 \psi + V\psi = E\psi \tag{8.162}$$

Now, as observable quantities are represented by mathematical operators, using the Hamiltonian operator (operator for the total energy consisting of kinetic and energy parts), the Schrodinger equation (8.162) can be written as follows

$$H\psi = E\psi \tag{8.163}$$

where the 3-D Hamiltonian operator is given by

$$H = -\frac{\hbar^2}{2m} \nabla^2 + U \tag{8.164}$$

Note that the wave function ψ entirely describes the state of a quantum system. Since the particle must be found somewhere in space, it follows

$$\int_\Omega \psi\psi^* d\Omega = \int_\Omega |\psi|^2 d\Omega = 1 \tag{8.165}$$

Functions satisfying (8.165) are referred to as eigenfunctions of operator H, while the corresponding values of the quantities (energies) are eigenvalues. Thus, if a certain wave function is an eigenfunction of H, then it follows

$$H\psi_n = E_n\psi_n \tag{8.166}$$

with the related eigenvalue E_n.

As the mean value of an observable is the expectation value of the related operator, according to the notation from statistical analysis for a certain wavefunction ψ_n, the expectation of an operator A is given by

$$\langle A_n \rangle = \frac{\int_\Omega \psi^* A \psi \, d\Omega}{\int_\Omega \psi^* \, \psi \, d\Omega} \tag{8.167}$$

A simplified case of 1-D time independent Schrodinger equation for a single non-relativistic particle can then be written in the following manner

$$H\psi(x) = E\psi(x) \tag{8.168}$$

where the corresponding Hamiltonian operator is given by

$$H = -\frac{\hbar^2}{2m} \frac{d^2}{dx^2} + U(x) \tag{8.169}$$

The squared wave function, $|\psi|^2$, is the PDF of the particle over the x-axis in the solution domain. The actual position is not known due to the Heisenberg uncertainty principle; rather one can describe the uncertainty of finding the particle for position in the observation interval. The sum of probabilities over the entire solution domain is 1. Therefore, wave function can be viewed as a mathematical tool to calculate the probability of observing the particle in a specific position, or simply the quantity whose squared value yields the probability of observing the corresponding outcome. Physically observable system is presented in a measurable and meaningful fashion, provided the wave function satisfies the following set of constraints:

- The wave function must be a solution to the Schrodinger equation;
- The wave function and the first derivative of the wave function must be continuous;
- The wave function must be normalizable – the wave function value approaches zero as x approaches infinity.

Considering the particle-in-a-box problem, an electron of the mass m_e confined to 1-D rigid, infinite potential well of width L and following the previously outlined constraints is observed and it can be shown that the wave function at the ground state is defined using the expression:

$$\psi(x) = \sqrt{\frac{2}{L}} \sin\left(\frac{n\pi}{L}x\right) \tag{8.170}$$

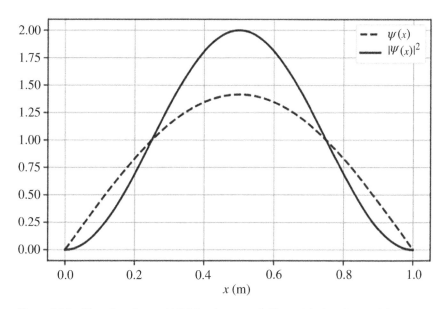

Figure 8.20 Wave function and PDF for the case of 1D particle-in-a-box problem.

An electron is able to move along x-axis and the collisions with the walls are considered perfectly elastic. The full mathematical description can be found elsewhere, e.g. in [39]. Figure 8.20 shows the analytic solution of the wave function and the PDF of an electron in infinite potential well.

The highest point of the PDF represents the most probable position of an electron at any given moment in time. Factor $\sqrt{\frac{2}{L}}$ in (8.170) represents the amplitude of the wave function and if it decreases, the probability density peak will decrease, which will subsequently cause the PDF curve to be flattened.

Finally, the family of wave functions obtained via FDM, FEM, and the neural network is shown in Figs. 8.21–8.23, for 10, 50, and 100 points, respectively.

It is worth noting that a single important advantage of the neural network approach are the interpolation capabilities inherently embedded in the method itself. Both FEM and FDM, and any other mesh-based numerical method, require additional interpolation computations in order to find the value of the solution at an arbitrary point in the domain. More mathematical details can be found in [39].

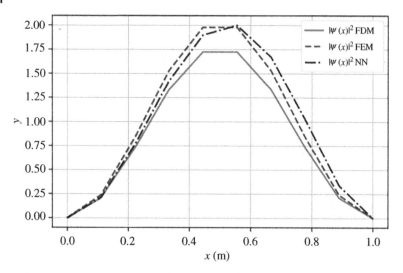

Figure 8.21 Wave functions approximated with 10 points using FDM, FEM, and NN approaches.

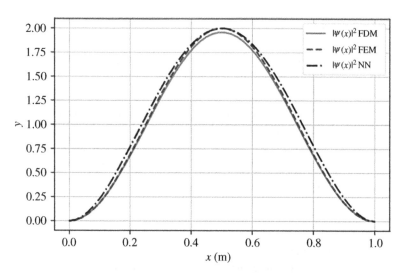

Figure 8.22 Wave functions approximated with 50 points using FDM, FEM, and NN approaches.

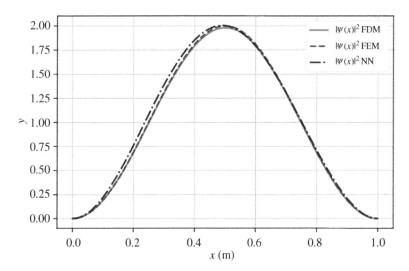

Figure 8.23 Wave functions approximated with 100 points using FDM, FEM, and NN approaches.

References

1 J. M. Jin and S. Jan, "Multiphysics modeling in electromagnetics," *IEEE Antennas and Propagation Magazine*, vol. 61, no. 2, pp. 14–26, 2019.

2 M. Celuch, M. Soltysiakand U. Erle, "Computer simulations of microwave heating with coupled electromagnetic," *Thermal and Kinetic Phenomena, ACCES Journal*, vol. 26, no. 4, pp. 275–283, 2011.

3 D. Poljak, S. Sesnic, A. Rubic and E. Maze, "A note on the use of analytical and domain discretisation methods for the analysis of some phenomena in engineering physics," *International Journal for Engineering Modelling*, vol. 31, no. 1–2, pp. 43–60, 2018.

4 D. Poljak and V. Doric, "A note on FEM modeling of some magnetohydrodynamics phenomena for application in fusion related research,"in 2014 International Conference on Software, Telecommunications and Computer Networks. Split: FESB, University of Split, 2014.

5 D. Poljak and K. El Khamlichi Drissi, Computational Methods in Electromagnetic Compatibility: Antenna Theory Approach versus Transmission line Models, Hoboken, NJ: John Wiley & Sons, 2018.

6 D. Poljak, A. Šušnjara, S. Šesnić, M. Cvetković and H. Dodig, "On some multiphysics models for Electromagnetic-Thermal-Hydrodynamics (ETHD)," in *2019 23rd International Conference on Applied Electromagnetics and Communications (ICECOM)*, Dubrovnik, Croatia, 2019.

7 D. Poljak and M. Cvetkovic, Human Interaction with Electromagnetic Fields: Computational Models in Dosimetry, St Louis, USA: Elsevier, 2019.

8 I. Laakso, S. Tanaka, S. Koyama, V. De Santis and A. Hirata, "Inter-subject variability in electric fields of motor cortical tDCS," *Brain Stimulation*, vol. 8, no. 5, pp. 906–913, 2015.

9 M. Cvetkovic, D. Poljak and A. Hirata, "The electromagnetic-thermal dosimetry for the homogeneous human brain model," *Engineering Analysis with Boundary Elements*, 63, pp. 61–73, 2016.

10 IEEE, "IEEE Standard for safety levels with respect to human exposure to radio frequency electromagnetic fields, 3 kHz to 300 GHz," IEEE Std C95.1-2005 (Revision of IEEE Std C95.1-1991), pp. 1–238, 2006. http://dx.doi.org/10.1109/IEEESTD.2006.99501.

11 F. Felici, O. Sauter, S. Coda, B. P. Duval, T. P. Goodman and J. M. Moret, J. I. Paley and TCV Team, "Real-time physics-model-based simulation of the current density profile in tokamak plasmas," *Nuclear Fusion*, vol. 51, no. 8, p. 083052, 2011.

12 R. R. Khayrutdinov and V. E. Lukash, "Studies of plasma equilibrium and transport in a tokamak fusion device with the inverse-variable technique," *Journal of Computational Physics*, vol. 109, no. (2), pp.193–211, 1993.

13 S. H. Aydin and M. Tezer-Sezgin, "Numerical solution of Grad-Shafranov equation for the distribution of magnetic flux in nuclear fusion devices," *Turkish Journal of Engineering and Environmental Science*, vol. 32, no. (5) pp. 265–275, 2008.

14 R. Zanino, "Advanced finite element modeling of the tokamak plasma edge," *Journal of Computational Physics*, vol. 138, no. 2, pp. 881–906, 1997.

15 E. Fable, C. Angioni, A. A. Ivanov, K. Lackner, O. Maj, S. Yu, G. Pautasso, and G. V. Pereverzev, "A stable scheme for computation of coupled transport and equilibrium equations in tokamaks," *Nuclear Fusion*, vol. 53, no. 3, p. 033002, 2013.

16 S. Sesnić, D. Poljak and E. Sliškovic, "A review of some analytical solutions to the Grad-Shafranov equation," in 2014 International Conference on Software, Telecommunications and Computer Networks. Split: FESB, University of Split, 2014.

17 P. McCarthy, "Analytical solutions to the Grad–Shafranov equation for tokamak equilibrium with dissimilar source functions," *Physics of Plasmas*, vol. 6, no. 9, pp. 3554–3560, 1999.

18 S. Zheng, A. Wootton and E. Solano, "Analytical tokamak equilibrium for shaped plasmas," *Physics of Plasmas*, vol. 3, no. 3, pp. 1176–1178, 1996.

19 C. Atanasiu, S. Gunter, K. Lackner and I. Miron, "Analytical solutions to the Grad–Shafranov equation," *Physics of Plasmas*, vol. 11, no. 7, pp. 3510–3518, 2004.

20 G. Falchetto, D. Coster, R. Coelho, B. D. Scott, L. Figini, D. Kalupin, E. Nardon, S. Nowak, L. L. Alves, J. F. Artaud and V. Basiuk, "The European Integrated Tokamak Modelling (ITM) effort: achievements and first physics results," *Nuclear Fusion*, vol. 54, 4, p. 043018, 2014.

21 V. S. Mukhovatov and V. D. Shafranov, "Plasma equilibrium in a tokamak," *Nuclear Fusion*, vol. 11, pp. 605–633, 1971.

22 F. Hinton and R. Hazeltine, "Theory of plasma transport in toroidal confinement systems," *Reviews of Modern Physics*, vol. 48, no. 2, pp. 239–308, 1976.

23 S. Sesnic, V. Doric, D. Poljak, A. Susnjara, A. Jean-Francois, J. Urban, "A finite element versus analytical approach to the solution of the current diffusion equation in tokamaks," *IEEE Transactions on Plasma Science*, vol. 46, no. 4, pp. 1027–1034, 2018.

24 D. P. Coster, V. Basiuk, G. Pereverzev, D. Kalupin, R. Zagórski, R. Stankiewicz, P. Huynh, F. Imbeaux, and Members of the Task Force on Integrated Tokamak Modelling, "The European transport solver," *IEEE Transactions on Plasma Science*, vol. 38, No. 9, pp. 2085–2092, 2010.

25 J. F. Artaud, V. Basiuk, F. Imbeaux, M. Schneider, J. Garcia, G. Giruzzi, P. Huynh, T. Aniel, F. Albajar, J. M. Ané and A. Bécoulet, "The CRONOS suite for integrated tokamak modelling," *Nuclear Fusion*, vol. 50, no. 4, pp. 1–25, 2010.

26 R. J. Hawryluk (Ed.), "An empirical approach to TOKAMAK transport," in Physics of Plasmas Close to Thermonuclear Conditions, Elsevier Ltd., pp. 19–46, 1981.

27 R. Budny, M. G. Bell, H. Biglari, M. Bitter, C. E. Bush, C. Z. Cheng, E. D. Fredrickson, B. Grek, K. W. Hill, H. Hsuan and A. C. Janos, "Simulations of deuterium-tritium experiments in TFTR," *Nuclear Fusion*, vol. 32, no. 3, pp. 429–448, 1992.

28 H. Grad and J. Hogan, "Classical diffusion in a tokamak," *Physical Review Letters*, vol. 24, no. 24, pp. 1337–1340, 1970.

29 A. D. Polyanin, Handbook of Linear Partial Differential Equations for Engineers and Scientists, Boca Raton, London, New York, Washington, DC: Chapman & Hall/ CRC, 2002.

30 D. Poljak, Advanced Modeling in Computational Electromagnetic Compatibility, New York: Wiley Interscience, 2007.

31 P. Holoborodko, "Smooth Noise Robust Differentiators," 2008. [Online]. Available: http://www.holoborodko.com/pavel/numerical-methods/numerical-derivative/ smooth-low-noise-differentiators (accessed 30 November 2015).

32 V. Parail, R. Albanese, R. Ambrosino, J. F. Artaud, K. Besseghir, M. Cavinato, G. Corrigan, J. Garcia, L. Garzotti, Y. Gribov and F. Imbeaux, "Self-consistent simulation of plasma scenarios for ITER using a combination of 1.5D transport codes and free-boundary equilibrium codes," *Nuclear Fusion*, vol. 53, no. 11, p. 113002, 2013.

33 G. Li and N. R. Aluru, "Hybrid Techniques for Electrostatic Analysis of Nanowires," ICCAD '04: Proceedings of the 2004 IEEE/ACM International Conference on Computer-Aided Design, 2004.

34 A. Udal, R. Reeder, E. Velmre and P. Harrison, "Comparison of methods for solving the Schrödinger equation for multiquantum well heterostructure applications," *Proceedings of the Estonian Academy of Sciences*, vol. 12, no. 3–2, pp. 246–261, 2006.

35 F. Gelbard and K. J. Malloy, "Modeling quantum structures with the boundary element method," *Journal of Computational Physics*, vol. 172, no. 1, pp. 19–39, 2001.

36 W. Moy, M. A. Carignano and S. Kais, "Finite element method for finite-size scaling in quantum mechanics," *Journal of Physical Chemistry A*, vol. 112, no. 24, pp. 5448–5452, 2008.

37 J. Baggot, The Meaning of Quantum Theory, New York, USA: Oxford University Press, 1992.

38 D. Poljak, A. Rubic, E. Maze, A Note on the Use of Domain Discretization Methods in Modeling of Some Phenomena in Engineering Physics, SpliTECH 2017.

39 A. Lojic Kapetanovic and D. Poljak, "Numerical solution of the Schrödinger equation using a neural network approach," 2020 International Conference on Software, Telecommunications and Computer Networks (SoftCOM), 2020.

40 Y. Shirvany, M. Hayati and R. Moradian, "Numerical solution of the nonlinear Schrödinger equation by feedforward neural networks,"*Communications in Nonlinear Science and Numerical Simulation*, vol. 13, no. 10, pp. 2132–2145, 2008.

Part III

Stochastic Modeling

9

Methods for Stochastic Analysis

The process of modeling the electromagnetic system can be viewed as a mathematical idealization of the physical processes governing its evolution. Building a mathematical model requires the definition of the system geometry, material properties, and the relationships between various quantities of interest. Rapid development of computer science enabled the numerical evaluation of approximated solutions of the mathematically described electromagnetic systems. However, it is important to question how confident we are about the models. Two independent procedures are used together for checking that a product, service, or system meets requirements and specifications and that it fulfils its intended purpose: the processes of verification and validation (V&V) [1]. The definitions for the V&V process taken from the "*Guide for the Verification and Validation (V&V) of Computational Fluid Dynamics Simulations*" by The American Institute for Aeronautics and Astronautics (AIAA) in 1998 are as follows [2]:

Verification: the process of determining that a model implementation accurately represents the developer's conceptual description of the model.
Validation: the process of determining the degree to which a model is an accurate representation of the real world for the intended uses of the model.

Obviously, the verification answers the question: "*Are we solving the equations correctly?*" stemming from the mathematics point of view [1]. The deficiencies between the mathematical idealization or algorithms and the actual physical process they represent can be recognized and measured as errors. The sources of errors are, for example, round-off, limited convergence of iterative algorithms, implementation mistakes, etc.

Contrarily, the validation is a process that answers the question: "*Are we solving the correct equations?*" related to the physics point of view [1]. In this case, the deficiency originates from our lack of knowledge about the input physical

Deterministic and Stochastic Modeling in Computational Electromagnetics: Integral and Differential Equation Approaches, First Edition. Dragan Poljak and Anna Šušnjara.
© 2024 The Institute of Electrical and Electronics Engineers, Inc.
Published 2024 by John Wiley & Sons, Inc.

parameters required for performing the analysis. As such, the uncertainties can be divided into reducible and irreducible uncertainties. The uncertainty that can be reduced by increasing our knowledge, for example, by performing more experimental investigations and/or developing new physical models is called *epistemic* or *systematic uncertainty*. On the other hand, the *aleatory* or *statistical uncertainty* cannot be reduced as it rises naturally from the observations of the system. Some additional experiments can be used only to better characterize the variability.

The usual practice in electromagnetic engineering is to use average values of input parameters, thus leading to a rough representation of reality. This practice is sometimes referred to as "deterministic modeling." However, uncertainty present in input parameters can be not only quantified by using the statistical/stochastic tools but also propagated to the output parameter of interest. In this chapter, the uncertainty quantification (UQ) framework is outlined with four main steps, i.e. step 1 is UQ of the model input, step 2 is uncertainty propagation (UP) from input to the output parameter setup, step 3 is UQ of output parameters, and step 4 is the sensitivity analysis (SA) of input parameters according to their impact on output variation. The chapter further outlines the four steps in more detail and emphasizes the stochastic collocation (SC) method which is extensively used by the authors in their applications. In the last part, two different approaches for SA are described.

9.1 Uncertainty Quantification Framework

The four steps of the UQ framework are depicted in Fig. 9.1. Starting with the choice of a mathematical/computational model that allows the calculation of a certain quantity of interest (mathematical description in the form of equations), what follows is the identification of input parameters that exhibit random nature. In step 2, the uncertainty is propagated to the output of interest via a suitable UP method. The next step, step 3, is the quantification of the uncertainty present in the output parameter. Finally, step 4 is the SA of the input parameters regarding the impact of their variability on the output variance. The following text describes these steps in more detail. Also, a traditional and robust method for UP, Monte Carlo (MC), is briefly outlined.

9.1.1 Uncertainty Quantification (UQ) of Model Input Parameters

The parameters with uncertainties are identified and represented in terms of random variables (RVs). Based on the available information, the UQ of input parameters can be carried out by using direct methods such as experimental observations, theoretical arguments, and expert opinions, or by using inverse methods such as inference, calibration, etc. [1]. The UQ of input parameters is a demanding task and the detailed discussion related to its theory and approaches

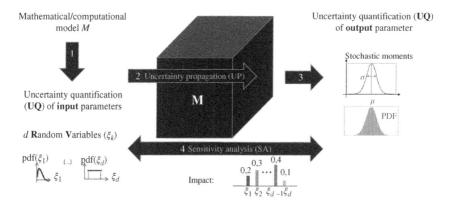

Figure 9.1 The uncertainty quantification (UQ) framework. M is a mathematical/computational model that enables computation of the output parameter. ξ_k is the k-th out of d input parameters with standard probability density function, pdf(ξ_k). σ and μ denote standard deviation and stochastic mean of the output of interest.

is out of the scope of this book. A review of available tools can be found in, for example, [3, 4].

For most of the scenarios encountered in computational electromagnetics (CEM) input parameters are modeled as RVs attributed with the corresponding standard probability density function (pdf). Furthermore, in many CEM applications, the variables are assumed to be independent. In this work, an input parameter modeled as RV is denoted by X. If there are more than one input parameters modeled as RVs, a vector of d input parameters is formed as $X = [X_1, ..., X_d]$. Since, in practice, RVs which represent the input parameters are usually not standardized, the input vector X is often transformed into a set of reduced random variables through the isoprobabilistic transforms [5]. Depending on the marginal distribution of each input parameter X_k, $k = 1, ..., d$, the associated reduced variable may be standard normal: $\xi \sim N(0,1)$, standard uniform: $\xi \sim U(-1,1)$, or some other variables with standard distribution. In this case, the vector of input parameters is denoted by $\boldsymbol{\xi} = [\xi_1, ..., \xi_d]$. Fig. 9.2 shows three types of RVs with typical standard distributions: uniform, normal, and beta, respectively.

9.1.2 Uncertainty Propagation (UP)

UP refers to the propagation of uncertainties from the input parameter set to the output parameter of interest. Given the model M, we seek to represent the output Y as a function of input RVs. Different methods exist and they can be classified in several ways. The general classification is into the statistical and non-statistical methods.

The statistical methods are straightforward to implement as they are all sampling-based methods relying on the statistical analysis of a large set of output

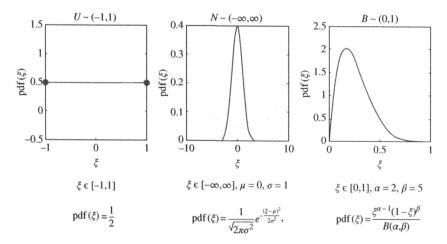

Figure 9.2 Probability density functions (pdf) of random variables (RVs) with uniform, standard normal and beta distribution, from left to right, respectively.

realizations. Therefore, their accuracy depends on the sample size which usually turns out to be computationally very expensive. However, the sample size does not depend on the stochastic dimensionality of a problem, i.e. the number of random input parameters showing uncertain nature, which means that these methods do not suffer from the *"curse of the dimensionality."* Some representative methods in this group are MC sampling, Latin Hypercube Sampling (LHS), etc.

The aim of non-statistical methods is to avoid the problem of large sample sizes present in traditional statistical techniques. Such methods exploit different approaches such as functional approximation theory or novel algorithms for optimization for "smarter" non-statistical sampling, or even the combination of two approaches.

The first group of non-statistical methods aims to represent the unknown stochastic solution as a polynomial in the stochastic space of input parameters. Among various techniques available in the literature, two emerged as the most often used approaches in the stochastic CEM. The first one is a spectral discretization-based technique, known as the generalized polynomial chaos expansion (gPCE). The gPCE framework implies intrusive approach to dealing with the uncertainties in the input and output parameters which leads to a mandatory change of governing equations [6, 7]. Once the equations are adapted to a stochastic point of view, it takes only one run to obtain the desired stochastic output. However, the procedure can be very challenging when the governing equations take complicated forms.

The second non-statistical approach based on the polynomial representation of the output value relies on the high-order SC techniques which are based on deterministic sampling [6, 8]. Its non-intrusive nature enables the use of previously validated deterministic models as black boxes in stochastic computations; therefore, the use of SC is not affected by the complexity or nonlinearity of the original problem [6]. The SC method combines the sampling nature of MC method and the polynomial approximation of the output from the gPCE. Although the total number of samples required for stochastic analysis is lower than in case of MC, the SC method suffers from the *"curse of the dimensionality"* for a large number of input RVs. This problem is alleviated to certain extent by use of high-order methods such as sparse grids and cubature rules. Both gPCE and SC approaches exhibit fast convergence and high accuracy under different conditions which has been reported in many CEM applications. However, it can be stated that gPCE offers the best accuracy in multidimensional random spaces and it should always be used when the coupling of gPCE does not involve additional computational cost or when efficient solvers for decoupling the Galerkin system of equations exist [6]. The comparison of the two approaches is documented in [9, 10]. Both are well-established UP methods, and their respective variants are still being developed.

Finally, in order to outperform the well-established methods such as gPCE and SC, in recent years, a different type of non-statistical methods emerged: the methods based on the optimization and pattern classification algorithms [11, 12] and machine learning [13].

9.1.3 Monte Carlo Method

One of the oldest and the most used methods for the stochastic simulation of an arbitrary system affected by random variations is MC method or one of its variants. MC is a statistical sampling method popularized by physicists from Los Alamos National Laboratory in the United States in the 1940s. The MC algorithm is robust and relatively simple. Due to its intuitiveness and versatility, it is often used for design purposes and as a reference tool for other methodologies. The details about the MC-based methods are out of the scope of this work, and a complete definition with thorough discussion can be found elsewhere, for example [14, 15]. Here, the MC is considered a computational algorithm relying on repeated random sampling to obtain statistical information.

The general MC procedure can be outlined as follows:

1) Determine input parameters exhibiting random nature and their distributions.
2) According to the distributions, generate a set of random samples $\{\mathbf{X}^{(k)}\}$ with $k = 1, ..., N_{MC}$. For $d > 1$, each sample is d-dimensional, i.e. $\mathbf{X}^{(k)} = [X^{(k)}_1 ... X^{(k)}_d]$. It is worth noting that in practical applications, pseudorandom sequences are used [14].
3) The deterministic solver is run repeatedly for each input sample. The output of a model $Y^{(k)}$ is the model realization for the k-th input sample: $Y^{(k)} = M(\mathbf{X}^{(k)})$. For N_{MC} input samples, the MC simulation results in a set of output samples $\{Y^{(k)}\}$, $k = 1 ... N_{MC}$.

4) The results, i.e. N_{MC} outputs of deterministic solver, are gathered and analyzed in order to obtain the statistical information.

The MC estimators of stochastic moments and pdf are given as follows. The mean value of the output variable can be estimated by using the following expression:

$$\mu(Y) \approx \hat{\mu}(Y) = \frac{1}{N_{MC}} \sum_{k=1}^{N_{MC}} Y^{(k)} \tag{9.1}$$

Different MC simulations for the same system and with the same number of samples N_{MC} will produce different values of estimated mean value; therefore, $\hat{\mu}$ itself is an RV. The error estimate for the MC method follows directly from the central limit theorem. The set of $\{Y^{(k)}\}$ represents the set of vectors with independent and identically distributed RVs. Therefore, the distribution function $\hat{\mu}$ converges in the limit of $N_{MC} \to \infty$, to a Gaussian distribution and the widely adopted concept is that the error convergence rate of MC is inversely proportional to the square root of the number of realizations.

The MC estimator of the variance is given as:

$$Var(Y) \approx \frac{1}{N_{MC}-1} \sum_{k=1}^{N_{MC}} \left(Y^{(k)} - \hat{\mu} \right)^2 \tag{9.2}$$

Besides some specific situations, there is no explicit expression for the variance of the estimator in Eq. (9.2). Nevertheless, some general qualitative conclusion can be drawn that the convergence rate of the variance estimator is similar to the convergence rate of the mean estimator, but its fluctuation will be, in general, larger.

The pdf of output $Y, f_y(Y)$, is constructed at a discrete set of points $\{Y^{(j)}\}, j = 1 \dots B$, from the samples $\{Y^{(k)}\}, k = 1 \dots N_{MC}$. The points $Y^{(j)}$ are considered to be equally spaced by $\Delta y = (Y^{(B)} - Y^{(1)})/(B-1)$. The value of the pdf at the point $Y^{(j)}$ can be estimated as:

$$\hat{f}_y \left(Y^{(j)} \right) = \frac{n_j}{N_{MC} \Delta y} \tag{9.3}$$

where n_j is the number of samples $Y^{(j)}$ from the interval $[Y^{(j)} - \Delta y/2, Y^{(j)} + \Delta y/2]$. The pdf of variable Y is approximated by a staircase function, or histogram, where a bin is associated to each point $Y^{(j)}$. The accuracy of the pdf estimator can be improved with large values of B and N_{MC}.

The accuracy of the stochastic moments and the pdf depends directly on the total number of samples, N_{MC} and not on the number of random input variables. Thus, the method does not suffer from the *"curse of the dimensionality"*. However, desired accuracy can be accomplished only for a very large N_{MC} which leads to a certain computational burden.

9.2 Stochastic Collocation Method

The fundamental principle of SC lies in the polynomial approximation of the considered output Y for d random input parameters [6]. The expansion coefficients for the SC are the deterministic outputs of the considered model, calculated at N_{sc} predetermined input points, also called the collocation points. However, not all classical sampling methods are automatically labelled as SC. Instead, the term *"stochastic collocation"* is reserved for the type of collocation methods that result in a strong convergence, e.g. mean-square convergence to the true solution [16]. This is typically achieved by utilizing the classical multivariate approximation theory to strategically locate the collocation nodes in order to construct a polynomial approximation to the solution.

Hence, the considered output Y is approximated in the following way [16]:

$$Y \approx \hat{Y}(\xi) = \sum_{i=1}^{N} L_i(\xi) \cdot Y^{(i)} \tag{9.4}$$

where $L_i(\xi)$ is the multivariate basis function and $Y^{(i)}$ is the output realization for the i-th multidimensional input point $\xi^{(i)}$. Note that the transformation of the vector of random input variables to a vector of standardized random input variables is assumed ($X^{(i)} = [X_1^{(i)}, ..., X_d^{(i)}]$ into $\xi^{(i)} = [\xi_1^{(i)}, ..., \xi_d^{(i)}]$). Equation (9.4) is often referred to as *surrogate model*. Namely, to compute the output quantity of interest as a function of input parameters defined as RVs with prescribed pdfs, and thus the specified domains, it is computationally less demanding to run equation (9.4), i.e. compute the surrogate $\hat{Y}(\xi)$ at defined input parameters' values, than the full model Y. Moreover, this is also convenient when it is of interest to construct the *pdf* of output Y. Again, it is computationally less expensive to use MC sampling of Eq. (9.4) than the original model.

When constructing the *surrogate model* in Eq. (9.4), there are two issues to be considered. The first refers to the choice of the basis functions $L_i(\xi)$, while the second refers to the choice of collocation points $Y^{(i)} = M[X^{(i)}]$. Again, different approaches exist to answer these questions. The basis functions can be, for example, Lagrange type [8], wavelet basis [17], or piecewise constant [18], while the choice of collocation points follows different quadrature [8] and cubature rules [19] as it will be presented in the further text.

9.2.1 Computation of Stochastic Moments

To address the questions pertaining to the choice of basis functions and collocation points, it is useful to first recall the definitions of stochastic moments. According to the statistics theory, the first two moments are defined as follows:

$$\mu(Y(\xi)) = \int_{\Gamma} Y(\xi) p(\xi) d\xi \tag{9.5a}$$

$$Var(Y) = E\left[(Y - E[Y])^2\right] = E\left[Y^2 - 2YE[Y] + E[Y]^2\right]$$
$$= E\left[Y^2\right] - 2E[Y]E[Y] + E[Y]^2 = E\left[Y^2\right] - E[Y]^2 \tag{9.5b}$$

$$Var(Y(\boldsymbol{\xi})) = \int_\Gamma (Y(\boldsymbol{\xi}) - \mu(Y(\boldsymbol{\xi})))^2 p(\boldsymbol{\xi})d\boldsymbol{\xi} \tag{9.5c}$$

where $p(\boldsymbol{\xi})$ is the joint pdf defined as follows:

$$p(\boldsymbol{\xi}) = \prod_{k=1}^{d} p(\xi_k) \tag{9.6}$$

Inserting (9.4) into (9.5), it follows that

$$\mu(Y(\boldsymbol{\xi})) \approx \int_\Gamma \sum_{i=1}^{N} L_i(\boldsymbol{\xi}) \cdot Y^{(i)} p(\boldsymbol{\xi})d\boldsymbol{\xi} = \sum_{i=1}^{N} Y^{(i)} \int_\Gamma L_i(\boldsymbol{\xi}) \cdot p(\boldsymbol{\xi})d\boldsymbol{\xi} \tag{9.7}$$

and

$$Var(Y(\boldsymbol{\xi})) = \int_\Gamma (Y(\boldsymbol{\xi}) - \mu(Y(\boldsymbol{\xi})))^2 p(\boldsymbol{\xi})d\boldsymbol{\xi}$$

$$Var(Y(\boldsymbol{\xi})) = \sum_{i=1}^{N} \left(Y^{(i)}\right)^2 \int_\Gamma L_i(\boldsymbol{\xi}) \cdot p(\boldsymbol{\xi})d\boldsymbol{\xi} - \mu\left(\hat{Y}(\boldsymbol{\xi})\right)^2 \tag{9.8}$$

Given that basis functions $L_i(\boldsymbol{\xi})$ and the joint pdf $p(\boldsymbol{\xi})$ are known, the integral over the space Γ can be precomputed and its value is called the weight of i-th collocation point, w_i:

$$w_i = \int_\Gamma L_i(\boldsymbol{\xi})p(\boldsymbol{\xi})d\boldsymbol{\xi} \tag{9.9}$$

Consequently, stochastic moments can be expressed in a very simple form. The expressions for the first four stochastic moments are given in Table 9.1. The higher orders of stochastic moments are computed accordingly [16].

In many applications it is more convenient to display the standard deviation instead of variance. The standard deviation is computed as the square root of variance. Additionally, confidence intervals (*CI*) are convenient metric to estimate the dispersion of the output of interest around its expected value and are computed as mean $\pm K*$ standard deviation, where K is usually 1, 2 or 3.

9.2.2 Interpolation Approaches

The most popular approach to choosing the basis function is by following the well-developed and extensive classical theory of univariate Lagrange polynomial

Table 9.1 The expressions for the first four stochastic moments.

Stochastic moment	Expression:
1 Mean	$\mu(Y(\xi)) \approx \sum\limits_{i=1}^{N} Y^{(i)} w_i$
2 Variance	$\sigma_Y^2 = Var(Y) \approx \sum\limits_{i=1}^{N} \left(Y^{(i)}\right)^2 w_i - \mu^2$
3 Skewness	$skew(Y) \approx \dfrac{\sum\limits_{i=1}^{N} \left(Y^{(i)}\right)^3 w_i - 3\mu\sigma_Y^2 - \mu^3}{\sigma_Y^3}$
4 Kurtosis	$kurt(Y) \approx \dfrac{\sum\limits_{i=1}^{N} \left(Y^{(i)}\right)^4 w_i - 4\mu \cdot skew \cdot \sigma_Y^3 - 6\mu^2 \cdot \sigma_Y^2 - \mu^4}{\sigma_Y^4}$

interpolation [16]. Hence, for the univariate case, i.e. $d = 1$ and $\boldsymbol{\xi}^{(i)} = [\xi_1^{(i)}]$ and the total of m_k collocation points in the k-th dimension, the Lagrange basis function is given as:

$$l_i(\xi_k) = \prod_{j=0,\, j\neq i}^{m_k} \frac{\xi_k - \xi_k^{(j)}}{\xi_k^{(i)} - \xi_k^{(j)}} \quad i = 1, ..., m_k \tag{9.10}$$

with the property

$$l_i(\xi_j) = \delta_{ij} \tag{9.11}$$

where δ_{ij} denotes Kronecker symbol.

One alternative approach is to use piecewise linear functions defined as follows:

$$l(\xi_k)_j^{(u)} = \begin{cases} 1 - \left(m_k^{(u)} - 1\right) \cdot \left|\xi_k - \xi_k^{(u)}{}_j\right|, & \text{if } \left|\xi_k - \xi_k^{(u)}{}_j\right| < \dfrac{1}{m_k^{(u)} - 1} \\ 0 & \text{otherwise} \end{cases} \tag{9.12}$$

where the total number of m_k collocation points in the k-th dimension is defined as:

$$m_k^{(u)} = \begin{cases} 1, & \text{for } u = 1 \\ 2^{u-1} + 1, & \text{for } u > 1 \end{cases} \tag{9.13}$$

where u is the level of supporting point.

Lagrange polynomials have the character of locally global basis functions, while piecewise linear basis functions are used when it is important to capture the discontinuous issues in stochastic solutions. Since the manipulation of Lagrange or other types of basis functions may be a cumbersome procedure, a pseudospectral collocation approach as a part of gPC theory is proposed in [20].

9.2.3 Collocation Points Selection

The choice of the collocation points is essential part of any collocation-based method. The aim of SC method is to approximate the integral in Eq. (9.9) as accurately as possible:

$$w_i = \int_\Gamma L_i(\xi)p(\xi)d\xi$$

The weights w_i are computed numerically. If the stochastic dimension is equal to 1 ($d = 1$), the points selection is straightforward. There are numerous numerical studies proposing a wide range of quadrature rules to deal with the one-dimensional integral evaluation and the optimal choice is Gauss quadrature [6]. Depending on the pdf of the input RV, one can choose between different Gauss quadrature rules, e.g. Gauss–Hermite, Gauss–Legendre, or Gauss–Jacobi for variables with normal, uniform, or beta distributions, respectively [6].

As an example, for parameter $X = 5$, uniformly distributed in the interval $U = [2, 8]$, the selection of collocation points follows Gauss–Legendre quadrature rule which belongs to interval $[-1, 1]$. To estimate the stochastic moments given by expressions in Table 9.1, the mathematical model will be run N times, where N corresponds to the number of collocation points. The accuracy is increased with the selected number of collocation points needed for the computation of the integral (9.9). Gauss–Legendre quadrature points are given in Table 9.2 for odd values of N along with the corresponding collocation points of parameter X and their respective weights. Basis functions used in this example are of Lagrange type.

Table 9.2 Collocation points for random variable uniformly distributed in range [2, 8].

N	$\xi = [-1, 1]$	$X = [2, 8]$	$w_i = \int_\Gamma L_i(\xi)p(\xi)d\xi$
3	−0.7746	2.6762	0.2778
	0	5.0000	0.4444
	0.7746	7.3238	0.2778
5	−0.9062	2.2815	0.1185
	−0.5385	3.3846	0.2393
	0	5.0000	0.2844
	0.5385	6.6154	0.2393
	0.9062	7.7185	0.1185

It is worth noting that the computation of weights does not depend on the boundary values of parameter X. Therefore, they can be precomputed and stored for further use.

Other integration rules may be used as well. Very often instead of Gauss quadrature, a Clenshaw–Curtis quadrature rule is used, especially when sparse grid multivariate interpolation is employed [16], which will be further discussed in the following subsections. Additionally, it is a practice to use equidistant points when the interpolation is done by using the piecewise linear basis functions.

9.2.4 Multidimensional Stochastic Problems

The real challenge, however, is numerical computation of multidimensional integral for $d > 1$, especially for $d >> 1$. For extremely large dimensions in realistic scenarios, the MC method and its variants are still the only applicable approaches. However, if the dimension is not too large, there are quicker approaches which are also used in the framework of the SC method.

9.2.4.1 Tensor Product

The most natural approach to multidimensional integration is the tensor product of one-dimensional quadrature rules leading to a relatively simple generalization of integration properties from one-dimensional to d-dimensional case [16]. The idea of tensor product has been introduced in [8, 21] while the errors are analyzed in [22].

The multidimensional integral (9.9) is thus given as:

$$w_i = \int_{\Gamma_1} l\left(\xi_1^{(i)}\right) p(\xi_1) d\xi \cdot \int_{\Gamma_2} l\left(\xi_2^{(i)}\right) p(\xi_2) \xi \cdot \ldots \cdot \int_{\Gamma_d} l\left(\xi_d^{(i)}\right) p(\xi_d) d\xi$$

$$w_i = w_{1(j)}^{(i)} \cdot w_{2(j)}^{(i)} \cdot \ldots \cdot w_{d(j)}^{(i)} \quad i = 1, \ldots, N_{SC}$$

(9.14)

where $l(\xi)$ is a one-dimensional basis function. The multivariate basis functions $L_i(\boldsymbol{\xi})$ from Eq. (9.4) are also formed by means of a tensor product of univariate basis functions in each dimension:

$$L_i(\boldsymbol{\xi}) = l\left(\xi_1^{(i)}\right) \otimes l\left(\xi_2^{(i)}\right) \otimes \ldots \otimes l\left(\xi_d^{(i)}\right)$$

(9.15)

The total number of simulation points is thus:

$$N_{SC} = \prod_{k=1}^{d} m_k$$

(9.16)

Table 9.3 Two-dimensional collocation points for two input parameters represented as random variables uniformly distributed in ranges [2, 8] and [2, 4], respectively.

i	$X_i = [X_1, X_2]^{(i)}$	w_i
1	[2.6762, 2.2254]	0.0772
2	[2.6762, 3.0000]	0.1235
3	[2.6762, 3.7746]	0.0772
4	[5.0000, 2.2254]	0.1235
5	[5.0000, 3.0000]	0.1975
6	[5.0000, 3.7746]	0.1235
7	[7.3238, 2.2254]	0.0772
8	[7.3238, 3.0000]	0.1235
9	[7.3238, 3.7746]	0.0772

In most of the applications, the number of collocation points in each dimension is equal, thus $N_{SC} = m_k^d$. Obviously, the number of simulation points grows exponentially with the number of input RVs. Therefore, the tensor product is mostly used at lower dimensions. The generally accepted limitation is $d \leq 5$ [16].

As an example, for a two-dimensional stochastic problem with parameter $X_1 = 5$ and $X_2 = 3$, uniformly distributed in intervals $U_1 = [2, 8]$ and $U_2 = [2, 4]$, respectively, the selection of three collocation points following Gauss–Legendre quadrature in each dimension results in a total of $N = 9$ two-dimensional input points. The points and the corresponding weights are given in Table 9.3.

9.2.4.2 Sparse Grids

The idea behind the sparse grids is to alleviate the problem of a *"curse of dimensionality"* present in the tensor product by using a sparse, instead of a tensorized, grid of points. The approach was first proposed by Smolyak in [23] and it has been widely used and improved ever since in the context of multivariate integration and interpolation [24, 25]. The sparse grids were introduced to SC framework in [16].

The sparse grid represents the subset of the full tensor product grid. The basic idea behind the sparse grid is an optimal linear approximation of the low-level tensor products such that an integration property for *d = 1* is preserved as much as possible for *d > 1* [16]. Consequently, only those products with a relatively small number of points are used, thus reducing the total number of simulation points. Lots of prominent researchers investigated different algorithms for sparse grid construction and perhaps the most famous works are those of Griebel and Zenger [26, 27].

The classical sparse-grid approach applied to the construction of multivariate basis function $L_i(\xi)$ can be expressed in the following way [16]:

$$L_i(\xi) = \sum_{q+1 \leq |\vec{h}| \leq q+d} (-1)^{q+d-|\vec{h}|} \cdot \binom{d-1}{q+d-|\vec{h}|}$$
$$\cdot \left(l\left(\xi_1^{(i)}, h_1\right) \otimes ... \otimes l\left(\xi_d^{(i)}, h_d\right) \right) \tag{9.17}$$

where q is a sparseness parameter or the sparseness level and h denotes the depth coordinate for each dimension: $k = 1, ..., d$. The vector $|h| = h_1 + h_2 + ... h_d$ lists the levels of the rules used by each component. The algorithms for a sparse grid construction may be found in [26].

It is worth noting that there is no mandatory choice of the one-dimensional quadrature rules used in the sparse grid algorithm. However, in order to reuse the collocation points from lower levels h, in the higher levels as well, it is a good practice to use those quadrature rules that result in a set of nested points. As an example, Fig. 9.3 illustrates the sequence of Gauss–Hermite collocation points and the sequence of Clenshaw–Curtis points computed as the extrema of the Chebyshev polynomials. As it is obvious from the figure, at the individual level h, number of Gauss rule results in a smaller number of points with respect to Clenshaw–Curtis rule. Nevertheless, due to the nesting property of Clenshaw–Curtis set of points, the total number of 17 points is less than the total of 21 Gauss points.

To illustrate the sparse grids, a simple example with three input parameters uniformly distributed in the interval $[-1, 1]$ is depicted in Fig. 9.4. The sparseness level for sparse grids is set to $q = 2$ with Clenshaw–Curtis quadrature. The grids of points are also presented for MC simulation tensor grid of points with Gauss–Legendre quadrature. The total number of points for different techniques is $N_{MC} = 1000$, $N_{SC-SG} = 69$, and $N_{SC-TG} = 343$, respectively.

The dependence of the total number of simulation points of the sparse grid products on the dimension is much weaker than in case of tensor product with the reduction from $N_{SC} = m_k^d$ to approximately N_{SC-SG} $(2^* m_k)^d / d!$ simulation points. The sparse grid approximation is accurate for $d > 5$. Some of the first sparse grid SC computations in [16] went to 50-dimensional stochastic space.

9.2.4.3 Stroud's Cubature Rules

The integration rules for the evaluation of multidimensional integrals, also known as cubature rules, are a part of active research. Cools has published a collection of available cubature formulas for the approximation of multivariate integrals over

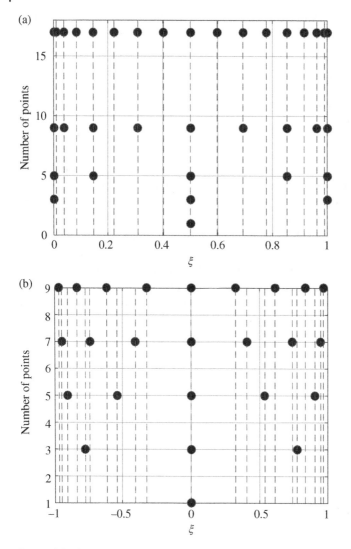

Figure 9.3 The (a) Clenshaw–Curtis versus (b) Gauss–Hermite sequence of points for five levels of interpolation.

some standard regions in [28]. As is the case for every numerical integration rule, the idea is to represent the multidimensional integral as a sum of products of weights ω_i and the function evaluations at a carefully chosen point $\xi^{(i)}$:

$$\int_{[-1,1]^d} f(\xi)d\xi = \sum_{i=1}^{M} \omega_i f\left(\xi^{(i)}\right) \tag{9.18}$$

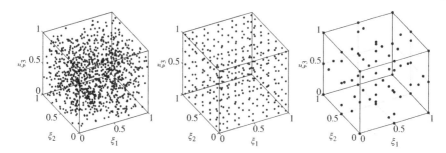

Figure 9.4 Three-dimensional grid of points for Monte Carlo, tensor grid, and sparse grid techniques (left to right).

where $[-1,1]^d$ is the hypercube space.

A large collection of cubature rules can be a good option for SC technique and the most used ones are Stroud's cubatures [16]. Stroud proposed two sets of cubature points, one accurate for multiple integrals of polynomials of degree 2 and the other for polynomials of degree 3, the Stroud-2 and Stroud-3 formulas, respectively [29].

The Stroud-2 formula is given as follows:

$$\xi_{2r-1}^{(i)} = \sqrt{\frac{2}{3}}\cos\left(\frac{2r(i-1)\pi}{d+1}\right) \text{ and } \xi_{2r}^{(i)} = \sqrt{\frac{2}{3}}\sin\left(\frac{2r(i-1)\pi}{d+1}\right), \omega_i = \frac{1}{d+1}$$

$$r = 1, 2, ..., \text{floor}\left(\frac{d}{2}\right) \tag{9.19}$$

provided that d is an even number. If number d is odd, the location of the i-th node for the d-th dimension is given by the following expression:

$$\xi_d^{(i)} = \frac{(-1)^{i-1}}{\sqrt{3}} \tag{9.20}$$

while the formula for weights ω_i remains the same.

The Stroud-3 formula is defined with:

$$\xi_{2r-1}^{(i)} = \sqrt{\frac{2}{3}}\cos\left(\frac{(2r-1)i\pi}{d}\right) \& \xi_{2r}^{(i)} = \sqrt{\frac{2}{3}}\sin\left(\frac{(2r-1)i\pi}{d+1}\right), \omega_i = \frac{1}{2d}$$

$$r = 1, 2, ..., \text{floor}\left(\frac{d}{2}\right) \tag{9.21}$$

again with the additional expression for the i-th node along the last dimension:

$$\xi_d^{(i)} = \frac{(-1)^i}{\sqrt{3}} \tag{9.22}$$

The total number of collocation points given by the Stroud-2 rule is $M = (d + 1)$ while in case of the Stroud-3 rule, $M = 2^*d$. The number M is fixed. It depends only on the dimensionality of a stochastic model, and it presents a minimal number of points necessary for the desired integration accuracy of polynomials with the corresponding degree. It is worth noting that Stroud rules present a good choice when the demand on accuracy is not very strict in case of high-dimensional stochastic spaces $(d \gg 1)$ [16].

9.3 Sensitivity Analysis

One of the definitions for the SA, adopted in this work as well, is the one describing it as the study of how the uncertainty in the output of a mathematical model or system (numerical or otherwise) can be apportioned to different sources of uncertainty in its inputs [30]. There are many SA methods available in literature and the particular choice depends on actual purpose. The extensive study on different SA approaches can be found elsewhere, for example, in [30]. In this chapter the two approaches are described: the so-called "one-at-a-time" approach and the approach based on the variance analysis known as the ANOVA SA.

It is worth noting that the ideal way for carrying out the SA assumes running both UQ and SA in the same stochastic framework, usually UQ preceding the SA, thus minimizing the computational burden as much as possible. Some effort has been made regarding the coupling of the UQ and SA in the same SC framework. Buzzard and Xiu exploited the nature of sparse grid interpolation and cubature methods of Smolyak together with combinatorial analysis to give a computationally efficient method for computing the global sensitivity values of Sobol [31]. Furthermore, Tang and Iaccarino showed the accuracy and efficiency of Sobol-like indices computed by using the SC method in [32]. SC methodology has also been used for inverse problems under the Bayesian approach. The aim of inverse problems is to reduce the uncertainty of the input parameters in order to make predictions, and even to devise control strategies based on the predictions [33]. The Bayesian inference in the SC framework has been reported in [34, 35]. The coupling of SC and ANOVA-based SA has been used extensively in [36].

9.3.1 "One-at-a-Time" (OAT) Approach

The idea behind the "one-at-a-time" approach, as the name itself says, is to change the input parameter one at a time while the others are kept at some nominal value. The sensitivity is then measured by monitoring the changes in the output which can be done in different ways, e.g. partial derivatives or linear regression [30]. Within the stochastic framework presented in this work, the sensitivity is measured by monitoring the change in the variance of the output after computing the variance for d univariate cases.

Thus, the impact factor of each input parameter is given by the following formula:

$$I_i = Var_i(Y)/Var(Y) \tag{9.23}$$

where $Var(Y)$ is the total variance when vector of input parameters is d-dimensional $X = [X_1, X_2, ..., X_d]$ and $Var_i(Y)$ is the variance of i-th one-dimensional stochastic problem, i.e. when vector of input parameters is $X = [X_i]$.

The alternative approach would be to simply compare the variances of the output value for each univariate case without normalization in Eq. (9.23).

Although, in this way, any change observed in the output is unambiguously prescribed to the single variable changed, the approach does not fully explore the input space since it does not take into account the simultaneous variation of input variables. The OAT approach cannot detect the presence of interactions between input variables.

9.3.2 ANalysis Of VAriance (ANOVA)-Based Method

The ANOVA is an approach originating from the work of Sobol [37] and it is based on variance decomposition within the probabilistic framework. The total variance of a model output is decomposed into terms depending on the input factors and their mutual interactions [30]:

$$Var(Y) = \sum_k Var_k + \sum_k \sum_{j>k} Var_{kj} + ... + Var_{12...d} \tag{9.24}$$

where $Var(Y)$ is the output variance when $X = [X_1, ..., X_d]$, while the other terms are defined as follows:

$$Var_k = Var(f_k(X_k)) = Var_{X_k}[E_{X_{\sim k}}(Y|X_k)] \tag{9.25a}$$

$$Var_{ij} = Var\left(f_{ij}(X_i, X_j)\right) \tag{9.25b}$$

$$Var\left(f_{ij}(X_i, X_j)\right) = Var_{X_i X_j}\left[E_{X_{\sim ij}}(Y|X_i, X_j)\right] - Var_{X_i}[E_{X_{\sim i}}(Y|X_i)] \\ - Var_{X_i}\left[E_{X_{\sim j}}(Y|X_{ij})\right] \tag{9.26}$$

Here E denotes the expected value operator and notation tilde ($\sim$) denotes "all but" operator, i.e. $E_{X_\sim}(Y|X_k)$ means the expected value of Y calculated by varying all input variables except X_k. The sign "|" denotes the conditional expectation or variance. The computational or mathematical model for computation of Y is denoted with f.

Normalizing the above expression by total variance $Var(Y)$, the sensitivity indices are obtained as follows:

$$1 = \sum_k S_k + \sum_k \sum_{j>i} S_{kj} + \dots + S_{12\dots d} \tag{9.27}$$

where the first-order sensitivity indices measuring the effect of only the k-th random input parameter, without any interaction with other input parameters exhibiting random nature, is given by the following expression:

$$S_k = \frac{Var_{X_k}[E_{X_{\sim k}}(Y|X_k)]}{Var(Y)}, k = 1, \dots, d \tag{9.28}$$

The second- and high-order sensitivity indices, S_{ij} and $S_{12\dots d}$, give information about the effect that the interaction of two or more random input parameters has with respect to the output variance.

The computational burden may become very prohibitive when all groups of sensitivity indices need to be computed. Therefore, very often only first-order sensitivity index is computed. To still obtain information about the potential significant interactions between the variables, a total effect sensitivity index is defined as:

$$S_{T_k} = \frac{E_{X_{\sim k}}[Var_{X_k}(Y|X_{\sim k})]}{Var(Y)} = 1 - \frac{Var_{X_{\sim k}}[E_{X_k}(Y|X_{\sim k})]}{Var(Y)} \tag{9.29}$$

The total effect index measures the contribution to the output variance of X_k, including all variances caused by its interactions, of any order, with any other input parameter exhibiting random nature.

References

1 G. Iaccarino, Uncertainty Quantification in Simulations of Reactive Flows, Marseille, 2012.

2 American Institute of Aeronautics and Astronautics, Guide for the Verification and Validation of Computational Fluid Dynamics Simulations, Reston, VA, USA: American Institute of Aeronautics and Astronautics, 1988,.

3 A. Papoulis, Probability, Random Variables and Stochastic Processes, New York: McGraw-Hill, Inc, 1991.

4 B. Sudret and A. Der Kiureghian, Stochastic Finite Element Methods and Reliabiltiy, Report No. UCB/SWMM-2000/08, Berkley: Department of Civil & Environmental Engineering, University of California, 2000.

5 B. Sudret, "Polynomial Chaos Expansions and Stochastic Finite Element Methods," Risk and Reliability in Geotechnical Engineering (Chap. 6), K.-K. Phoon & J. Ching (Eds.), pp. 265–300, CRC Press, Zurich, 2014.

6 D. Xiu, "Fat numerical methods for stochastic computations: a review," *Communications in Computational Physics*, vol. 5, no. 2-4, pp. 242–272, 2009.

7 D. Xiu and G. E. Karniadakis, "The Wiener–Askey polynomial chaos for stochastic differential equations," *SIAM Journal on Scientific Computing*, vol. 24, no. 2, pp. 614–644, 2002.

8 L. Mathelin and M. Y. Hussaini, "A Stochastic Collocation Algorithm fo Uncertainty Analysis,"Hanover: NASA Center for AeroSpace Information, 2003.

9 M. S. Eldred and J. Bukardt, "Comparison of Non-Intrusive Polynomial Chaos and Stochastic Collocation Methods for Uncertainty Quantification," in *47th AIAA Aerospace Sciences Meeting including The New Horizons Forum and Aerospace Exposition*, Orlando, Florida, 2009.

10 P. Manfredi, D. De Zutter and D. Vande Ginste, "On the relationship between the stochastic Galerkin method and the pseudo-spectral collocation method for linear differential algebraic equations," *Journal of Engineering Mathematics*, vol. 108, no. 1, pp. 73–90, 2018.

11 M. Grigoriu, "Reduced order models for random functions. Application to stochastic probelms," *Applied Mathematical Modelling*, vol. 33, pp. 161–175, 2009.

12 R. V. Field, Jr., M. Grigoriu and J. M. Emery, "On the efficacy of stochastic collocation, stochastc Galerkin, and stochastic reduced order models for solving stochastic problems," *Probabilistic Engineering Mechanics*, vol. 41, 2015, pp. 60–72.

13 R. Trinchero, M. Larbi, H. M. Torun, F. G. Canavero and M. Swaminathan, "Machine learning and uncertainty quantification for surrogate models of integrated devices with a large number of parameters," *IEEE Access*, vol. 7, pp. 1–12, 2018, doi: 10.1109/ACCESS.2018.2888903.

14 I. M. Sobol, A Primer for the Monte Carlo Method, Boca Raton, FL, USA: CRC Press, Inc., 1994.

15 M. H. Kalos and P. A. Whitlock, The Monte Carlo Methods, Hoboken, NJ: Wiley, 2008.

16 D. Xiu and J. S. Hesthaven, "High-order collocation methods for differential equations with random inputs," *SIAM Journal on Scientific Computing*, vol. 27, no. 3, pp. 1118–1139, 2005.

17 M. Gunzburger, C. G. Webster and G. Zhang, An Adaptive Wavelet Stochastic Collocation Method for Irregular Solutions of Stochastic Partial Differential Equations, Oak Ridge, TN: Oak Ridge National Laborator, 2012.

18 N. Agarwal and N. R. Aluru, "Stochastic analysis of electrostatic MEMS subjected to parameter variations," *Journal of Microelectromechanical Systems*, vol. 18, no. 6, pp. 1454–1468, 2009.

19 C. Chauvire, J. S. Hesthaven and L. C. Wilcox, "Efficient computation of RCS from scatterers of uncertain shapes," *IEEE Transactions on Antennas and Propagation*, vol. 55, no. 5, pp. 1437–1448, 2007.

20 D. Xiu, "Efficient collocation approach for parametric uncertainty analysis," *Communications in Computational Physics*, vol. 2, no. 2, pp. 293–309, 2007.

21 M. A. Tatang, W. Pan, R. G. Prinn and J. McRae, "An efficient method for parametric uncertainty analysis of numerical geophysical models," *Journal of Geophysical Research*, vol. 102, no. D18, pp. 21925–21932, 1997.

22 I. Babuška, F. Nobile and R. Tempone, "A stochastic collocation method for elliptic partial differential equations with random input data," *SIAM Journal on Numerical Analysis*, vol. 45, no. 3, pp. 1005–1034, 2007.

23 S. A. Smolyak, "Quadrature and interpolation formulas for tensor products of certain classes of functions," *Doklady Akademii Nauk SSSR*, vol. 148, no. 5, pp. 1042–10445, 1963.

24 V. Barthelmann, E. Novak and K. Ritter, "High dimensional polynomial interpolation on sparse grids," *Advances in Computational Mathematics*, vol. 12, no. 4, pp. 273–288, 2000.

25 E. Novak and K. Ritter, "Simple cubature formulas with high polynomial exactness," *Constructive Approximation*, vol. 15, no. 4, pp. 499–522, 1999.

26 J. Garcke and M. Griebel, Sparse Grids and Applications, Bonn: Springer Science & Business Media, 2012.

27 C. Zenger, Sparse grids, W. Hackbusch (Eds.), Parallel Algorithms for Partial Differential Equations, *Proceedings of the Sixth GAMM-Seminar*, Kiel, 1991.

28 R. Cools, "An encyclopaedia of cubature formulas," *Journal of Complexity*, vol. 19, no. 3, pp. 445–453, 2003.

29 A. H. Stroud, Approximate Calculation of Multiple Integrals, Englewood Cliffs, NJ: Prentice–Hall, 1971.

30 A. Saltelli, M. Ratto, T. Andres, F. Campologno, F. Cariboni, D. Gatelli, M. Saisana and S. Tarantola, Global Sensitivity Analysis: The Primer, West Susex, England: John Wiley & Sons, Ltd, 2008.

31 G. T. Buzzard and D. Xiu, "Variance-based global sensitivity analysis via sparse-grid interpolation and cubature," *Communications in Computational Physics*, vol. 9, no. 3, pp. 542–567, 2011.

32 G. Tang, G. Iaccarino and M. S. Eldred, "Global Sensitivity Analysis for Stochastic Collocation Expansion," in *Proceedings of the 51st AIAA/ASME/ASCE/AHS/ASC Structures, Structural Dynamics, and Materials Conference*, Orlando, Florida, 2010.

33 M. Iglesias and A. M. Stuart, "Inverse Problems and Uncertainty Quantification," *SIAM NEWS,* July/August 2014.

34 X. Ma and N. Zabaras, "An efficient Bayesian inference approach to inverse problems based on an adaptive sparse grid collocation method," *Inverse Problems*, vol. 25, no. 2009, pp. 1–27, 2009.

35 L. Yan and L. Guo, "Stochastic colloation algorithms using L1-minimization for Bayesian solution of inverse problems," *SIAM Journal on Scientific Computing*, vol. 37, no. 3, p. 26, 2015.

36 D. Poljak, S. Šesnić, M. Cvetković, A. Šušnjara, H. Dodig, S. Sebastien and K. El Khamlichi Drissi, "Stochastic collocation applications in computational electromagnetics," *Mathematical Problems in Engineering*, vol. 2018, pp. 1–13, 2018.

37 I. M. Sobol, "Sensitivity estimates for nonlinear mathematical models," *Matematicheskoe Modelirovanie*, 2, pp. 112–118, 1990.

10

Stochastic–Deterministic Electromagnetic Dosimetry

An important aspect to be considered in electromagnetic (EM) dosimetry is the stochastic nature of the model input parameters related both to the source of radiation and electromagnetic characteristics of human body tissues. Deterministic modeling assumes fixed value of input parameter which corresponds to an average of a set of values resulting from various measurements, observations, and/or computations. Unfortunately, in the case of internal field dosimetry, these data sets are limited because direct measurements of human beings are not ethically acceptable. Namely, tissue-relative electric permittivity and electric conductivity depend on age, health, and gender of each individual person. Furthermore, their values are obtained from in vitro measurements on human and animal tissues which fail to accurately represent realistic scenarios. Consequently, the relevant literature and databases list a set of values for these properties, scattered around some expected mean values [1, 2].

Besides the tissue parameters, the sources of radiation also exhibit uncertain behavior. Even in 2G/3G/4G communication systems, such uncertainties are noticed, e.g. it has been demonstrated in [3, 4] that antenna position and orientation as well as specific probability distribution of incidence angle and magnitude of electromagnetic waves have different impacts on the resulting radiated and thus internal fields. Additionally, new features of 5G and 6G technologies such as beam forming assume that the radiation pattern is not fixed [5]. Accordingly, in the last decade, some efforts have been undertaken in order to incorporate stochastic approaches in the field of electromagnetic dosimetry, thus resulting in a novel area, i.e. stochastic dosimetry [6].

This chapter examines the application of the stochastic collocation method (SCM) in electromagnetic dosimetry. The examples start with internal field dosimetry and cylinder representation of human body. It is important to consider canonical geometries in order to test the SCM convergence on simple geometries.

Deterministic and Stochastic Modeling in Computational Electromagnetics: Integral and Differential Equation Approaches, First Edition. Dragan Poljak and Anna Šušnjara.
© 2024 The Institute of Electrical and Electronics Engineers, Inc.
Published 2024 by John Wiley & Sons, Inc.

Namely, human body is an extremely heterogeneous electromagnetic media with a complex, i.e. highly irregular, and inhomogeneous geometry. This fact *per se* presents a rather demanding task when posing a physical formulation and its related mathematical solution/numerical implementation, and finally in terms of a required computer capacity. By introducing the stochastic dimensions into the model via random input parameters, the problem becomes even more complex due to growing model dimensionality. Therefore, the chapter starts with two examples of stochastic–deterministic modeling with cylindrical body model; in the first one, the body is exposed to a low-frequency plane wave, while the second example features the exposure to electromagnetic pulse. In both cases, the output of interest is the current induced in the body. Then, SCM is coupled with deterministic approach to computing the electric field induced in a realistic human head with three compartments exposed to high-frequency plane waves. Finally, in the third example, the incident field dosimetry is considered featuring the stochastic analysis of the field radiated by a base station antenna (BSA). Therefore, the presented examples cover all aspects of EM dosimetry; LF versus HF dosimetry, incident versus internal field dosimetry, and simple versus realistic body models.

10.1 Internal Stochastic Dosimetry for a Simple Body Model Exposed to Low-Frequency Field

The following example features the stochastic-deterministic model of a human body exposed to low-frequency (LF) electric field [7]. The model is depicted in Fig. 10.1 along with the values of input parameters, both deterministic and stochastic.

Human body is represented as a thick-wire cylindrical antenna illuminated by a low-frequency electric field, $f = 60$ Hz. At such low frequency, human body is considered a conductive medium. Thus, its dielectric properties are characterized by electric conductivity, σ. The cylinder geometry is defined with its length L corresponding to the body height and diameter $2a$ representing the width of the body. The outputs of interest are the induced current and electric field. Namely, the results for induced current along the human body starting from the feet to the top of the head are in consistency with the International Commission on Non-Ionizing Radiation Protection (ICNIRP)'s basic restriction on exposure of humans to extremely low-frequency (ELF) electric fields from 1998 which is induced current density [8]. However, international standards [9, 10], also recommended by the World Health Organization (WHO), report the change in the basic restriction

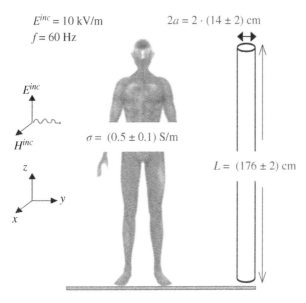

Figure 10.1 Stochastic–deterministic model of a human body (a thick wire representation) exposed to LF electric field. Both deterministic and stochastic input parameters are denoted.

parameter on ELF electric field exposure from induced current into induced electric field inside the body.

The current induced in the body is governed by Pocklington's integro-differential equation as follows [7]:

$$E_z^{inc} = \frac{1}{j4\pi\omega\varepsilon_0} \int_{-L}^{L} \left[\frac{\partial^2}{\partial z^2} + k^2 \right] g_E(z, z') I(z') dz' + I(z) Z_L(z) \tag{10.1}$$

where $I(z)$ is the unknown axial current, k is a free space phase constant computed as $k^2 = 4\pi^2 f^2 \mu_0 \varepsilon_0$, the term $g_E(z, z')$ is Green's function, and $Z_L(z)$ is the impedance of the human body. The equation is solved analytically by R. W. P. King in [11] and numerically via Galerkin–Bubnov scheme of Indirect Boundary Element Method [12].

Once the axial current is determined, it is possible to calculate current density, electric field, and other related parameters. The induced electric field $E^{ind}(z)$ is given as [12]:

$$E^{ind}(z) = \frac{1}{\sigma} \cdot \frac{I(z)}{a^2 \pi} \tag{10.2}$$

A new perspective on the problem is introduced in [7]. The variability inherent to input parameters such as body height (L), shape (a), and conductivity (σ) is propagated to the output-induced axial current and electric field, $I(z)$ and E^{int}, respectively. The three input parameters are modeled as uniformly distributed random variables as depicted in Fig. 10.1. Lagrange SCM with Gauss–Legendre (GL) quadrature is used for uncertainty propagation. The three-dimensional stochastic problem is solved with full-tensor approach.

The convergence of the SC method in computation of output mean and variance is tested by successive increases of SC points in each dimension and by comparison with the Monte Carlo (MC) method. The number of collocation points in each dimension is 3, 5, 7, and 9, thus resulting in a total of 27, 125, 343 and 729 samples, respectively. It is assumed that the exact mean and variance are obtained by MC simulation with $N = 10{,}000$ samples. A sample is the i-th variant of total N three-dimensional vectors of input parameters $X_i = [L, a, \sigma]^{(i)}$. Besides the exact solution, another 50 different MC simulations are carried out, each with a different set of $N = 50$ samples. Since electric field is computed after current as in Eq. (10.2), Fig. 10.2 depicts the convergence results only for the induced field.

It can be observed that the solutions of 50 MC simulations form an envelope, thus indicating that MC does not converge with $N = 50$. The envelope is even wider

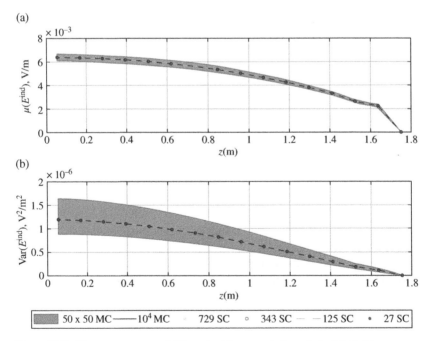

Figure 10.2 The convergence of SC method in computation of electric field mean (a) and variance (b). SC method with 27, 125, 343, and 729 samples is compared with MC method with 10,000 samples and another 50 MC simulations, each with 50 samples.

for the variance than the mean of the induced field. The SCM solution with $N = 27$ samples converges to exact solution both for mean and variance. The same conclusion is valid for the induced current. The convergence is accomplished for 3 points in each dimension both for mean and variance, i.e. only 27 simulations are necessary to obtain the mean and variance of the induced current.

The induced current and electric field expectations with their respective confidence intervals (CIs) are depicted in Fig. 10.3. CIs are computed for 99.7% certainty of coverage, i.e. CI is mean ± 3 standard deviations. The extreme values are included inside the confidence margins and they are not sufficient to predict the dispersion of results around the mean electric field since lower extreme values are much higher than the lower interval margin. The mean electric field computed stochastically is higher than the deterministic one.

Finally, sensitivity analysis is carried out in one-at-a-time (OAT) approach. CIs of the induced current are depicted for three one-dimensional cases in Fig. 10.4. The CIs are practically the same for three-dimensional case (Fig. 10.3a) and

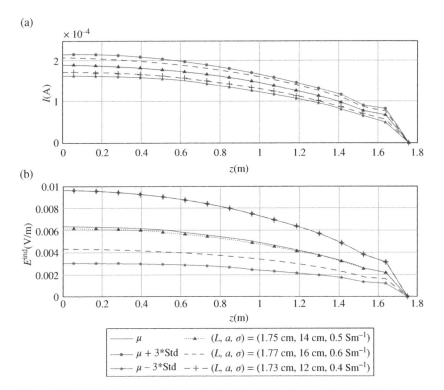

Figure 10.3 Induced current (a) and electric field (b) in a human body with body length, radius, and electric conductivity as RVs. μ and *Std* stand for mean and standard deviation. Different variants of L, a, and σ stand for deterministic values of length, radius, and conductivity in three cases: average, maximum, and minimum, respectively.

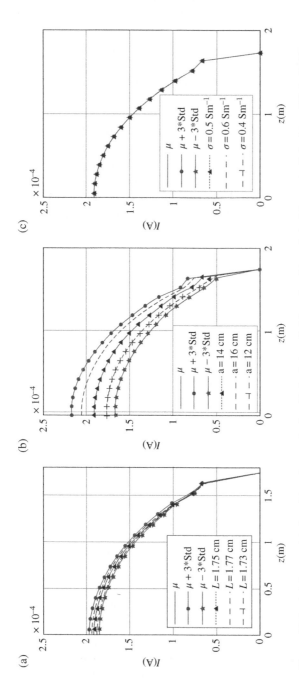

Figure 10.4 Axial current induced in a human body with body length (a), radius (b), and electric conductivity (c) as RVs in OAT manner. μ and Std denote mean and standard deviation. L, a, and σ in panel legends are extreme values of the corresponding input parameter modeled as RV.

one-dimensional case when only antenna radius is RV (Fig. 10.4b). Therefore, antenna radius is the input parameter with the highest impact on induced current. The resulting current variance is practically independent of conductivity which is in accordance with the analytic prediction that the current does not vary with the change in conductivity [11].

10.2 Internal Stochastic Dosimetry for a Simple Body Model Exposed to Electromagnetic Pulse

In the past two decades, some studies have provided evidence of greater effect on biological systems when exposed to pulsed signals such as radar or mobile radio telephones than to non-pulsed signals [13]. The following example presents the uncertainty quantification of the transient axial current induced along the human body exposed to electromagnetic pulse radiation [14]. The body is modeled as a straight wire antenna with length and radius as uniformly distributed random variables as depicted in Fig. 10.5.

A human equivalent thin-wire antenna model valid in the frequency region 50–110 MHz has been proposed for experimental dosimetry in [15] while a time domain (TD) simulation of the body based on the human-equivalent antenna

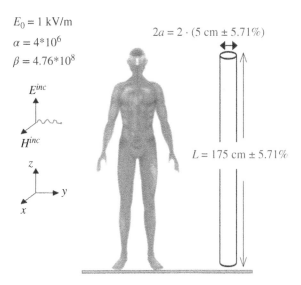

Figure 10.5 Stochastic–deterministic model of a human body (a thin wire representation) exposed to electromagnetic pulse. Both deterministic and stochastic input parameters are denoted.

model is proposed in [16]. The TD formulation is based on Hallen's type of integral equation solved via the GB-IBEM. The solution has been validated providing stable numerical results. Thus, the deterministic part of the proposed stochastic–deterministic model is based on Hallen's equation governing the distribution of the induced axial current in [16]:

$$\int_0^L \frac{I\left(z',t-\frac{R}{c}\right)}{4\pi R}dz' = F_0\left(t-\frac{z}{c}\right) + F_L\left(t-\frac{L-z}{c}\right) + \frac{1}{2Z_0}\int E_z^{inc}\left(z',t-\frac{|z-z'|}{c}\right)dz'$$

(10.3)

where $I(z', t\text{-}R/c)$ is the unknown space-time dependent current, E_z^{inc} is the incident electric field, c is the light velocity, and Z_0 is the free-space wave impedance. Multiple reflections of the current wave from the wire-free ends are considered via time signals $F_0(t)$ and $F_L(t)$. A detailed derivation can be found elsewhere, e.g. in [16].

The excitation field is given as the standard double exponential EMP waveform [16]:

$$E_z^{inc}(t) = E_0\left(e^{-\alpha t} - e^{-\beta t}\right)$$

(10.4)

where $E_0 = 1\,\text{kV/m}$, $\alpha = 4{*}10^6\,\text{s}^{-1}$, and $\beta = 4.76{*}10^8\,\text{s}^{-1}$. The frequency range of the chosen EMP corresponds to the frequency range of the human equivalent antenna presented in [15].

SCM with Lagrange interpolation and GL quadrature is used to propagate the uncertainties from the input antenna length (L) and radius (a) to the induced transient current. Both input parameters are uniformly distributed around their respective average values with the coefficient of variation, CF as depicted in Fig. 10.5. Since antenna is considered a perfect conductor, the output of interest is the induced current instead of the induced electric field. Although the latter would be more appropriate to use according to the latest updates in international standards and guidelines [17, 18], the results are still valid and can be readily interpreted.

To test the convergence of the SCM, the full tensor model was built with 3×3, 5×5, 7×7, and 9×9 simulation points. CV is changed for both parameters from 5.71% to 12%. The results presented in Fig. 10.6 exhibit a satisfactory convergence. SC method converges even with three collocation points in each dimension. Changing the CF for input parameters does not deteriorate the convergence of the SCM. The chosen observation point is the middle of the antenna, thus representing the body waist. Similar results are obtained for other space coordinates. The size of confidence margins doubles when CF is changed from 5.71% to 12%.

Furthermore, it is interesting to analyze the stochastic response of the maximal possible current value denoted here as I_{max}. Observing Fig. 10.6, the maximal

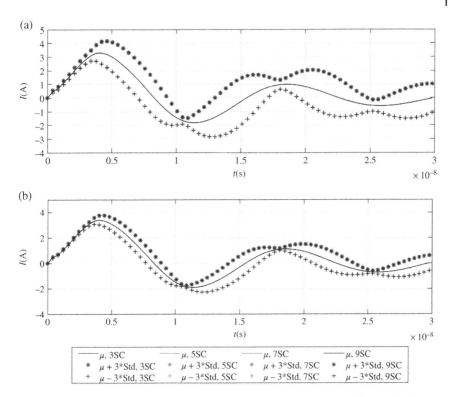

Figure 10.6 Confidence intervals for the transient current induced in a human body exposed to the double-exponential EMP. The observation point is the middle point of the antenna. Stochastic mean is denoted with μ and *Std* stands for standard deviation. The top figure (a) corresponds to the coefficient of variation CF = 12% while the bottom figure (b) corresponds to CF = 5.71%.

current value is expected in the first 5 ns. The stochastic analysis is done in the following way. First, the surrogate model for the maximal current is built according to Eq. (10.1):

$$I_{max} \approx \hat{I}_{max}(L, a) = \sum_{i=1}^{N} L_i(L, a) \cdot I_{max}^{(i)} \tag{10.5}$$

where $I_{max}^{(i)}$ is the i-th maximal current from the $N = 3 \times 3$ SC simulations. Once the surrogate is constructed, new design of experiment is defined with 1,000,000 MC samples. This time the surrogate model from Eq. (10.5) is run 1,000,000 times, thus resulting in 1,000,000 outputs. Finally, the quantiles are obtained from the resulting 1,000,000 values for the maximal current. Hence, Table 10.1 lists the estimated CIs based on quantile estimation obtained from MC simulation of the surrogate with 1,000,000 samples.

Table 10.1 The estimate of the confidence intervals for maximal possible current value based on quantile computation.

	I_{max} (A)	
Quantile range	CF = 5.71%	CF = 12%
[10.0%, 90%]	[3.23, 3.54]	[3.03, 3.65]
[5.00%, 95%]	[3.20, 3.57]	[2.96, 3.71]
[1.00%, 99%]	[3.15, 3.61]	[2.87, 3.80]

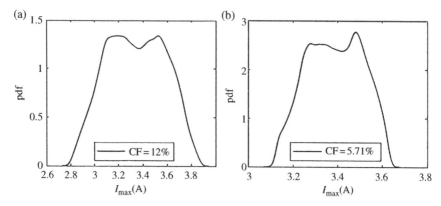

Figure 10.7 Probability density function of the maximal possible current value in case of CF = 12% (a) and CF = 5.71% (b). The results are based on MC sampling of surrogate model with 1,000,000 samples.

Surrogate model from Eq. (10.5) is used to obtain the probability density function for the I_{max} depicted in Fig. 10.7 for CF = 5.71% and 12%. The maximal possible value of the current appears within 5 ns which is the early time of response. It is within the recommended intervals.

Finally, impact factors for the antenna length and radius, respectively, are calculated in OAT manner as depicted in Fig. 10.8. Body height has higher impact throughout the simulation period. However, body width exhibits much higher influence in the early time behavior of the transient response. This is exactly the time period in which the current reaches its maximal possible value. Hence, the radius variability has high influence on confidence margin width of the maximum current. The mutual interaction between the two parameters is obviously weak, as the summation of the $I(L)$ and $I(a)$ adds up to 1 at almost every time instant.

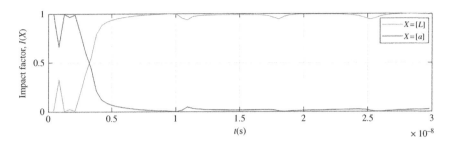

Figure 10.8 Impact factors of antenna parameter.

The modeling carried out in the presented stochastic-deterministic approach may be useful in designing such antennas like in Gandhi and Aslan patent in [15].

10.3 Internal Stochastic Dosimetry for a Realistic Three-Compartment Human Head Exposed to High-Frequency Plane Wave

Moving on from EM dosimetry at low frequencies and in TD to high-frequency EM dosimetry, as well as from canonical body models to anatomically realistic ones, the following SCM+EM coupling example features the stochastic dosimetry of induced electric field in a three-compartment human head [19]. The head model is depicted in Fig. 10.9.

The head model consisting of brain, skull, and scalp tissues is illuminated by a plane wave with incident amplitude of 1 V/m and frequency $f = 900$ MHz. Relative electric permittivity and electric conductivity of each tissue are modeled as uniformly distributed RVs, thus forming a six-dimensional stochastic model $X = [X_1, X_2, X_3, X_4, X_5, X_6]$ or $X = [\varepsilon_{r1}, \sigma_1, \varepsilon_{r2}, \sigma_2, \varepsilon_{r3}, \sigma_3]$ where subscripts 1, 2, and 3 denote scalp, brain, and skull tissues. According to [1, 20], average values of relative permittivities and electric conductivities for scalp, brain, and skull at 900 MHz are as follows: $\varepsilon_{r1} = 40.936$, $\sigma_1 = 0.899$ S/m, $\varepsilon_{r2} = 52.258$, $\sigma_2 = 0.985$ S/m, $\varepsilon_{r3} = 20.584$, and $\sigma_3 = 0.364$ S/m. Uniform distributions are formed by varying each parameter $\pm20\%$ around its respective average.

In the deterministic part, the incident plane wave is treated as an unbounded scattering problem formulated via the Stratton–Chu equation, i.e. the time-harmonic electric field at the exterior domain is expressed by the boundary integral equation as follows [21]:

$$\alpha \vec{E}'_{ext} = \vec{E}'_{inc} + \oint_{\partial V} \hat{n}\left(\nabla \times \vec{E}_{ext}\right)G dS + \oint_{\partial V}\left[\left(\hat{n} \times \vec{E}_{ext}\right) \times \nabla G + \left(\hat{n} \cdot \vec{E}_{ext}\right)\nabla G\right]dS$$

$$(10.6)$$

where E_i is the incident electric field and G is free space Green's function.

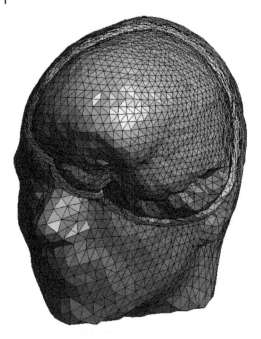

Figure 10.9 A three-compartment human head model consisting of scalp, skull, and brain tissues.

The interior domain is governed by the vector Helmholtz equation [21]:

$$\nabla \times \left(\frac{1}{k_B} \nabla \times \vec{E} \right) - k_A \vec{E} = 0 \tag{10.7}$$

with subscripts A and B denoting the exterior and interior regions, respectively.

The system of coupled Eqs. (10.6) and (10.7) is solved numerically via hybrid finite element method/boundary element method (FEM/BEM) [21]. The interior domain of the three-compartment model is discretized using 1,201,246 tetrahedral elements. Computations are executed by out-of-core Intel Pardiso matrix solver running as single core solver on Intel i7 processor with 32 GB RAM and 1TB SSD disk RAM available for the solver. A single deterministic simulation requires approximately 75 minutes for execution. Evidently, the traditional MC approach to stochastic analysis is impractical due to a large number of simulations needed for convergence.

SCM based on sparse grid (SG) interpolation is employed to assess stochastic moments of the induced electric field. Since MC simulations are not convenient to obtain the reference solution, the convergence is tested by gradually increasing the total number of simulation points of SC method. Clenshaw–Curtis formula up to order 4 is utilized to generate input samples resulting in maximum of 389 deterministic simulations. In addition, GL quadrature serves as a benchmark for the convergence in 6 univariate cases, i.e. when vector of input parameters is $X = [X_1]$, $X = [X_2]$, $X = [X_3]$, $X = [X_4]$, $X = [X_5]$, and $X = [X_6]$. Hence, induced field mean and variance are calculated for six univariate cases by using 3, 5, and 9 Chebyshev points obtained from Clenshaw–Curtis formula with the order 2, 3, and 4, respectively, and 9 collocation points from GL quadrature rule. The consecutive integration rules are denoted with CH2, CH3, CH4, and GL5, respectively.

The absolute relative error between the two consecutive integration rules is computed as follows:

$$Err = \left| \frac{V_i - V_j}{V_i} \right| \cdot 100 \qquad i,j = \text{CH2, CH3, CH4 or GL5} \qquad (10.8)$$

where V stands for variance while subscripts i and j denote the integration rule, j being the one with the expected higher accuracy. The absolute relative error is calculated at each point and the results are summarized in Table 10.2 in the form of the maximal gap between the integration rules for variance computation in brain tissue. The percentage, $P[\%]$, of the total number of observation points in the brain tissue for which the gap between the CH4 and GL5 is below 0.5% and 0.005% are given in Table 10.2.

The SCM convergence is satisfactory and computation of stochastic moments in six-dimensional case is further carried out with the CH4 SG rule. Note that the convergence results for the induced electric field mean are omitted here. However, numerical errors propagate from the lower to the higher level of the stochastic moment, e.g. from mean to variance according to expressions in Table 10.1.

Table 10.2 The maximal absolute relative error, Err for variance of incident field between different integration rules.

	$X = [X_1]$	$X = [X_2]$	$X = [X_3]$	$X = [X_4]$	$X = [X_5]$	$X = [X_6]$
Err [%]						
CH2–CH3	11575.79	4679.17	378069.3	4327.76	498.91	60.59
CH3–CH4	86.95	25.78	294.21	38.08	47.28	10.51
CH4–GL5	7.03	2.30	22.61	3.89	4.79	0.18
P [%]						
Err < 0.5 %	98.72	99.99	98.60	99.95	99.96	100
Err < 0.005 %	89.73	99.30	91.71	94.89	96.04	95.95

Therefore, as expected, the convergence is even better in case of the stochastic mean than in case of the presented stochastic variance.

Furthermore, distributions of the induced electric field mean and variance on the tissue surfaces are depicted in Fig. 10.10 for six-dimensional case. The mean (μ) and standard deviation (*Std*) of the induced electric field are given for the two characteristic points: the point with the maximal mean and variance, M_{max} and V_{max}, respectively. For brain tissue, the upper limit of *CI*, calculated as mean $\pm$ 3 standard deviations, is larger for the point V_{max} than M_{max} (269.76 mV/m > 279.43 mV/m). The distribution of the induced field's mean value is inhomogeneous and more pronounced in the ventral parts.

ANOVA sensitivity analysis is carried out at the characteristic points M_{max} and V_{max}. The first-order sensitivity indices are depicted in Fig. 10.11. The sum of the

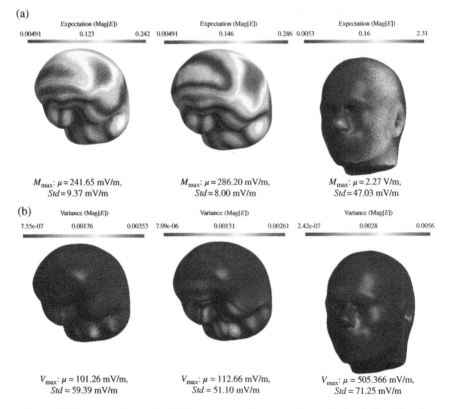

Figure 10.10 Induced electric field mean (a) and variance (b) on brain, skull, and scalp tissue surface (from left to right) for six-dimensional case. Points with the highest mean or variance value are denoted with M_{max} and V_{max}, respectively.

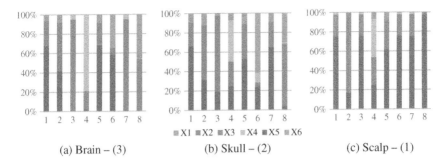

(a) Brain – (3) (b) Skull – (2) (c) Scalp – (1)

■X1 ■X2 ■X3 ■X4 ■X5 ■X6

Figure 10.11 The first-order sensitivity indices at eight points. Points numbered 1–6 correspond to V_{max} in one-dimensional cases; points number 7 and 8 correspond to M_{max} and V_{max} in six-dimensional case. Point M_{max} has the same coordinate in every case, regardless of dimension.

first-order sensitivity indices is equal or very close to 1 which indicates that no mutual interactions between the parameters have a significant impact on the field values in the characteristic points M_{max} and V_{max}.

The first-order sensitivity indices at points M_{max} and V_{max} in the brain tissue are given in Fig. 10.11a. Brain permittivity is the most influential parameter at points 1, 2, 5 and 6. However, skull permittivity has a strong impact at points 3 and 7 and skull conductivity has the strongest impact at point 4. Sensitivity indices for point M_{max} from six-dimensional case have similar values as for the point of maximal variance when only brain permittivity is RV. However, for point V_{max} in six-dimensional case, four input parameters have significant impact starting from skull permittivity, followed by brain permittivity, scalp conductivity, and skull conductivity, the last three having a similar impact.

Sensitivity indices at the surface of skull tissue are depicted in Fig. 10.11b. Overall, the influence of scalp's permittivity is the strongest. According to Fig. 10.11c, the impact of scalp permittivity on scalp tissue surface is the strongest. Scalp conductivity does not play an important role in the given coefficient of variation.

Next, CIs of the induced electric field and impact factor of input parameters are depicted in Figs. 10.13 and 10.14 along the sagittal axis denoted with a solid line in Fig. 10.12.

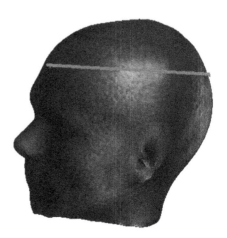

Figure 10.12 The sagittal axis of the head model denoted with a solid line.

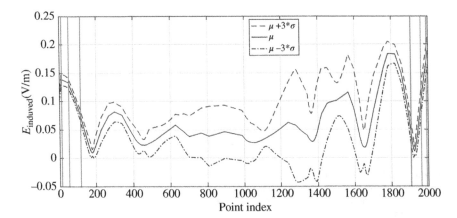

Figure 10.13 Confidence interval of the induced electric field along the sagittal axis of head model computed as $\mu \pm 3*Std$. μ and Std stand for mean and standard deviation. Vertical lines denote the interfaces between air-scalp-skull-brain on the left and brain-skull-scalp-air on the right.

Stochastic mean with CIs is presented in Fig. 10.13. CIs are wider in the interior part of the brain than in the other areas.

Impact factors of input parameters computed according to the OAT approach are depicted in Fig. 10.14. Brain permittivity and conductivity have much higher impacts compared to impact of other input parameters in the interior part of brain. Scalp and skull relative permittivities have a noticable influence outside the brain region. Overall, the difference between impact factors decreases in the vicinity of brain–skull interface.

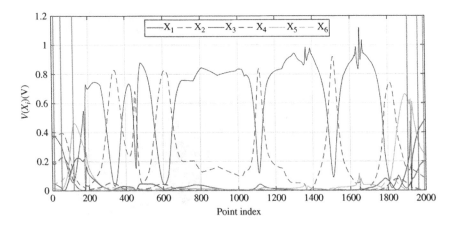

Figure 10.14 Impact of input parameters along the sagittal axis of head model. Vertical lines denote the interfaces between air-scalp-skull-brain on the left and brain-skull-scalp-air on the right.

10.4 Incident Field Stochastic Dosimetry for Base Station Antenna Radiation

Finally, the last example of stochastic EM dosimetry pertains to incident EM dosimetry [22]. The output of interest is the total electric field radiated by a BSA placed at height A_h above a two-layered lossy ground. Both incident and reflected rays in the far field zone are considered, which is depicted in Fig. 10.15.

The ground consists of two layers each, first of which has thickness of d_1 while the other has an infinite thickness, thus forming a half-space. Layers are characterized by the corresponding electric conductivities and relative electric permittivities, σ_1, σ_2, ε_{r1}, and ε_{r2}, respectively. Transmitting antenna is modeled as a panel antenna with horizontal and vertical radiation patterns obtained by software package NEC (Numerical Electromagnetic Code) [23]. The output antenna power is set to 58.45 dBmW and operating frequency is from GSM frequency range, $f = 936.8$ MHz.

Although different approaches can be used to compute the radiated electric field at high frequencies [24], according to regulations in [25], approximation of antenna insulated in free space is officially used to obtain radiated electric field while reflected field components are neglected. Other techniques include reflections with different approximations of the reflected coefficient. Here, the deterministic computation of total field assumes the summation of the incident and reflected field as follows [26]:

$$E_{TOT} = \sqrt{E_i^2 + E_r^2} = \sqrt{E^2 + (E \cdot \Gamma_r)^2} \tag{10.9}$$

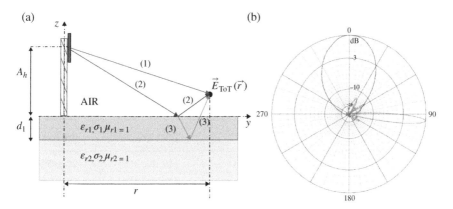

Figure 10.15 (a) Base station antenna above a two-layered lossy ground. Incident ray, rays reflected from the first and second ground layers are represented by arrows numbered (1), (2), and (3), respectively. (b) Horizontal and vertical antenna patterns.

where Γ_r is reflection coefficient and E is computed by means of an approximate formula [27]:

$$E = \frac{\sqrt{30 \cdot N \cdot EIRP \cdot 10^{\frac{G}{10}}}}{r} \tag{10.10}$$

Effective Isotropic Radiated Power (EIRP) is the output antenna power set to 58.45 dBmW, G is antenna gain in dB, N is the number of active channels set to 1, and r is the distance from the antenna. This formula is the official relation used from farfield computation in practical engineering studies such as in studies [28].

The reflection coefficient for the two-layered media, Γ_r is given by the following expression [29]:

$$\Gamma_r = \frac{R_{02} + R_{12} \cdot e^{-j2\beta d_1 \cdot \cos(\vartheta_{i-01})}}{1 + R_{02} \cdot R_{12} \cdot e^{-j2\beta d_1 \cdot \cos(\vartheta_{i-01})}} \tag{10.11}$$

where R_{mn} stands for the reflection coefficient between the two media indexed by m and n, each time computed either with Fresnel reflection coefficient model (FRM) or Modified Image Theory method (MIT) [26]. Indices 0, 1, and 2 denote air and first and second ground layers, respectively. The incidence angle between the air and first layer is denoted by θ_{i-01} while β is the phase propagation constant for air, $\beta^2 = \omega^2 \mu_0 \varepsilon_0$. The angular frequency is denoted by ω.

In the stochastic part, antenna height, conductivities, relative permittivities, and upper ground layer thickness are modeled as random variables as it is defined in Table 10.3.

Uncertainty from the input parameter set is propagated to the output of interest by means of both MC and SC methods. SC approach is based on SGs with Clenshaw–Curtis quadrature and Lagrange interpolation.

First, the convergence is tested by changing the size of input sets; $N = N_{MC}$ or $N = N_{SC}$ for MC or SC method, respectively. Input simulation points are

Table 10.3 Input parameters with the lower and upper boundaries of uniform distributions denoted as min and max, respectively. avg and CV stand for the average and coefficient of variation.

		[min, avg, max]	avg ± CV (%)
Antenna height	A_h (m)	[19, 20, 21]	20 ± 5%
First layer thickness	d_1 (cm)	[15, 20, 25]	20 ± 25%
First layer relative permittivity	ε_{r1}	[12, 15, 18]	15 ± 20%
First layer electric conductivity	σ_1 (mS/m)	[0.8, 1, 1.2]	1 ± 20%
Second layer relative permittivity	ε_{r2}	[3.2, 4, 4.8]	4 ± 20%
Second layer electric conductivity	σ_2 (S/m)	[0.04, 0.05, 0.06]	50 ± 20%

d-dimensional. The set sizes for MC method are $N_{MC} = 1{,}000$, 10,000, and 100,000 both for six-dimensional and one-dimensional stochastic models. The reference stochastic mean and standard deviation are obtained with $N_{MC} = 100{,}000$. While MC convergence depends only on N_{MC}, the convergence of the SC method depends both on dimensionality and on number of collocation points used in each dimension. Hence, for one-dimensional cases, SC set size is set to 3, 5, 9, and 17 and for six-dimensional model, $N_{SC} = 13$, 85, and 1457. The convergence of MC and SC methods in computation of stochastic mean and standard deviation for six-dimensional model is depicted in Figs. 10.16 and 10.17 for Fresnel reflection coefficient and Modified Image Theory approaches (FRM and MIT, respectively).

(a)

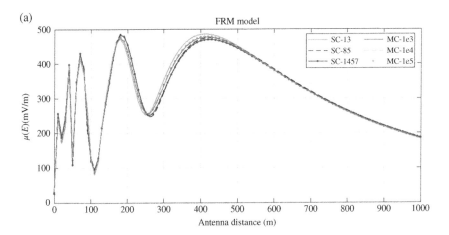

(b)

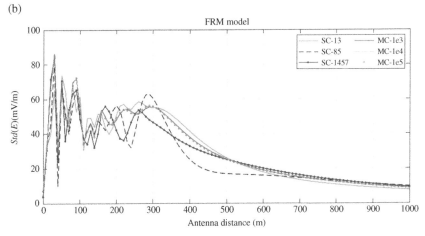

Figure 10.16 Electric field mean (a) and standard deviation (b) obtained with MC and SC methods with different number of simulation points. The deterministic computation is based on Fresnel reflection coefficient approach (FRM).

(a)

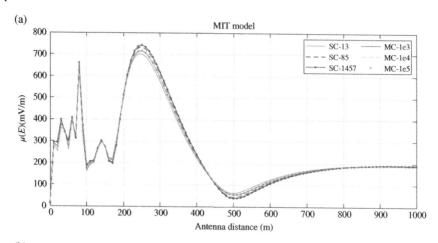

(b)

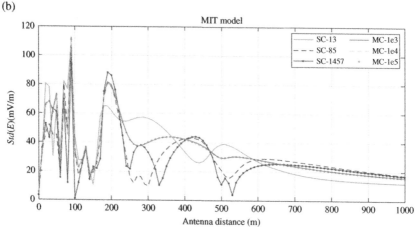

Figure 10.17 Electric field mean (a) and standard deviation (b) obtained with MC and SC methods with different number of simulation points. The deterministic computation is based on Modified Image Theory approach (MIT).

Although the convergence of SC method is satisfactory for mean electric field, it is rather poor when computing the standard deviation. To improve the convergence, N_{SC} should be increased over 1457 which is no longer efficient compared to MC method which converges at $N_{MC} = 1,000$ for computation of both stochastic mean and standard deviation in FRM and MIT models. Additionally, Fig. 10.18 depicts the convergence of standard deviation for six 1-dimensional cases for FRM model. The MC method, again, shows good results at $N_{MC} = 1,000$.

The electric field standard deviation obtained by SCM when antenna height, first-layer conductivity, second-layer relative permittivity, and second-layer

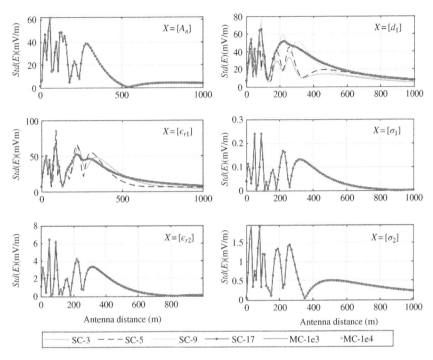

Figure 10.18 Convergence of MC and SC methods in computation of electric field standard deviation, *Std(E)* for six 1-dimensional cases. Deterministic computation is based on FRM model.

conductivity, respectively, are set as RVs one-at-a-time, exhibits good convergence even at lowest number of simulation points, $N_{SC} = 3$. The variation of these input parameters does not disturb the convergence of the standard deviation of six-dimensional stochastic model. When only first-layer relative permittivity is RV, SCM convergence is accomplished at $N_{SC} = 9$ which is relatively small set size. However, when nine collocation points are used in each of six dimensions, the total N_{SC} is 1,457, which is more than $N_{MC} = 1,000$. Finally, when the first-layer thickness is considered RV, the convergence is accomplished at $N_{SC} = 17$ which is efficient only for one-dimensional stochastic model. The same analysis is valid for MIT model, too.

Results presented in Fig. 10.18 contain information about the OAT sensitivity analysis for FRM model. Comparing the standard deviations, it can be concluded that antenna height, first-layer depth, and first-layer relative permittivity are significant parameters. Their ranking in top 3 depends on the observation point of interest. However, OAT cannot answer if the impact originates from the input parameter's sole variability or its interaction with another parameter. For that purpose, ANOVA analysis is carried out as depicted later in Figs. 10.20 and 10.21.

Based on the convergence results, the electric field expectation and CIs are computed by using MC method with 1,000 samples which are depicted in Fig. 10.19. It is of particular interest to investigate to which extent the CI obtained with FRM, and MIT models overlap since deterministic values of electric field obtained by using FRM and MIT models differ to a certain extent as reported in [29, 30].

Electric field mean is between 100 and 500 mV/m for FRM model and 50 and 700 mV/m for MIT model while CI width is practically constant for both models. CI stemming from FRM and MIT approaches overlap at distances shorter than 200 m. As the distance increases up to 750–800 m, FRM CI and MIT CI differ. Finally, at distances above 750–800 m, CIs overlap, while mean electric field obtained by FRM and MIT converges to the same value.

Sensitivity analysis based on ANOVA approach for FRM and MIT models is depicted in Figs. 10.20 and 10.21, respectively. The first-order sensitivity index of antenna height is higher than the first-order sensitivity index of other input parameters for most of the observation points of interest ($S1$ in Figs. 10.20 and 10.21a). According to Eq. (11.5), the summation of all indices must be equal to 1 or 100% which is not the case for all observation points of interest. Therefore, as the first-order sensitivity indices are not sufficient for complete sensitivity analysis, the total effect sensitivity indices are computed (ST in Figs. 10.20 and 10.21b).

Antenna height first-order sensitivity index and its total effect sensitivity index have approximately the same value. Hence, the interactions of antenna height with other input parameters do not significantly impact the output field variance. On the contrary, first-layer thickness and its relative permittivity total effect indices differ from the corresponding first-order sensitivity indices. Moreover, total

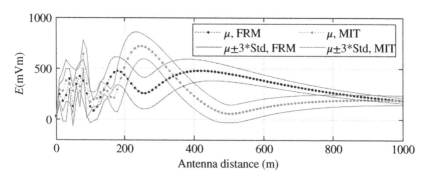

Figure 10.19 Electric field confidence intervals with FRM and MIT calculation model, respectively.

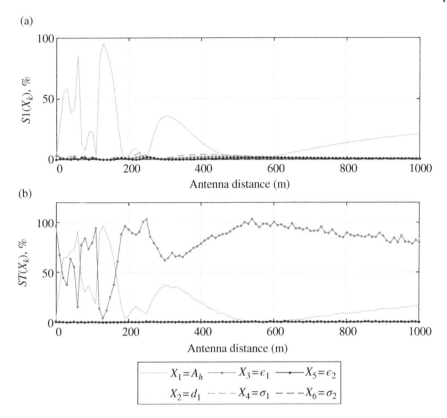

Figure 10.20 First-order and total effect sensitivity indices (a) and (b) for FRM model.

effect indices of the first-layer relative permittivity and total effect indices of the first-layer thickness are approximately the same. This leads to the conclusion that although the first-layer thickness and its relative permittivity do not directly impact the output variance, their interactions in the mathematical model cause variability in the electric field output. Therefore, their uncertainties should not be neglected. The impact of the antenna height alone and the impact originating from the interaction of the first-layer thickness and its permittivity changes from one observation point to another. For smaller distances, the impact switches from antenna height to combined effect of the first-layer thickness and its relative permittivity, but after approximately 500 m, the combined impact becomes completely dominant while antenna height loses its significance.

Since their impact is less than 5% for all observation points of interest, the rest of the input parameters do not impact the output variance significantly.

(a)

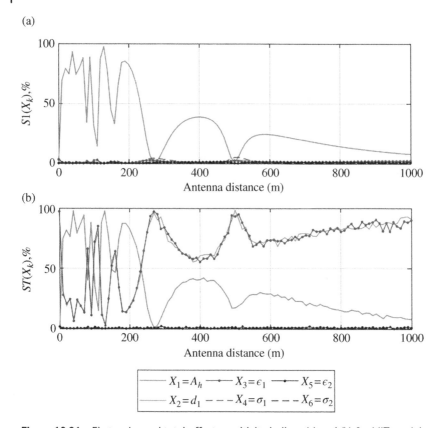

(b)

Figure 10.21 First-order and total effect sensitivity indices (a) and (b) for MIT model.

References

1 C. Gabriel, "Compilation of the dielectric properties of body tissues at RF and microwave frequencies," Technical Report: AL/OE-TR-1996-0037, TX: Brooks Air Force Base; 1, 1996.

2 C. Gabriel and A. Peyman, "Dielectric measurement: error analysis and assessment of uncertainty," *Physics in Medicine & Biology*, vol. 51, no. 23, pp. 6033–6046, 2006.

3 E. Chiaramello, M. Parazzini, S. Fiocchi, P. Ravazzani and J. Wiart, "Assessment of fetal exposure to 4G LTE tablet in realistic scenarios: effect of position, gestational age, and frequency," *IEEE Journal of Electromagnetics, RF and Microwaves in Medicine and Biology*, vol. 1, no. 1, pp. 26–33, 2017.

4 O. Jawad, D. Lautru, J. M. Dricot, F. Horlin, A. Benlarbi-Delaï and P. De Doncker, "Estimation of specific absorption rate with kriging method," in *USNC-URSI Radio Science Meeting*, Lake Buena Vista, FL, USA, 2013.

5 I. Ahmed, H. Khammari, A. Shahid, A. Musa, K. Soon Kim, E. De Poorter and I. Moerman, "A survey on hybrid beamforming techniques in 5G: architecture and system model perspectives," *IEEE Communications Surveys & Tutorials*, vol. 20, no. 4, pp. 3060–3097, 2018.

6 J. Wiart, Radio-Frequency Human Exposure Assessment: From Deterministic to Stochastic Methods, London, UK; Hoboken, NJ, USA: John Wiley & Sons, Inc., 2016.

7 A. Šušnjara and D. Poljak, "An efficient deterministic–stochastic model of the human body exposed to ELF electric field," *International Journal of Antennas and Propagation*, vol. 2016, Article ID 6153620, p. 8, 2016.

8 ICNIRP, "ICNIRP guidelines for limiting exposure to time-varying electric, magnetic fields and electromagnetic fields (up to 300 GHz)," *Health Physics*, vol. 74, no. 4, pp. 494–522, 1998.

9 ICNIRP, "ICNIRP guidelines for limiting exposure to electromagnetic fields (100 kHz to 300 GHz)," *Health Physics*, vol. 118, no. 5, p. 483–524, 2020.

10 IEEE Std 2889™-2021, IEEE Guide for the Definition of Incident Power Density to Correlate Surface Temperature Elevation, 2021.

11 R. W. P. King, "Fields and currents in the organs of the human body when exposed to power lines and VLF transmitters," *IEEE Transactions on Biomedical Engineering*, vol. 45, no. 4, pp. 520–530, 1998.

12 D. Poljak and Y. F. Rashed, "The boundary element modelling of the human body exposed to the ELF electromagnetic fields," *Engineering Analysis with Boundary Elements*, vol. 26, no. 2002, pp. 871–875, 2002.

13 J. Y. Chen and O. P. Gandhi, "Currents induced in an anatomically based model of a human exposure to vertically polarized electromagnetic pulses," *IEEE Trans. on Microwave Theory and Techniques*, vol. 39, no. 1, pp. 31–39, 1991.

14 A. Šušnjara and D. Poljak, "Uncertainty quantification for the transient response of human equivalent antenna using the stochastic collocation approach," *International Journal of Antennas and Propagation*, vol. 2019, Article ID 4640925, p. 7, 2019.

15 O. P. Gandhi and E. Aslan, "Human equivalent antenna for electromagnetic fields". USA Patent 5394164, 28 February 1995.

16 D. Poljak, C. Y. Tham, O. Gandhi and A. Sarolic, "Human equivalent antenna model for transient electromagnetic radiation exposure," *IEEE Transactions on Electromagnetic Compatibility*, vol. 45, no. 1, pp. 141–145, 2003.

17 ICNIRP, ICNIRP guidelines for limiting exposure to time-varying electric and magnetic fields (1 Hz–100 kHz), *Health Physics* 99(6):818–836; 201, 2010.

18 IEEE Standard for Safety Levels with Respect to Human Exposure to Electromagnetic Fields, 0–3 kHz, New York, NY, USA: Standard IEEE-C95.1, 10016-5997, 2019.

19 A. Šušnjara, H. Dodig, M. Cvetković and D. Poljak, "Stochastic dosimetry of a three compartment head model," *Engineering Analysis with Boundary Elements*, vol. 117, pp. 332–345, 2020.

20 ITIS Foundation, "Copyright © 2010–2021 IT'IS Foundation," June 2019. Available: [Online]. https://itis.swiss/virtual-population/tissue-properties/database/

21 M. Cvetkovic, H. Dodig and D. Poljak, "A study on the use of compound and extracted models in the high frequency electromagnetic exposure assessment," *Mathematical Problems in Engineering*, vol. 2017, Article ID 7932604, p. 12, 2017.

22 M. Galić, A. Šušnjara and D. Poljak, "Stochastic-deterministic assessment of electric field radiated by base station antenna above a two-layered ground," *Mathematical Problems in Engineering*, vol. 2022, Article ID 1833748, p. 15, 2022.

23 Numerical electromagnics code, https://www.nec2.org/ [Accessed October 26, 2021].

24 D. Poljak and M. Cvetković, Human Interaction with Electromagnetic Fields: Computational Models in Dosimetry, St. Louis, Misouri: Elsevier, 2019.

25 EN 50413, "Basic standard on measurement and calculation procedures for human exposure to electric, magnetic and electromagnetic fields (0 Hz - 300 GHz)," 2019.

26 M. Galic, D. Poljak and V. Doric, "Simple analytical models for the calculation of the electric field radiated by the base station antenna," *International Journal for Engineering Modelling*, vol. 31, no. 1–2, pp. 31–42, 2018.

27 EN 62232, "Determination of RF field strength, power density and SAR in the vicinity of radiocommunication base stations for the purpose of evaluating human exposure," 2017.

28 EN 61566, "Measurement of exposure to radio-frequency electromagnetic fields - field strength in the frequency range 100 kHz to 1 GHz (IEC 61566:1997; EN 61566:1997)," 2001.

29 M. Galic, D. Poljak and V. Doric, "Analytical technique to determine the electric field above a two-layered medium," in *2018 2nd URSI Atlantic Radio Science Meeting (AT-RASC)*, pp. 1–4, 2018.

30 M. Galić and D. Poljak, "Theoretical and experimental incident field dosimetry for GSM base stations," in *2019 URSI International Symposium on Electromagnetic Theory (EMTS)*, San Diego, CA, USA, 2019.

11

Stochastic–Deterministic Thermal Dosimetry

Analogously to internal electromagnetic dosimetry, thermal dosimetry is greatly dependent on accurate and trustworthy computational models. Namely, it is ethically unacceptable to measure the temperature distribution inside the human body. The only exception is the superficial temperature which can be measured by thermometers or infrared thermal cameras.

All computational models in thermal dosimetry are based on Pennes' bioheat equation which governs temperature distribution in biological tissue. In this chapter, the steady-state heat transfer in biological tissue is considered. The governing equation and the corresponding boundary condition at air–skin interface are given in the following form [1]:

$$\nabla(\lambda \nabla T) + w_b(T_{art} - T) + Q_m + Q_{ext} = 0 \tag{11.1a}$$

$$-\lambda \frac{\partial T}{\partial \hat{n}} = h_{eff}(T - T_{amb}) \tag{11.1b}$$

Thermal parameters λ, w_b, Q_m, T_{art}, T_{amb}, and h_{eff} are defined in Table 11.1. The heat transfer by the radiation and the forced convection are neglected.

Equation (11.1a) includes several mechanisms which are responsible for heat transfer in biological tissue. The mechanisms are represented by thermal parameters such as arterial blood temperature and tissue-dependent parameters such as thermal conductivity, volumetric perfusion blood rate, and heat source due to metabolic processes. Additional term Q_{ext} represents the external heat source. In thermal dosimetry, the external heat source corresponds to the generated EM energy in the tissue due to induced EM field which is a consequence of the external EM source. Of course, this quantity is an output from the EM internal dosimetry.

The analysis of thermal processes inside the human organism is complex due to many factors that need to be included, most important of which are the

Deterministic and Stochastic Modeling in Computational Electromagnetics: Integral and Differential Equation Approaches, First Edition. Dragan Poljak and Anna Šušnjara.
© 2024 The Institute of Electrical and Electronics Engineers, Inc.
Published 2024 by John Wiley & Sons, Inc.

Table 11.1 Thermal parameters from Pennes' bioheat equation and boundary condition.

Thermal parameter		
Thermal conductivity	λ	[W/m °C]
Volumetric perfusion blood rate	w_b	[kg/(sm^3)]
Arterial blood temperature	T_{art}	[°C]
Heat source due to metabolic processes	Q_m	[W/m^3]
Convection coefficient	h_{eff}	[W/(m^2 °C)]
Ambient temperature	T_{amb}	[°C]

The parameters λ, w_b, Q_m, and h_{eff} are tissue dependent.

metabolism and the blood flow [1]. As is the case with tissue relative electric permittivity and conductivity, the tissue dependent thermal parameters also depend on age, health, and gender of each individual person. Accordingly, their values obtained from *in vitro* measurements on human and animal tissues are listed in the relevant literature and databases as a set of values, scattered around average value [2]. Therefore, stochastic dosimetry comes in handy to consider the uncertain nature of thermal parameters and obtain the expected temperature and temperature elevation along with the corresponding confidence intervals (CIs). Also, it is desirable to rank the thermal parameters according to their impact on the CI width of the output of interest.

The rest of the chapter is organized as follows. First, stochastic analysis of Pennes' bioheat transfer equation is given for simplified tissue geometry without external EM source [3]. Both stochastic collocation and Monte Carlo methods are utilized for uncertainty quantification and sensitivity analysis. Then, stochastic dosimetry of temperature in homogeneous human brain and the three-compartment human head are given, [4, 5], respectively. In both examples, the EM source is a plane wave at 900 MHz. Finally, even more realistic human head model is utilized for stochastic thermal dosimetry at 3.5 GHz [6].

11.1 Stochastic Sensitivity Analysis of Bioheat Transfer Equation

The geometry of interest is a one-dimensional skin tissue with thickness L depicted in Fig. 11.1.

Deterministic one-dimensional bioheat equation governing the temperature distribution in the human skin is given as [3]:

$$\lambda \frac{d^2 T(x)}{dx^2} - w_b \cdot T(x) + w_b \cdot T_{art} + Q_m + Q_{ext} = 0 \qquad (11.2a)$$

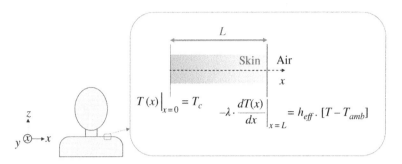

Figure 11.1 One-dimensional skin model. T_c is core temperature, T_a is ambient temperature, λ is tissue thermal conductivity, and h_{eff} is the convection coefficient.

$$T|_{x=0} = T_c \tag{11.2b}$$

$$-\lambda \cdot \frac{dT(x)}{dx}\bigg|_{x=L} = h_{\mathit{eff}} \cdot [T - T_{amb}] \tag{11.2c}$$

Expressions (11.2b) and (11.2c) are boundary conditions where T_c denotes the core temperature and $x = L$ corresponds to skin–air interface. Thermal parameters λ, T_{art}, T_{amb}, h_{eff}, w_b, and Q_m are presented in Table 11.1. The deterministic bio-heat Eq. (11.2a) is solved by means of FEM and the output of interest is the temperature distribution [3].

The Lagrange SC method is used as a wrapper around the deterministic code. Collocation points are generated from Gauss–Legendre and Clenshaw–Curtis quadrature rules for tensor and sparse grid SC, respectively [3]. Six thermal parameters, i.e. λ, w_b, T_{art}, O_m, h_{eff} and T_{amb} are modeled as random variables (RVs) uniformly distributed in the range computed as average $\pm 20\%$. The random input vector is: $X = [\lambda, w_b, T_{art}, Q_m, h_{\mathit{eff}}, T_{amb}]$. Average values of input parameters are as follows: $\lambda = 0.5$ W/m°C, $w_b = 2100$ kg/(sm³), $T_{art} = 37\,°C$, $Q_m = 33\,800$ W/m³, $h_{\mathit{eff}} = 10$ W/(m² °C) and $T_{amb} = 25\,°C$. The skin depth is set to $L = 3$ cm.

Six-dimensional stochastic problem with up to nine points in each dimension resulted in total of 729 and 389 deterministic simulations for full-tensor + Gauss–Legendre quadrature and sparse-grid + Clenshaw–Curtis quadrature, respectively. The reference stochastic moments are obtained with MC method with 1 000 000 samples. Additional 100 MC sets, each with 1000 samples are used to test the MC convergence.

Convergence of MC and SC methods in computation of stochastic moments; mean, variance, skewness and kurtosis of the output temperature is depicted in Fig. 11.2.

Compared to MC method, SC converges more quickly. Namely, 1000 samples are not enough for MC method to ensure the accurate solution for the stochastic moments. However, both full-tensor and sparse-grid SC provide a

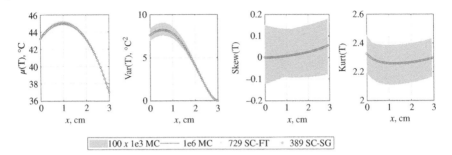

Figure 11.2 From left to right: stochastic mean, variance, skewness, and kurtosis of temperature distribution in the skin tissue obtained by MC, SC full-tensor (SC-FT), and an SC sparse-grid (SC-SG) method.

good approximation of the stochastic moments with 729 and 389 simulations, respectively.

Furthermore, CIs computed as mean ± 1 standard deviation are depicted in Fig. 11.3. The maximal deviation from the mean is 6.43%.

Variations of individual input parameters have different impact on the resulting temperature variation. Total effect sensitivity indices obtained by means of ANOVA approach combined with sparse-grid SC with 389 simulations and MC with 1e6 samples are depicted in Fig. 11.4.

Sensitivity analysis demonstrates the overwhelming impact of arterial temperature over the whole domain followed by the influence of metabolic processes, both in a homogenous trend. Other parameters exhibit nonhomogeneous influence, however, rather small. The accuracy of the ANOVA based approach within the SC framework is validated against the Sobol's sensitivity indices computed via

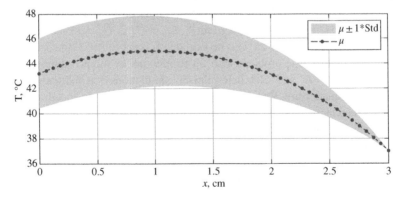

Figure 11.3 Confidence intervals computed as mean temperature ± 1 standard deviation, i.e. $\mu \pm 1 {}^* \mathrm{Std}$.

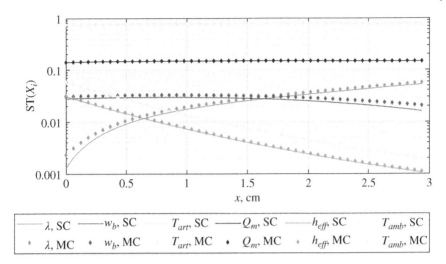

Figure 11.4 entries:

——— λ, SC	——— w_b, SC	T_{art}, SC	——— Q_m, SC	——— h_{eff}, SC	T_{amb}, SC
◆ λ, MC	◆ w_b, MC	T_{art}, MC	◆ Q_m, MC	◆ h_{eff}, MC	T_{amb}, MC

Figure 11.4 Total effect sensitivity indices obtained with variances computed with SC and MC method and 389 and 1 000 000 simulations, respectively.

MC based sampling. The values of the total effect and first order indices are almost the same for each input parameter, thus proving that mutual interactions have insignificant impact on the temperature distribution.

11.2 Stochastic Thermal Dosimetry for Homogeneous Human Brain

The first realistic example of stochastic thermal dosimetry refers to human brain exposed to incident plane wave with power density $P = 5\,\mathrm{mW/cm^2}$ and operating frequency $f = 900$ MHz [4]. The EM part necessary for computation of the induced electric field and thus the specific absorption rate (SAR) inside the human brain is formulated as a surface integral equation solved by Method of Moments [7]. Deterministic EM part is followed by Pennes' bioheat equation in from of Eq. (11.1). The SAR and induced electric field (E_{ind}) are related to the temperature distribution through the external heat source in Pennes' bioheat equation as follows [4]:

$$Q_{ext} = \rho \cdot SAR = \frac{\sigma}{2} \cdot |E_{ind}|^2 \tag{11.3}$$

where ρ is tissue density (kg/m³). Equation (11.1) is solved by means of Finite Element Method (FEM) [4, 7].

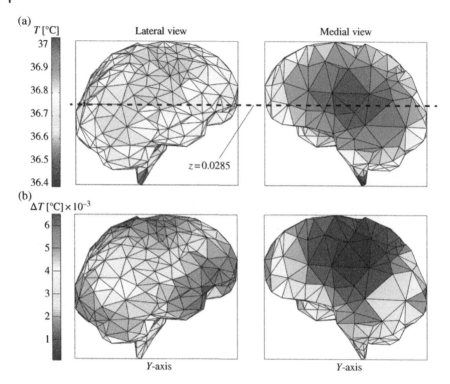

Figure 11.5 (a) Temperature distribution and (b) temperature elevation when human brain is exposed to the incident electric field. Thermal parameters have average values.

Thermal parameters presented in Table 11.1 vary significantly and are very difficult to find. Therefore, they are modeled as random variables with uniform probability distributions as follows: $\lambda = 0.513 \pm 0.1$ W/m °C, $w_b = 33\,287 \pm 6000$ kg/(sm³), $T_{art} = 37 \pm 0.5$ °C, $Q_m = 6385 \pm 3192.5$ W/m³, $h_{eff} = 12 \pm 10$ W/(m² °C), and $T_{amb} = 20 \pm 10$ °C. For illustration, Fig. 11.5 depicts the temperature distribution in the human brain of the adult for average values of thermal parameters after the exposure.

Lagrange full-tensor stochastic collocation with Gauss–Legendre quadrature is used to propagate the uncertainty to the output of interest. Beside the temperature (T) and temperature elevation (ΔT) distribution, some other specific parameters are observed such as maximal temperature (T_{max}), minimal temperature (T_{min}), and average temperature (T_{avg}). The convergence is tested by increasing the number of collocation points, 3, 5 and 7 in one-dimensional cases and 729, 15.625 and 117.649 deterministic simulations in six-dimensional case. Stochastic analysis is carried out with $Q_{ext} = 0$ and $Q_{ext} \neq 0$, i.e. the steady state case and HF exposure case, respectively.

In Fig. 11.6 CIs for maximal, minimal and average temperature in steady state and temperature gradient in case of exposure are depicted for six one-dimensional

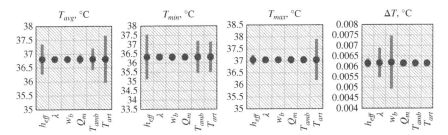

Figure 11.6 Confidence interval computed as mean ± 3 standard deviations for average, minimum, and maximum temperatures in steady state and temperature elevation in case of exposure. Six one-dimensional cases are depicted and denoted with the label of input parameter being the RV in OAT manner, h_{eff}, λ, w_b, Q_m, T_{amb}, and T_{art}, respectively.

cases. Note that the CI widths for maximal, minimal and average temperature are independent on Q_{ext}, i.e. they are the same no matter if brain is exposed to incident field or not.

The stochastic mean and variance in all one-dimensional cases computed with 3, 5 and 7 collocation points resulted in the same values, hence SC method in one-dimensional cases has excellent convergence. The results in Fig. 11.6 are obtained with three collocation points.

Inspecting the results for CI in Fig. 11.6, some conclusion regarding the OAT sensitivity analysis can be drawn. The average temperature and the maximal temperature are mostly affected by the variation of the arterial blood temperature. Overall, the arterial blood temperature is expected to have the highest impact on the temperature distribution. The effective thermal convection coefficient and ambient temperature have significant influence on minimum and average temperature. The influence of the remaining thermal parameters is negligible. As for the temperature elevation, the most influential parameter is the volumetric perfusion blood rate followed by thermal conductivity. Other input parameters have small impact.

Furthermore, impact of input parameters' variations on the distribution of temperature elevation along and across the brain at certain fixed height are depicted in Fig. 11.7. Once again, the influence is given in form of CIs computed as mean temperature ± 3 standard deviations.

When only heat transfer coefficient is RV, the CI is rather narrow, and its trend is the same along the x and y axes (Fig. 11.7a). Its effect is slightly more significant on edges corresponding to the brain surface which is expected since this input parameter governs the heat exchange with the air. On the contrary, thermal conductivity affects the temperature elevation in the central part of the brain more than in the superficial layers (Fig. 11.7b). Finally, the most dominant thermal

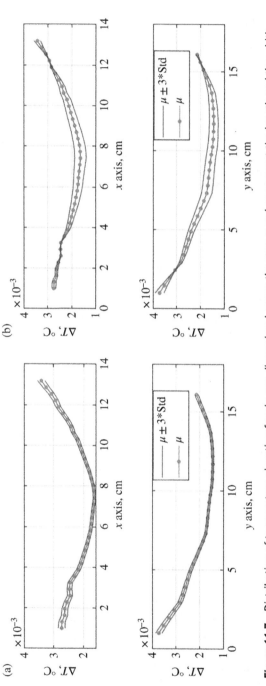

Figure 11.7 Distribution of temperature elevation for six one-dimensional cases; the axes x and y represent the length and the width of the brain, respectively; μ and Std are mean and standard deviation. Panels from (a) to (f) belong to one-dimensional cases when RV is (a) convection coefficient, h_{eff}, (b) thermal conductivity, λ, (c) volumetric perfusion, w_b, (d) heat due to metabolic processes, Q_m, (e) arterial blood temperature, T_{art}, (f) ambient temperature, T_{amb}.

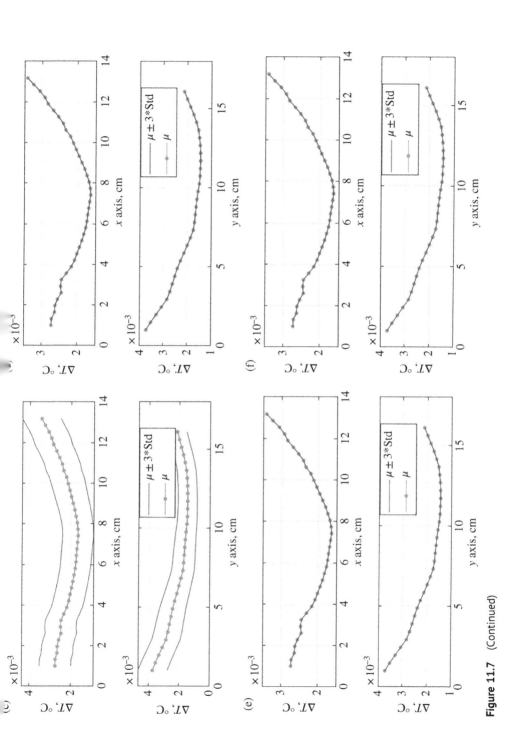

Figure 11.7 (Continued)

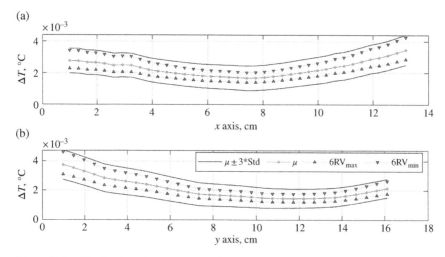

Figure 11.8 Distribution of temperature elevation for six-dimensional case along the axis penetrating the (a) brain width denoted as *x* axis and (b) brain length denoted as *y* axis; μ and Std are mean and standard deviation.

parameter is volumetric perfusion blood rate. Variation of this parameter has uniform effect on temperature elevation along both axes, (Fig. 11.7c). The remaining thermal parameters have negligible influence on temperature elevation which is obvious from Fig. 11.7e–f.

The distribution of the temperature elevation with CI computed as mean ± 3 standard deviations is depicted in Fig. 11.8. for six-dimensional case, i.e. all input parameters are modeled as random.

The results are similar to the one-dimensional case when only volumetric perfusion blood rate is RV. In addition to CIs, Fig. 11.8 depicts the temperature distribution when all input parameters are set to either their maximal or minimal values. Considering the general way of defining the confidence margins as the mean ± 3 standard deviations, extreme values are insufficient when making predictions on the dispersion of results around the mean.

Finally, a detailed sensitivity analysis (SA) in ANOVA fashion is carried out. Not only the first order and total effect sensitivity indices, but also the higher order sensitivity indices are computed and depicted in Figs. 11.9–11.14 both for steady state and case of exposure. In addition to minimal, maximal, and average temperatures, the temperature gradient $T_{min-max}$ is also considered as output parameter of interest.

The steady state first order sensitivity indices and total effect sensitivity indices for temperature gradient and minimal, maximal and average temperatures are depicted in Fig. 11.9. Once again, it is demonstrated that convection coefficient

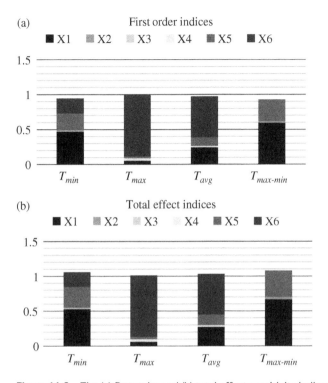

Figure 11.9 The (a) first order and (b) total effect sensitivity indices of thermal parameters for steady state.

and ambient temperature have high influence on the minimal temperature and the temperature gradient. These two input parameters are the key parameters in boundary condition which describes the heat exchange between the brain and the environment. Since temperature of superficial layers is lower than in the brain parenchyma the obtained results are justified. The average temperature and the maximum temperature are greatly affected by the arterial blood temperature. The remaining thermal parameters have negligible influence on observed temperature outputs.

Furthermore, Figs. 11.10–11.12 depict the first, second, third, fourth and fifth order sensitivity indices and total effect sensitivity indices for minimal, maximal and average temperatures, temperature gradient and temperature elevation in case of exposure.

The influence of input parameters on minimal/maximal/average temperatures and temperature gradient is the same for steady state and in case of exposure. Higher order indices are very small indicating that the mutual interactions between the input parameters have negligible impact on output parameters of

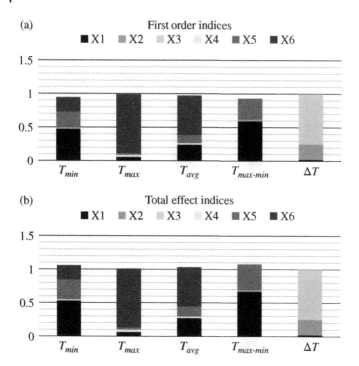

Figure 11.10 The (a) first order and (b) total effect sensitivity indices of thermal parameters for case of exposure.

interest. The only somewhat significant impact is the one from the second order index S_{ae} (Fig. 11.11a) which represents the mutual interaction between the convection coefficient and ambient temperature. The values of the fifth order sensitivity indices are basically zeros as depicted in Fig. 11.12b.

The total effect sensitivity indices of input thermal parameters are depicted in Fig. 11.10b. Their values are very similar to the first order sensitivity indices which explains why the higher order sensitivity indices have small values.

Considering the temperature elevation, the most influential thermal parameter is the volumetric perfusion rate of blood, thus confirming previous findings from [7]. The second influential thermal parameter is the thermal conductivity, followed by the effective convection coefficient which has small impact on temperature elevation. The remaining parameters are of minimal importance in this case.

Additional analysis is carried out regarding the influence of the input RVs on the distribution of temperature elevation along the two brain axes. The total effect sensitivity indices are depicted on Figs. 11.13 and 11.14 for x and y axis, respectively.

The mutual interactions are weak and have no impact on the temperature elevation. The influence of input parameters is quite similar along both axes. The

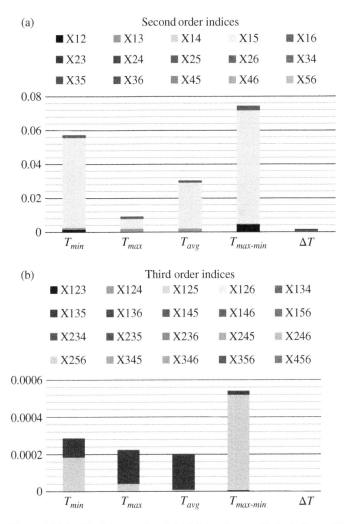

Figure 11.11 The (a) second and (b) third order sensitivity indices of thermal parameters for case of exposure.

effects of the effective heat transfer coefficient are hardly noticeable on the edges of axis which corresponds to the surface of the brain. On the contrary, thermal conductivity affects the temperature rise in the central part of the brain more than in the superficial layers. Finally, the most dominant thermal parameter is volumetric perfusion blood rate which affects the temperature elevation uniformly along both axes. It is interesting to notice how the temperature elevation is not significantly influenced by the thermal parameters whose impact is very high in case of other

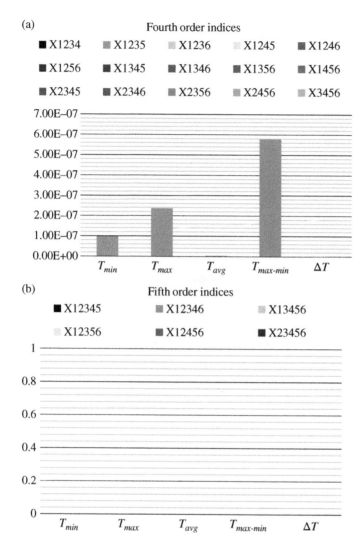

Figure 11.12 The (a) fourth and (b) fifth order sensitivity indices of thermal parameters for case of exposure.

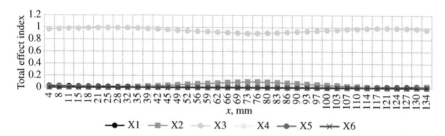

Figure 11.13 Total effect sensitivity indices of thermal parameters along the brain *x* axis (brain length).

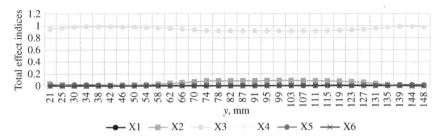

Figure 11.14 Total effect sensitivity indices of thermal parameters along the brain y axis (brain width).

temperature output values such as average, minimum and maximum temperature inside the brain.

11.3 Stochastic Thermal Dosimetry for Three-Compartment Human Head

Next, stochastic collocation method is applied to thermal dosimetry of a three-compartment human head [5]. The human head consisting of scalp, skull and brain and the electromagnetic part of the problem are the same as in Section 10.3. However, in this chapter only the thermal part is stochastic while the EM part is deterministic meaning that tissues' relative electric permittivities and electric conductivities are set to their respective nominal values. The Pennes' bioheat equation and the corresponding boundary condition are in the form of Eq. (11.1). Deterministic Pennes' equation is solved via FEM with Intel Pardiso matrix solver on a computer with Intel i7 processor with 32 GB RAM. One deterministic simulation is executed in approximately 75 min. The details on FEM solution of Pennes' bioheat equation for the described model are available elsewhere, for example [7].

Based on the study presented in Section 13.2, in the stochastic part, only the volumetric perfusion blood rate and tissue thermal conductivity of scalp, skull, and brain are modeled as random variables, thus creating the six-dimensional vector of input parameters $X = [\lambda_1, w_{b1}, \lambda_2, w_{b2}, \lambda_3, w_{b3}]$. Average values are given as follows: (i) scalp parameters $\lambda_1 = 0.498$ W/m °C and $w_{b1} = 2700$ kg/(sm^3), (ii) brain parameters $\lambda_2 = 0.534$ W/m°C and $w_{b2} = 3500$ kg/(sm^3), and (iii) scull parameters $\lambda_3 = 0.3$ W/m °C and $w_{b3} = 1000$ kg/(sm^3). The coefficient of variation (CV) used to construct the uniform distributions is $\pm 20\%$.

The sparse grid stochastic collocation with Lagrange interpolation and Clenshaw–Curtis quadrature points is used for uncertainty propagation. Traditional Monte Carlo approach would require at least 10,000 simulations to guarantee the exact solutions for stochastic mean and variance, while the full

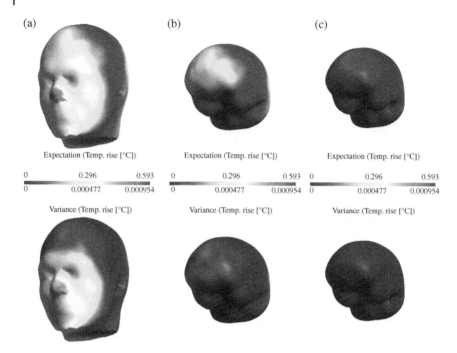

Figure 11.15 Stochastic mean (or expectation, top) and variance (bottom) of the temperature elevation on the surface of (a) scalp, (b) scull, and (c) brain.

tensor SC model requires at least 117.649 simulations. Both alternatives are highly impractical since one deterministic simulation requires approximately 75 min. Therefore, the convergence of the SC method is tested by inspecting the results for the consecutive orders of the Clenshaw–Curtis formula up to order 4, which leads to the maximal number of the resulting simulation points equal to 389.

Stochastic mean and variance of the temperature elevation on the surface of the three tissues are presented in Fig. 11.15 for six-dimensional case.

The mean temperature elevation distribution is non-homogeneous for the scalp with the maximum value in the nose area. The temperature variance is also non-homogeneously distributed, but the highest variance value of approximately 0.6 °C is under the limitation of 1 °C. The distribution of the stochastic moments is homogeneous for the skull and brain with values around 0 °C.

Impact factor of input thermal parameters obtained in OAT manner on temperature elevation variance along the scalp surface is depicted in Fig. 11.16. The results for the skull and brain tissues are omitted here since the values are around 0 °C.

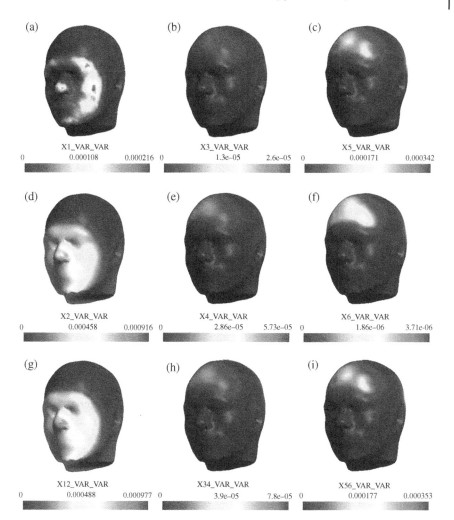

Figure 11.16 Thermal parameters' impact factors on temperature elevation on the scalp surface: (a)–(f) correspond to impact of λ_1, w_{b1}, λ_2, w_{b2}, λ_3, and w_{b3}, respectively, according to expression $V(X_i)/V(X)$ while (g)–(i) correspond to impact of the interaction between the two parameters $[\lambda_1, W_{b1}]$, $[\lambda_2, W_{b2}]$, and $[\lambda_3, W_{b3}]$ respectively, according to expression $V(X_iX_j)/V(X)$.

Scalp thermal conductivity and scalp volumetric perfusion have higher impact than the rest of the input parameters (Fig. 11.16g). The impact of volumetric perfusion blood rate is greater than the impact of thermal conductivity. Additionally, volumetric perfusion affects wider area which is in accordance with the results from [4]. Some smaller areas such as the forehead are weakly influenced by the

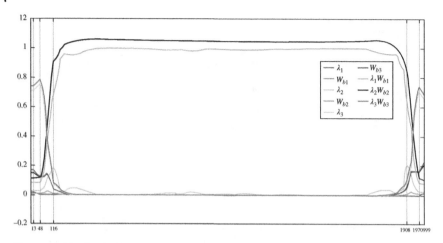

Figure 11.17 The influence of the input parameters on temperature elevation along the head axis. The axis is depicted in Fig. 10.13.

variations in skull thermal parameters. The variation of brain thermal parameters has no effect at all.

Finally, the impact of input parameters on the distribution of the temperature elevation along the head axis for the fixed height is depicted in Fig. 11.17. The axis chosen for the illustration is depicted in Fig. 10.13. Once again, the prevailing influence of the volumetric perfusion blood rate is noticeable. However, the impact of the tissue thermal conductivity is more pronounced on the interfaces between the tissues. For some areas, the influence of the other tissues' thermal parameters is more significant than of the thermal parameters of the given tissue. For example, skull thermal conductivity has a higher effect on the temperature elevation in the scalp area than the scalp thermal conductivity itself. This is in accordance with the results depicted in Fig. 11.16f.

11.4 Stochastic Thermal Dosimetry below 6 GHz for 5G Mobile Communication Systems

The last example of stochastic thermal dosimetry features the exposure of realistic human head to a plane wave radiation at 3.5 GHz [6]. This frequency belongs to the lower spectrum of 5G frequencies. As an example, 5G frequency range in Croatia is 3.4–3.6 GHz.

Figure 11.18 Plane wave incidence on the right side of human head model.

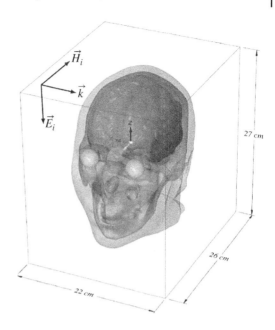

The geometry of interest is a realistic multi-layered human head depicted in Fig. 11.18. The head model is illuminated by an incident plane wave coming from the right side. The model consists of eyes, skull, mandible, cerebrospinal fluid (CSF), skin, gray matter, cerebellum and brain stem. Except for the eyes, tissues' geometry originates from MRI scans of 24-year-old male volunteer [8]. Since spatial resolution of most MRI scans is not sufficiently accurate to capture the sophisticated details of the human eye tissues, the geometric representation of the human eye obtained from available anatomical data is used instead. Thus, the 24 mm rotationally symmetric geometric eye consists of 12 tissues (ora serrata, ciliary body, ligaments, cornea, iris, anterior and posterior chambers, lens, sclera, choroid, retina, and vitreous body). The geometry of the eye lens surface is calculated using Gullstrand's schematic eye model [9]. The conformal mesh consisting of 8 885 916 tetrahedrons is obtained using AnSyS ICEM meshing software.

The deterministic EM part used for computation of the induced electric field and *SAR* is based on coupled Stratton–Chu time-harmonic boundary integral equation and Helmholtz equation that governs the electromagnetic field inside of the computational domain. The set of coupled equations is solved by means of hybrid BEM/FEM approach. Hence, the governing equations of the EM deterministic part are the same as in Sections 12.3 and 13.3. However, due to higher model complexity related to this EM-scattering problem, the conformal mesh contains 19 374 unknowns for BEM discretization and 10 387 137 unknowns for FEM part. The sparse matrix is solved using Intel Pardiso direct solver with total execution time of 22 h.

Instead of temperature elevation, the output of interest is heating factor defined as the ratio of the temperature elevation to SAR at each given point [6]:

$$\alpha = \frac{\Delta T}{SAR} \tag{11.4}$$

The temperature elevation is the difference in temperatures obtained when head is illuminated by incident plane wave and when there is no exposure at all. Of course, the temperature is governed by Pennes' bioheat equation with the corresponding boundary condition as in Eq. (11.1). The equation is solved via FEM [7]. For the computation of temperature elevation and thus the heating factor, the conformal mesh yielded 1 497 936 unknowns and 22 272 210 elements in the sparse matrix.

Based on the experience from previous studies [3–5, 7], which are partly outlined in this chapter, thermal parameters modeled as RVs are thermal conductivity and volumetric blood perfusion rate since their impact on temperature elevation is more significant compared to other thermal parameters. The importance of considering the variability of the volumetric blood perfusion rate is reported by other authors as well, e.g. in [10]. Furthermore, due to the incidence direction of choice (Fig. 11.18), only parameters of brain, skull, and scalp are considered RVs, thus resulting in six-dimensional stochastic problem, i.e. thermal conductivity and volumetric blood perfusion rate of brain, skull, and scalp are modeled as RVs. The corresponding uniform distributions of those parameters are given as follows: (i) brain parameters $w_{b\text{-}brain} = 50\,212.32 \pm 11.18\%$ W/(m^3 °C) and $\lambda_{brain} = 0.545 \pm 66.67\%$, (ii) skull parameters $w_{b\text{-}skull} = 1993.86 \pm 56.25\%$ W/(m^3 °C) and $\lambda_{skull} = 0.325 \pm 2.75\%$ W/(m °C), and (iii) scalp parameters $w_{b\text{-}scalp} = 7443.80 \pm 10.77\%$ W/(m^3 °C) and $\lambda_{scalp} = 0.41 \pm 21.95\%$ W/(m °C). The coefficients of variations are obtained from the databases in [2]. The vector of input parameters is $X = [w_{b\text{-}brain}, w_{b\text{-}skull}, w_{b\text{-}scalp}, \lambda_{brain}, \lambda_{skull}, \lambda_{scalp}]$.

Sparse-grid SC method with Lagrange interpolation and Clenshaw–Curtis quadrature points are used to obtain the stochastic mean and variance. To test the convergence of the SC method, collocation points are increased consecutively in each dimension: 3, 5, and 9, thus resulting in a total number of simulations of 13, 85, and 389. The convergence rate is depicted in Fig. 11.19. Observation points are set along

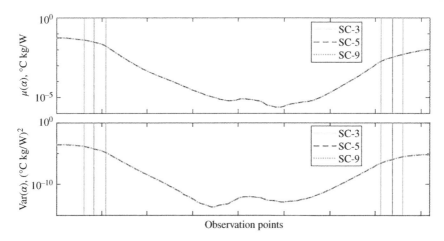

Figure 11.19 Heating factor expectation (top) and variance (bottom) for six-dimensional case computed with levels SC-3, SC-5, and SC-9, which means 13, 85, and 389 total number of simulation points. Observation points are chosen along the axis penetrating through head, skull, CSF, and brain tissues in the incident wave direction.

the axis penetrating the head model through scalp, skull, CSF, and brain tissues in the direction of the incident wave. The convergence for heating factor expectation is accomplished with 13 simulation points, while for variance in some observation points, 85 simulation points are needed.

Next, CIs for heating factor along the axis in the incidence direction are presented in Fig. 11.20. The CIs are defined as mean ± 3 standard deviations.

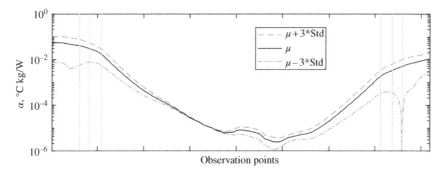

Figure 11.20 Heating factor confidence intervals obtained as $\mu \pm 3*$Std. μ and Std are mean and standard deviation.

The values of heating factor mean and its standard deviation are practically zero in the interior head parts. The width of CI is wider in the exterior parts. The confidence margins on the surface of scalp are $CI = [0.733, 10.101]^* 10^{-3}$ °C/(W/kg) while the largest spread of heating factor values in the brain region is $CI = [1.031, 3.777]^* 10^{-6}$. The temperature elevation is not even close to the basic restriction limitation even when the width of CI is defined using the three standard deviations rule at each observation point.

Distribution of heating factor stochastic mean and standard deviation on the surface of four head tissues is depicted in Fig. 11.21a–d.

Comparing the standard deviations from one-dimensional cases, scalp blood perfusion has the highest impact on six-dimensional standard deviation. The standard deviations from the rest of five one-dimensional cases are very small. Note that six one-dimensional cases are depicted in logarithmic scale while six-dimensional cases are shown in linear scale.

Sensitivity analysis carried out in OAT fashion is depicted in Fig. 11.22 for observation points along the axis of interest. The uncertainty of the scalp's volumetric blood perfusion rate is the major source of uncertainty for the output heating factor in the exterior parts of the brain, while the variance of brain perfusion blood rate has the highest impact in the interior part of head. The distribution of variances pertaining to cases when skull and scalp thermal parameters are RVs have the same character as the distribution of standard deviation from six-dimensional cases but with lower magnitude. When brain thermal parameters are RVs in a one-at-a-time manner, the corresponding variances tend to increase in the direction of wave propagation. It is worth noting that surface views presented in Fig. 11.21 are taken closer to the right head part where the impact of the brain perfusion is not high. For this reason, the impact of the brain perfusion blood rate is not visible in Fig. 11.21 as is the case in Fig. 11.22.

Results for ANOVA sensitivity analysis are depicted in Fig. 11.23. The first order sensitivity indices are close to 1 for the scalp and brain volumetric blood perfusion rate in different regions. Namely, the brain's volumetric blood perfusion rate is more significant in the interior brain region and scalp's volumetric blood perfusion rate prevails in the rest of the domain. Since the total sum of the first order sensitivity indices for all input parameters equals 1 at each observation point, there is no need for computation of total or higher-order effect indices. Interestingly, the CV for scalp and brain perfusion are 11.67% and 10.77%, respectively, which are lower than CV for some other input parameters. Nevertheless, the impact of other input parameters is not significant.

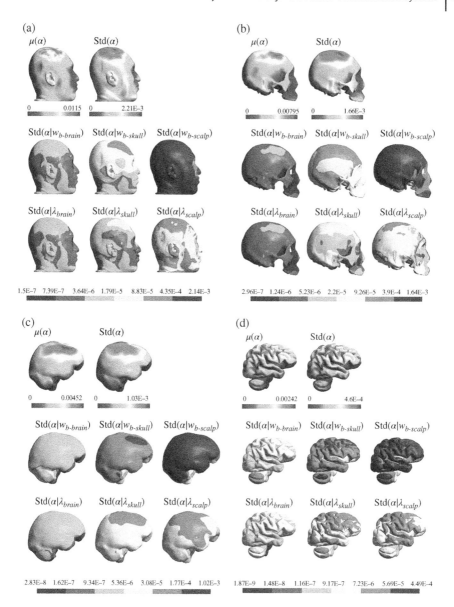

Figure 11.21 Distribution of heating factor mean and standard deviation on the surface of head tissues (a–d). $\mu(\alpha)$ (°Ckg/W) and Std(α) (°Ckg/W) are mean and standard deviation from six-dimensional case; Std($\alpha|X_i$) (°Ckg/W) is a standard deviation from i-th one-dimensional case, $X_i = w_{b\text{-}brain}$, $w_{b\text{-}skull}$, $w_{b\text{-}scalp}$, λ_{brain}, λ_{skull}, λ_{scalp}. (a) Scalp surface. (b) Skull surface. (c) CSF surface. (d) Brain surface.

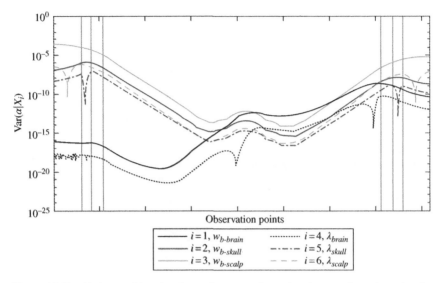

Figure 11.22 Variance of heating factor from one-dimensional cases along the axis of incidence; Var($\alpha|X_i$) ($^\circ$Ckg/W), i = 1,2,...,6.

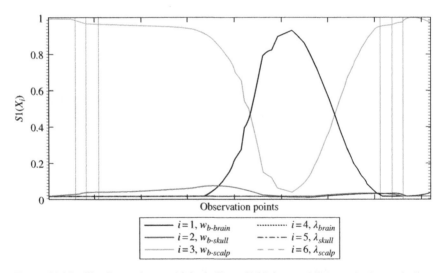

Figure 11.23 The first order sensitivity indices, $S1(X_i)$ from ANOVA sensitivity analysis, i = 1, 2,...,6.

References

1 D. Poljak and M. Cvetković, Human Interaction with Electromagnetic Fields: Computational Models in Dosimetry, St. Louis, Misouri: Elsevier, 2019.

2 ITIS Foundation, "Tissue properties: Dielectric Properties," [Online]. Available: https://itis.swiss/virtual-population/tissue-properties/database/dielectric-properties. [Accessed September 2021].

3 A. Šušnjara, D. Poljak, F. Rezo and J. Matković, "Stochastic Sensitivity Analysis of Bioheat Transfer Equation," in *URSI International Symposium on Electromagnetic Theory (EMTS)*, San Diego, CA, USA, 2019.

4 A. Šušnjara, M. Cvetković, D. Poljak, S. Lallechere and K. El Khamlichi Drissi, "Stochastic sensitivity in thermal dosimetry for the homogeneous human brain model," in *The Joint Annual Meeting of the Bioelectromagnetics Society and the European BioElectromagnetics Association – BioEM 2016*, Ghent, Belgium, 2016.

5 A. Šušnjara, M. Cvetković, H. Dodig and D. Poljak, "Stochastic Thermal Dosimetry for the Three Compartment Head Model," in *International Conference on Software, Telecommunications and Computer Networks*, Split, Croatia, 2018.

6 A. Šušnjara, H. Dodig, D. Poljak and M. Cvetković, "Stochastic-deterministic thermal dosimetry below 6 GHz for 5G Mobile communication systems," *IEEE Transactions on Electromagnetic Compatibility*, vol. 63, no. 5, pp. 1667–1679, 2021.

7 M. Cvetkovic, D. Poljak and A. Hirata, "The electromagnetic-thermal dosimetry for the homogeneous human brain model," *Engineering Analysis with Boundary Elements*, vol. 63, no. 2016, pp. 61–73, 2016.

8 I. Laakso, S. Tanaka, S. Koyama, V. D. Santis and A. Hirata, "Inter-subject variability in electric fields of motor cortical tDCS," *Brain Stimulation*, vol. 8, no. 5, pp. 906–913, 2015.

9 Gullstrand, A., "The optical system of the eye," *Physiological Optics*, vol. 1, pp. 35–358, 1909.

10 P. R. Wainwright, "Computational modelling of temperature rises in the eye in the near field of radiofrequency sources at 380, 900 and 1800 MHz," *Physics in Medicine and Biology*, vol. 52, no. 12, pp. 3335–50, 2007.

12

Stochastic–Deterministic Modeling in Biomedical Applications of Electromagnetic Fields

Biomedical applications of EM fields assume desired biological effects induced to help in either diagnosis or treatment of some disease. Again, computational modeling plays a crucial role in better understanding the underlying phenomena due to limitations of direct measurement procedures of physical quantities in a human body. Moreover, computational modeling is important tool for optimization and design of medical devices and protocols.

Depending on the type of biomedical application, the character of computational modeling may be completely electromagnetic (EM) or coupled EM-thermal. Just as is the case with EM and thermal dosimetry, input parameters such as tissue electric conductivity, relative permittivity, and thermal parameters in Pennes' bioheat equation exhibit uncertain nature [1] and [2]. Additionally, input parameters used to describe the source of EM radiation may also vary to a certain extent [3–5]. Hence, stochastic approach to modeling is necessary in order to provide confidence margins for the output parameters, whether they are induced EM or thermal values in the human body, or some specific features of EM source used for their final design.

The chapter consists of four sections. The first two sections are dedicated to stochastic modeling of transcranial magnetic [6, 7], and electric stimulation of human brain [8–11], respectively. Both techniques are used for stimulation or inhibition of certain brain regions, one using the principle of magnetic induction, the other by conducting the electric current through the head. The third section features the stochastic analysis of the EM induction effect on a neuron's action potential dynamics [12], while the final section presents stochastic analysis of radiation efficiency of implantable antennas [13, 14].

Deterministic and Stochastic Modeling in Computational Electromagnetics: Integral and Differential Equation Approaches, First Edition. Dragan Poljak and Anna Šušnjara.
© 2024 The Institute of Electrical and Electronics Engineers, Inc.
Published 2024 by John Wiley & Sons, Inc.

12.1 Transcranial Magnetic Stimulation

Transcranial magnetic stimulation (TMS) is a noninvasive and painless technique for stimulation or inhibition of certain brain regions used in diagnostic and therapeutic purposes as well as in studying the specific cortex regions. In TMS, a stimulation coil positioned over the head surface generates a time-varying high-intensity magnetic field, which penetrates the head tissues. Following the principle of Faraday's law, the time-varying magnetic field induces an electric field, resulting in the depolarization or hyperpolarization of the cortex neuronal cell membranes. The desired action potential is initiated when the membrane potential threshold value is reached. From the macroscopic point of view, brain activation happens in regions with the highest values of the induced electric field [15–17]. The efficiency of TMS stimulation varies due to differences in relevant stimulation parameters such as pulse waveform, frequency, intensity of TMS, etc. [18] and [19]. Furthermore, finding the optimal stimulation intensity is not an easy task. Unless using the precise robotic navigation system for the TMS coil placement, the coil orientation and positioning affect the misalignment from the targeted brain region. Additionally, since the optimization of the TMS setup is done by using computational modeling, the prescribed tissue EM parameters may introduce some uncertainties in the output as well.

This section presents the example of stochastic–deterministic modeling of TMS applied to human brain [6] and [7]. The output parameters of interest are induced electric field, induced current density, and induced magnetic flux. Human brain model and formulation for the induced electric field are the same as in Section 11.2. Namely, the deterministic EM part is formulated as surface integral equation derived by using the equivalence theorem and applying the boundary conditions on the brain surface [7]. The resulting set of coupled integral equations is solved by means of Method of Moments. The source is the magnetic field originating from TMS coil instead of a plane wave. Figure 12.1 depicts the human brain and the TMS stimulation coil. The three selected locations of points for visualization of the results are also depicted in the figure.

The brain model is obtained from the average adult human brain in [20] and it is linearly scaled to dimensions of a 10-year-old brain, similar to [21] with scaling factors 0.805 in the horizontal plane, and 0.782 in the vertical axis. The TMS figure-of-eight-coil depicted in Fig. 12.1 enables the focusing action, thus facilitating stimulation of a significantly smaller brain area. The coil parameters are as follows: operating frequency is 3 kHz, the radius of wings is 3.5 cm, the number of wire turns is 15, and the amplitude of the coil current is 2.843 kA. The separation between the coil and the brain surface (corresponding to primary motor cortex area) is 1 cm.

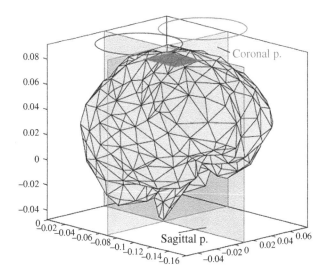

Figure 12.1 TMS figure-of-eight-coil configuration above the human brain model with denoted sagittal and coronal cross sections, respectively (coronal p. and saggital p. in the figure). The location of selected points is 1 cm under the brain surface denoted by horizontally placed square.

The stochastic response is obtained by means of stochastic collocation method with full-tensor grid, Lagrange interpolating polynomials, and Gauss–Legendre quadrature points. Two different stochastic models are constructed. The first model features brain permittivity and conductivity as RVs while coil displacements in both directions are set to their initial values. The initial value and variation range for brain conductivity is $\sigma = 0.1 \pm 0.02$ S/m and for brain relative permittivity is $\varepsilon_r = 46\,315 \pm 9000$. The brain relative permittivity and conductivity are taken from [22] as an average between the white matter and the gray matter. In the second stochastic model, the brain parameters are fixed to initial values while coil displacements are modeled as RVs. The initial value and variation range for coil displacement in x direction is $C_x = 0 \pm 1$ cm while coil displacement in y direction is $C_y = 0 \pm 1$ cm. Therefore, there are two two-dimensional stochastic problems denoted as $f(\varepsilon_r, \sigma)$ and $f(C_x, C_y)$, respectively.

The outputs of interest are as follows: the induced electric field (E) and the induced electric current density (J) in the brain tissue, the maximum of induced electric field (E_{max}), the maximum of induced magnetic flux (B_{max}), and the maximum of induced current density (J_{max}). The convergence of the SC method to obtain the stochastic mean and variance of the outputs of interest is tested by consecutive increase of collocation points in each dimension, 3, 5, and 7, thus leading to total of 9, 25, and 49 simulation points for two-dimensional problems.

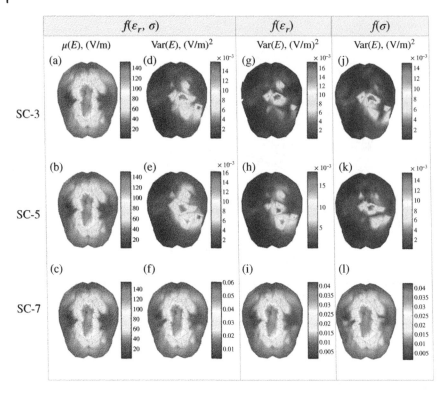

Figure 12.2 Electric field mean and variance on brain surface obtained with 9, 25, and 49 simulation points in two-dimensional case $f(\varepsilon_r, \sigma)$ and variances of electric field for two one-dimensional cases $f(\varepsilon_r)$ and $f(\sigma)$ obtained with 3, 5, and 7 simulation points.

Figure 12.2 depicts the stochastic mean and variance of the induced electric field on the brain surface in a two-dimensional case, $f(\varepsilon_r, \sigma)$, and variance in two one-dimensional cases $f(\varepsilon_r)$ and $f(\sigma)$.

The maximum values of the mean electric field and its variance are located directly under the geometric center of TMS coil. The convergence of the SC method in computation of mean electric field is good even at the lowest number of simulation points. On the contrary, the convergence of E_{max} variance is not that good and it is necessary to increase the number of SC points to show the same accuracy for the variance. Nonetheless, some rough estimates can still be made from the proposed results. Furthermore, the variance of the electric field from a two-dimensional problem is higher than for one-dimensional variances. Moreover, one-dimensional variances have the same values. This indicates that both brain permittivity and bran conductivity have a significant impact on the output field variations.

Another perspective is given in Fig. 12.3. Observation points are located on two cross sections depicted in Fig. 12.1. The region with the highest value of electric

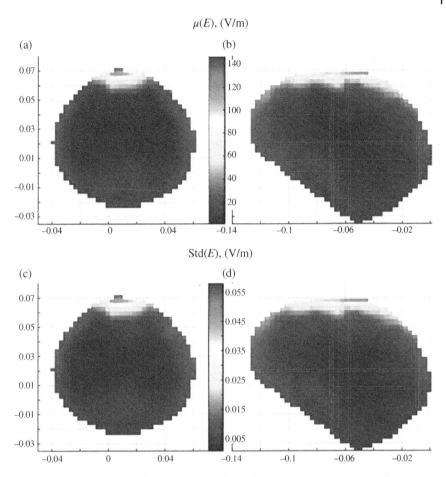

Figure 12.3 Induced electric field mean and standard deviation on the coronal (a and c) and sagittal (b and d) cross sections of the brain.

field mean and standard deviation is located directly under the coil center. As observation points move from the surface, both mean and standard deviation decrease more rapidly.

Furthermore, Fig. 12.4 illustrates the convergence of SC method in computation of mean and standard deviation of the maximum electric field (E_{max}), maximum magnetic flux density (B_{max}), and maximum induced current density (J_{max}). The results are depicted for two-dimensional and two one-dimensional cases. The convergence of the SC method is satisfactory for maximum magnetic flux density and maximum induced current density.

Confidence intervals computed as mean $\pm$ 3 standard deviations are computed for all three characteristic outputs. The maximum electric field confidence interval

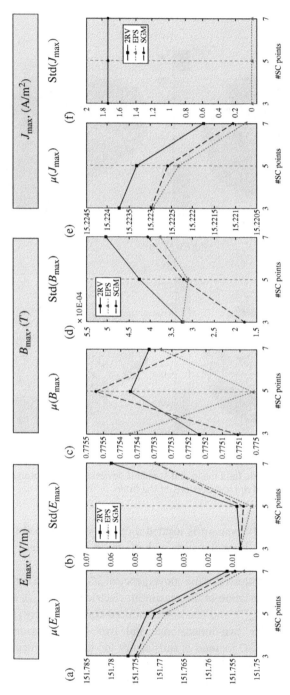

Figure 12.4 SC method convergence in computation of mean and standard deviation of the maximum electric field E_{max}: (a–b), maximum magnetic flux density B_{max}: (c–d), and maximum induced current density J_{max}: (e–f), with respect to number of SC points for two one-dimensional and one two-dimensional cases, denoted with EPS, SGM, and 2RV, respectively.

is $CI(E_{max}) = [151.5750, 151.9350]$ V/m. The standard deviation is 0.04% of the expected maximum value. Comparing the standard deviations of E_{max} for all three cases, i.e. $f(\varepsilon_r, \sigma)$, $f(\varepsilon_r)$, and $f(\sigma)$, it is obvious that permittivity and conductivity have the same impact on the maximal electric field.

The standard deviation of the maximum induced magnetic flux density is 0.06% of its expected value, thus the confidence margins are $CI(B_{max}) = [0.7738, 0.7768]$ T. The impact of the input permittivity and conductivity is practically equal.

As expected, permittivity has no significant impact on the maximum induced current density as all influence originates from conductivity variations. Confidence interval is $CI(J_{max}) = [9.9718, 20.4718]$ A/m^2 while the standard deviation is 11.5% of its expected value. Therefore, the maximum induced current density has the largest confidence margins.

Moving on to another stochastic problem denoted with $f(C_x, C_y)$, the effect of coil displacement in x and y directions is investigated. Ideally, the coil is centered 1 cm above the brain surface corresponding to the primary motor cortex area. However, it is possible for the coil positioning to dislocate. Therefore, the coil is uniformly displaced in both directions. The output-induced electric field is computed at equidistant points inside magenta square from Fig. 12.1. The square dimension is 1.5 cm $\times$ 1.5 cm and its perpendicular to the z-axis is 1 cm under the brain surface. Electric field mean and standard deviation are shown in Fig. 12.5.

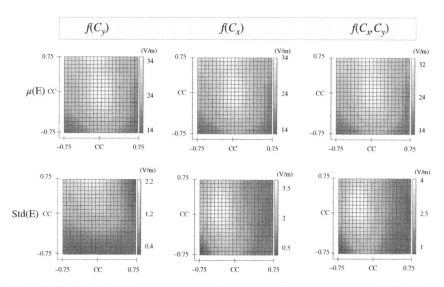

Figure 12.5 Electric field mean and standard deviation along magenta square in Fig. 12.1. Results for one-dimensional cases, $f(C_x)$ and $f(C_y)$ and two-dimensional case $f(C_x, C_y)$ are depicted.

The uncertainty in coil displacement in the x direction has more impact on the variance of the induced electric field than the displacement in y direction. Furthermore, in case of $f(C_x, C_y)$, the points with the highest variance are moved along both axes, while in $f(C_y)$ case, the points with the highest deviation are moved along y-axis. In the case $f(C_x)$, the points are moved along the x-axis. The reason might be the nonsymmetric nature of the utilized brain surface mesh. As the homogeneous model was obtained by smoothing the otherwise very complicated brain surface, at some points on the surface, greater density of triangular elements was achieved. These can contribute to higher values of the electric field, especially if very thin elements are obtained. These results indicate the importance of a careful preparation of model mesh. Furthermore, the $f(C_x, C_y)$ case showed the variability of the electric field value to be around 10% of the value at focal point.

12.2 Transcranial Electric Stimulation

Transcranial electric stimulation (TES) is another type of a noninvasive brain stimulation technique. Compared to TMS, the difference is in modulation of spontaneous neuronal activity instead of induced activity in the resting neuronal. Hence, the effects depend upon the previous physiological state of the target neural structures [23]. Another difference is related to the fact that TES electrodes are positioned on the head surface, while TMS coils do not have any contact with the scalp.

The low-intensity TES is based on a subthreshold modulation of neuronal membrane potentials, which alters the cortical excitability and the activity-dependent on the current flow direction through the target neurons [23] and [24]. It comprises several different techniques such as transcranial direct current stimulation (tDCS), transcranial alternating current stimulation (tACS), transcranial random noise stimulation (tRNS), and transcranial pulsed current stimulation (tPCS) [25]. TES techniques are applied according to pre-established protocols and the procedures are reported to be well tolerated [25]. Research indicates the important role of low-intensity TES in the treatment of various neurological and psychiatric disorders such as depression, anxiety, and Parkinson's disease [26, 27].

Even though the clinical experience and empirical research have led to the improvement of brain stimulation techniques, the underlying phenomenon, i.e. the character of the induced current in the human head, is still not entirely understood [24]. Computational modeling of TES can alleviate this problem [28, 29]. As an example, this tool is used for the design of electrode positions as different positions stimulate different parts of the cortex [30]. Additionally, due to differences in morphology of individual heads, the overall treatment effectiveness may be improved through the patient-specific modeling approach [31, 32].

Different choices of head models such as canonical (e.g. sphere, ellipsoid) and anatomically realistic high-resolution models, as well as different solution tools (analytical or numerical like FEM or MoM) lead to different approaches in computational modeling of TES. More details on such approaches can be found in [27].

A novel approach compared to previous ones features TES simulation based on the Boundary Element Method (BEM). The problem has been approached in three steps, each with different level of approximation for the human head model. Starting with simple cylindrical structure representing human head [8], the second step features the three-compartment human head consisting of scalp, skull, and brain tissues [9] and [10], and finally, a nine-subdomain human head is used in the third step [11].

In all presented examples, the formulation of the TES is based on a quasi-static approximation of the currents and voltages in living tissues [26]:

$$\nabla(-\sigma\nabla\varphi) = 0 \tag{12.1}$$

where σ is a tissue electric conductivity and φ is the unknown electric scalar potential. Boundary conditions are given as Dirichle's boundary condition, $\varphi = \pm 1$ V at the electrodes' area, while for the rest of the domain, the Neumann's boundary condition is prescribed, $\sigma(\partial\varphi/\partial n) = 0$. The deterministic solution of electric scalar potential is obtained via BEM. Once the potential is known, the resulting electric field and current are straightforward to obtain.

Furthermore, the three examples are treated as stochastic problems, i.e. input parameters are modeled as random variables (RVs). The output of interest is the electric scalar potential whose stochastic moments are computed by using stochastic collocation method.

12.2.1 Cylinder Representation of Human Head

The common models in analytical bioelectromagnetism are sphere, ellipsoid, and cylinder [33]. In the first example head is modeled as a homogeneous cylinder of length L and a base diameter $D = 2a$ as in Fig. 12.6. The conductivity σ is set to a constant value and the electrodes are circles of diameter d with the potential of ± 1 V.

The cylinder representing human head is dimensionless with $L = 3$, $D = 1$, and $d = 0.5$. The cylinder length and the electrode radius are varied 20% around their respective mean values, thus forming a vector of input parameters $X = [X_1, X_2]$, where $X_1 = L$ and $X_2 = d$. Stochastic collocation with Lagrange interpolation and Gauss–Legendre quadrature is used to obtain stochastic response. The number of collocation points in each dimension is 3, 5, 7, and 9, thus resulting in total of 9, 25, 49, and 81 deterministic simulations. The SC convergence is depicted in Fig. 12.7, and it is satisfactory both for potential mean and variance.

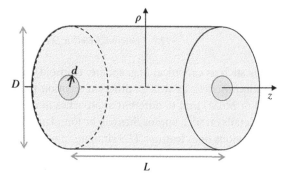

Figure 12.6 The cylinder model of human head.

(a)

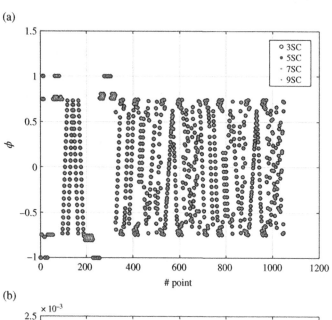

(b)

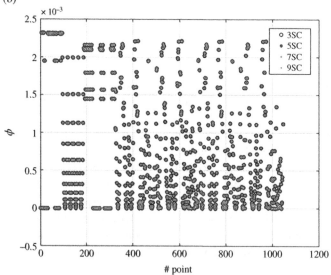

Figure 12.7 Potential mean (a) and variance (b) at every computation point.

(a)   (b)

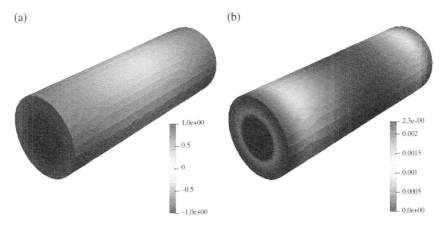

Figure 12.8 The distribution of the potential stochastic mean (a) and variance (b) on the surface of the cylinder.

The potential mean and the variance on the surface of the cylinder are depicted in Fig. 12.8. The only parts that exhibit variance greater than 0 are at circular edges of the cylinder bases.

Sensitivity analysis in ANOVA manner is depicted for z and ρ axes in Fig. 12.9. The impact of the electrode's radius variability is lower than that of the cylinder's length variability, except at the point $z = 0$. The mutual impact of the two RVs is negligible.

12.2.2 A Three-Compartment Human Head Model

A three-compartment head model commonly used in electroencephalography and magnetoencephalography consists of skin, skull, and brain compartments as depicted in Fig. 12.10. The circularly shaped electrodes of diameter d and potential of ± 1 V are placed according to Cz-Fpz electrode setup defined by 10/20 electro-encephalogram standard [34].

The conductivities of the three tissues are chosen after the examination of a large number of studies [35]. The values are spread around their respective mean. Hence, they are modeled as RVs uniformly distributed between the minimal and maximal reported values. Since only one study measured the skull conductivity, 20% uniform distribution around the published average value is considered instead. The uniform distributions for conductivities of the three tissues are as follows: skin conductivity is $\sigma \sim (0.09, 0.25)$ S/m; skull (cortical bone) conductivity is $\sigma \sim (0.256, 0.384)$ S/m; and brain (grey and white matter) conductivity is $\sigma \sim (0.0644, 1.28)$ S/m.

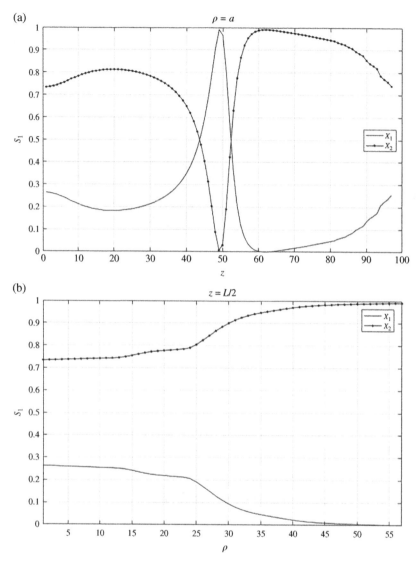

Figure 12.9 The first-order sensitivity indices along the axis z at point $\rho = a$ (a) and along the axis ρ at point $z = L/2$ (b); $X_1 = L$ and $X_2 = d$.

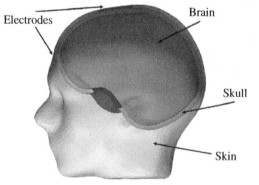

Electrodes

Brain

Skull

Skin

Figure 12.10 A three-compartment head model.

(a)

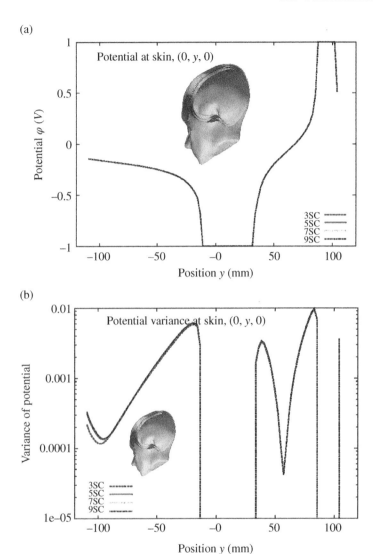

Figure 12.11 Distribution of scalar electric potential mean (a) and variance (b) along *y*-axis for fixed $(x,z) = (0,0)$ at skin surface.

Sparse-grid stochastic collocation with Lagrange interpolation and Chebyshev type of quadrature points is used for stochastic analysis. 3, 5, 7, and 9 collocation points in each dimension, i.e. total of 27, 125, 343, and 729 deterministic simulations are used. The convergence is depicted in Fig. 12.11 for the axis of choice. The convergence is satisfactory with three points in case of the stochastic mean and five points for variance.

(a)   (b)

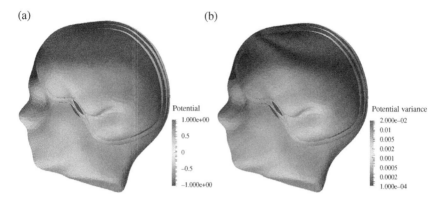

Figure 12.12 Distribution of the scalar electric potential mean (a) and variance (b) in the head tissues.

The distribution of the stochastic mean and the variance at the surface of the head tissues is further depicted in Fig. 12.12.

The ANOVA sensitivity analysis at all computation points is depicted in Fig. 12.13. Brain conductivity has the highest impact at most of the points, although skin conductivity is the most significant parameter at some other points of interest. Also, in minor part of the computation domain, the impact of the mutual interaction of the skin and brain conductivities has a very high impact on total variance of the resulting electric scalar potential.

Finally, in order to illustrate the impact of the tissue parameters through the computational domain, the observation points are chosen along the y-axis for fixed $(x, z) = (0\,\text{mm}, 90\,\text{mm})$. Fig. 12.14 depicts the confidence intervals in six-dimensional case, total sensitivity indices, and sensitivity indices of orders 1 and 2. Confidence margins are not homogeneous along the axis. The brain conductivity has the highest impact at most of the observation points, while in the smaller region in the interior scalp, conductivity is more significant. The values of second-order sensitivity index are too small. However, at the beginning of the computational domain, the mutual interaction between the scalp and brain conductivity has an impact of 5%.

12.2.3 A Nine-Compartment Human Head Model

A follow-up for three-compartment head model is a nine-compartment model consisting of scalp, cerebrospinal fluid (CSF), skull, brain gray matter, white matter part of cerebrum, cerebellum, ventricles, jaw, and tongue as depicted in Fig. 12.15. This is a model of a female from a Visible Human Project (VHP) of the US National Library of Medicine [36, 37]. The presented version is based on version given by

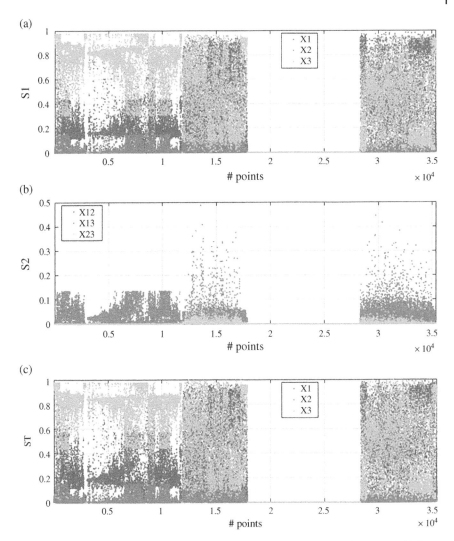

Figure 12.13 The first-order (a), the second-order (b) and the total effect (c) sensitivity indices at all computation points. X1, X2, and X3 represent the impact of tissue conductivities, while X12, X13, and X23 represent the impact of the mutual interaction. 1, 2, and 3 stand for the head, skull, and scalp.

Noetscher et al. in [27] and Elloian et al. in [38]. A detailed description can also be found in work by Makarov et al. in [39]. Electrodes are positioned on the top of the head and forehead, thus forming a Cz-Fpz setup defined by 10/20 electroenceph-alogram standard for electrode placements [34]; Cz is the central midline electrode

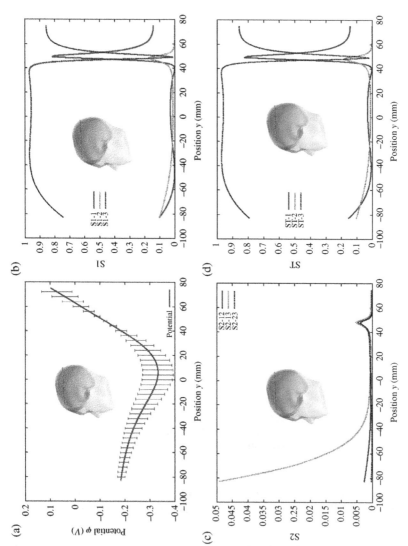

Figure 12.14 Stochastic analysis of electric scalar potential along y-axis for fixed $(x,z) = (0\,\text{mm}, 90\,\text{mm})$; (a) Confidence intervals computed as mean ± 1 standard deviation; Sensitivity indices: (b) S1, the first-order; (c) S2, the second-order and (d) ST, total effect. 1, 2, and 3 stand for the head, skull, and scalp.

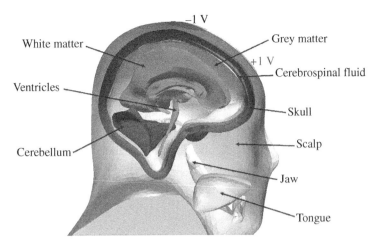

Figure 12.15 The nine-compartment head model with Cz-Fpz electrode position.

and Fpz is the frontopolar midline electrode. The electrodes are of circular shape with the applied potential of $\varphi = \pm 1$ V.

Tissues' conductivities are modeled as RVs uniformly distributed in the range defined by their maximal and minimal values as reported in [35] (Table 12.1).

Table 12.1 Conductivity values for nine subdomains; $\sigma_{avg} = (\sigma_{max} + \sigma_{min})/2$ and *CV* (%) = $100^*(\sigma_{max} - \sigma_{avg})/\sigma_{avg}$.

Subdomain		Conductivity			RV Index	
Tissue name	Index	$[\sigma_{min}, \sigma_{max}]$ (S/m)	σ_{avg} (S/m)	CV (%)	$i = 1...d;$ $d = 7$ or 8	
Cerebellum (WM)	I	[0.22, 1.31]	0.765	71.24	1	1
Ventricles	II	[1.59, 1.8]	1.695	6.19	2	2
Cerebrospinal fluid	III	[1.59, 1.8]	1.695	6.19	2	2
Gray matter (GM)	IV	[0.109, 0.481]	0.295	63.05	3	3
Jaw	V	[0.00185, 0.00588]	0.003865	52.13	4	(−)
Scalp (skin)	VI	[0.09, 0.25]	0.17	47.59	5	4
Tongue	VII	[0.02, 0.67]	0.345	94.20	6	5
Cerebrum (WM)	VIII	[0.0644, 1.20]	0.632	89.87	7	6
Skull	IX	[0.256, 0.384]	0.32	20	8	7

The Roman and Arabic numerals denote the anatomical subdomain and RV index, respectively. OAT analysis eliminates jaw's conductivity from stochastic dimensionality which reduces total RV index to $d = 7$.

Since only one study measured the skull conductivity, 20% uniform distribution around the published value is considered instead. The total number of input parameters is lower than nine because ventricles are filled with CSF; thus, the conductivity of both subdomains is treated as one parameter.

Firstly, eight-dimensional stochastic problem is analyzed in OAT manner using the stochastic collocation with Lagrange interpolation and Gauss–Legendre (GL) quadrature. The indexing of the "stochastic" subdomains in this case corresponds to $i = 1, 2, ...8$ as denoted in Table 12.1. Convergence is tested by changing the number of collocation points $n = 3, 5, 7$, and 9. To investigate the convergence in the whole computational domain, the error function E_r is computed as follows:

$$E_r = \left\|M_j^{i+1} - M_j^i\right\| / \left\|M_j^i\right\|, \left\|M_j^i\right\| = \sqrt{\sum_{j=1}^{J}\left|M_j^i\right|^2}, i = 1, 2, ...N - 1 \qquad (12.2)$$

where M corresponds to either $\mu(\varphi)$ or $Std(\varphi)$ at jth point, N stands for total number of different n sizes. Since here n is set to 3, 5, 7, and 9, the value of N is 4. $\|M\|$ is the Euclidean norm of M.

The resulting E_r is depicted in Fig. 12.16. It can be observed that the error for potential expectation is negligible for all sizes of n, while the error for standard deviation decreases to value of 0.0381.

Examining the variances from eight one-dimensional cases, it is found that the variance of jaw conductivity is very small. Therefore, d-dimensional stochastic problem is reduced from $d = 8$ to $d = 7$. Figure 12.17 depicts one-dimensional standard deviations (square roots of 1-dim variances) along the chosen x and y axes.

Furthermore, sparse-grid stochastic collocation with 3, 5, and 9 points originating from Clenshaw–Curtis 1-dim quadrature is used to obtain the stochastic response from the seven-dimensional stochastic problem. Thus, the total number of simulation points are 15, 113, and 589. Convergence is accomplished with 113 simulations as depicted in Figs. 12.18 and 12.19. Only two axes in head model are shown, but the conclusions are valid for all points of observation.

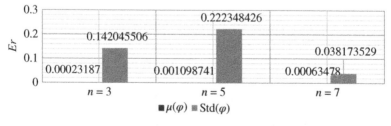

Figure 12.16 The error function values for different n sizes for potential expectation and standard deviation.

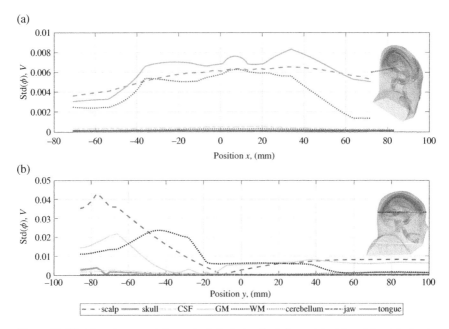

Figure 12.17 Electric potential standard deviation for left-to-right (a) and front-to-back (b) profiles for eight one-dimensional cases.

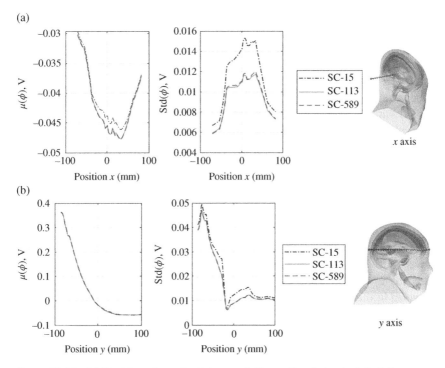

Figure 12.18 (a) Electric scalar potential mean (left panel) and standard deviation (right panel) along the x-axis for different n sizes. (b) Electric scalar potential mean (left panel) and standard deviation (right panel) along the y-axis for different n sizes.

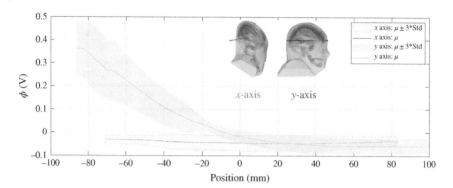

Figure 12.19 Confidence intervals along *x* and *y* axes in the human head computed as $CI = \mu(\varphi) \pm 3\text{Std}(\varphi)$.

Confidence intervals are computed as $CI = \mu(\varphi) \pm 3 \cdot \text{Std}(\varphi)$ and depicted in Fig. 12.19. The CI width is evenly distributed along *x*-axis which is oriented from the left to the right side of the head. On the other hand, since CI is the widest around the electrode area, the CI width along *y*-axis oriented front-to-back tends to decrease.

Next, the distributions of the electric scalar potential expectation and standard deviation on the surface of seven subdomains are depicted in Fig. 12.20. The uncertainty in the input parameter setup does not cause the points of maximal or minimum potential to change the position to some greater extent, i.e. the character of the potential distribution remains the same which is in accordance with results in [40]. Expectedly, the highest standard deviation is in the close area around the electrodes. Moreover, moving the observation points toward interior domains, the largest deviation is still observed under the electrode area. The maximal standard deviation is of order ~50 mV in scalp, skull, CSF, and brain, and of order ~10 mV in cerebellum, ventricles, jaw, and tongue.

Maximal potential expectation and maximal potential standard deviation are depicted in Fig. 12.21 and compared to the average value of input conductivities and their coefficient of variation. The conductivities of CSF and ventricles are represented by the same RV, but maximal standard deviation is much lower in the ventricles area due to their physical position with respect to the electrode placement. Potential expectation decreases toward the inner domains. Although standard deviation generally follows this trend, the largest standard deviation is observed in the skull. Interestingly, although tongue conductivity has the largest coefficient of variation, the standard deviation is small when compared to standard deviations in the first four domains.

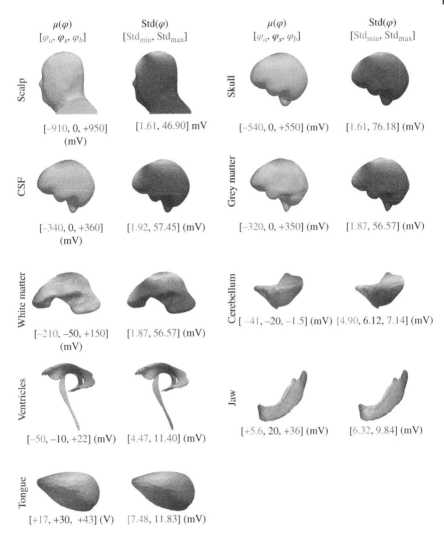

Figure 12.20 Side view of the electric scalar potential mean and standard deviation distribution on the surface of head subdomains. φ_a and φ_b stand for maximal negative and positive potential values, φ_s is minimal absolute potential value.

Finally, ANOVA sensitivity analysis is depicted in Fig. 12.22 for observation points along the chosen axes. Only the first-order sensitivity index (S1) is shown. Conductivities of scalp, gray matter, and cerebrum (white matter) have more significant impact on the output variance of electric scalar potential compared to conductivities of other tissues. The conclusion is the same for the observation points that are not depicted in Fig. 12.22. This is in accordance with results reported in [41]. Note that the impact of the skull conductivity is reported as

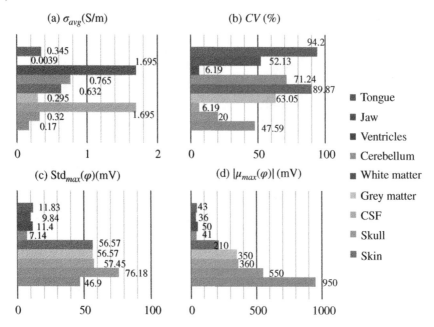

Figure 12.21 Average conductivity values (a), coefficient of variation (b), potential maximal standard deviation (c), and potential maximal mean (d) for each subdomain. Results in (c) and (d) are for seven-dimensional stochastic model.

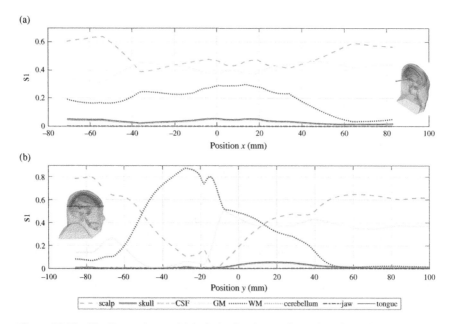

Figure 12.22 The first-order sensitivity index for observation points along the axes x and y, left-to-right and front-to-back axis, respectively.

"significant" in [42] and [40]. In this example, the skull conductivity is taken from [35] with the coefficient of variation $CV = 20\%$, while [42] and [40] have different nominal values for the skull's conductivity with CV up to 90.8% and 60%, respectively.

12.3 Neuron's Action Potential Dynamics

The exposure to time-varying magnetic fields may lead to neuron electrostimulation, i.e. electrical polarization of presynaptic processes, thus leading to a change in post-synaptic cell activity. The main interaction between the time-varying magnetic field and neuron action potential dynamics can be described by using the principle of Faraday induction of electrical fields and associated currents which depend on tissue morphology and electric conductivity. Therefore, action potential dynamics can be investigated by simulating electrophysiology of a single cortex neuron. Neuron numerical models range from simple ones such as integrate-and-fire [43, 44], over more biophysically realistic models such as Hodgkin–Huxley neuron model [45], and its respective simplifications [46, 47], to completely theoretical models [48, 49].

Hodgkin–Huxley model used as an equivalent electrical circuit of a single neuron cell is depicted in Fig. 12.23.

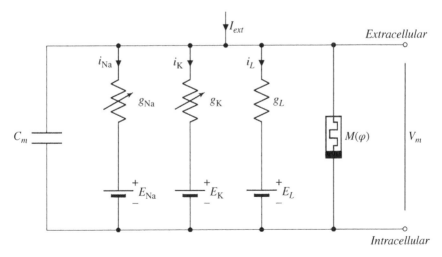

Figure 12.23 Hodgkin–Huxley neuron model equivalent electrical circuit [45] with the magnetic flux-controlled memristor, used to bridge the gap between the magnetic flux and the membrane potential and to enable the trans-membrane feedback current flow.

This bio-physically meaningful model describes the dynamics of membrane potential under the influence of some external stimulus, the dynamics of ionic currents, synaptic integration, etc. The model is measurable and suitable for bifurcation analysis [50]. It is given in the form of three coupled equations [51]. The first equation describes the change of the membrane potential, V_m, in time:

$$\frac{dV_m}{dt} = \frac{1}{C_m}\left(\hat{g}Kn^4(V_m - E_K) + \hat{g}N_am^4h(V_m - EN_a) + \hat{g}_L(V_m - E_L) + I_{ext} - k\rho(\varphi)V_m\right)$$

(12.3)

where C_m denotes the capacitance of the lipid bilayer, and $\hat{g}_K$, $\hat{g}_{Na}$, and $\hat{g}_L$ are potassium, sodium, and leakage ion channel conductivity maximum values, respectively. E_K, E_{Na}, and E_L are associated reversal potentials for potassium, sodium, and leakage ion channel, respectively. I_{ext} represents the external neuro-stimulus by means of the electric current source.

The second equation governs the dynamics of gating variables, denoted as y:

$$\frac{dy}{dT} = \alpha_y(1 - y) - \beta_y y, y = (n, m, h)$$

(12.4)

where α_y and β_y represent the closing rate of an ion channel gate's opening, respectively. Each gating variable has its activation and inactivation steady state represented by the Boltzmann equation as a function of the membrane potential, V_m, and the environmental temperature, T. The potassium, sodium and leakage ion channels are denoted by n, m and h.

The presence of a time-varying magnetic field across an electrical conductor will induce a spatially varying nonconservative electric field. This electric field causes a change in the membrane currents resulting in depolarization or hyperpolarization of neurons [52]. The feedback current of magnetic flux is realized by parallel placement of the memristor $M(\varphi)$ in the equivalent electric circuit as in Fig. 12.23. Memristor, i.e. memory resistor is a nonlinear resistor with memory introduced in [53] as fourth fundamental electric circuit element, which compensates for the missing link between the charge and the magnetic flux.

Finally, the EM induction in a neuron cell is represented through an additive variable in a membrane potential equation, thus extending the original model with an additional equation describing the change in the magnetic flux, φ.

$$\frac{d\varphi}{dt} = k_1 V_m - k_2 \varphi$$

(12.5)

where $k_1 V_m$ and $k_2 \varphi$ describe the membrane potential-induced flux changes and flux leakage, respectively.

The following example presents a so-called tonic spiking electrical mode from [50] where the action potential, represented as a time series of discrete, nearly identical membrane potential spikes, encodes information, and carries it throughout a neuron's cell. This kind of tonic spiking behavior represents the optimal behavior of a neuron exposed to the EM induction effect, where optimality refers to the ability to transfer maximum information, expressed through the entropy measure.

The effect of the EM induction is investigated via bifurcation analysis where the bifurcation parameter is the induction coefficient, k. Analysis consists of repeated simulations of a fixed configuration Hodgkin–Huxley neuron model at a constant temperature of $T = 10\,°C$ for the induction coefficient, k, ranging from 0 to 2. The rest of the input parameters and the initial condition setup of Hodgkin–Huxley neuron model are given in Table 12.2. The distribution of inter-spike intervals (ISIs) is close to an exponential distribution [54] where the total entropy is around 2. The external noisy direct current is set to $I_{ext}(t) = 10\,\mu A/cm^2$ for a complete simulation duration of $\Delta t_{sim} = 1500\,ms$. Furthermore, parameters k_1 and k_2 are set to 0.001 and 0.01, respectively.

The output of interest is the ISI but instead of displaying all ISIs, the expected value of ISI denoted with *mISI* is evaluated. Deterministic solution of Eqs. (12.3)–(12.5) is carried out by using the MATLAB implementation of nonstiff differential equation solver based on the fourth-order Runge–Kutta integration method [55]. The action potential dynamics along with the histogram of ISIs, captured post-simulation, are shown in Fig. 12.24.

Table 12.2 Hodgkin–Huxley neuron model setup.

Name or description	Label	Value
Parameters		
Lipid bilayer capacitance	C_m	$1\,\mu F/cm^2$
Potassium channel reversal potential	E_K	$-77\,mV$
Sodium channel reversal potential	E_{Na}	$50\,mV$
Leakage channel reversal potential	E_L	$-54.387\,mV$
Initial conditions		
Membrane potential values at $t = 0$	$V_{m,0}$	$-65\,mV$
Magnetic flux value at $t = 0$	φ_0	$0.1\,mV$
Potassium channel gating variable value at $t = 0$	m_0	$\dfrac{\alpha_m(V_{m,0}T)}{\alpha_m(V_{m,0}T) + \beta_m(V_{m,0}T)}$
Sodium channel gating variable value at $t = 0$	n_0	$\dfrac{\alpha_n(V_{m,0}T)}{\alpha_n(V_{m,0}T) + \beta_n(V_{m,0}T)}$
	h_0	$\dfrac{\alpha_h(V_{m,0}T)}{\alpha_h(V_{m,0}T) + \beta_h(V_{m,0}T)}$

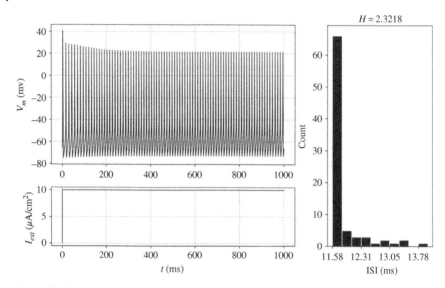

Figure 12.24 Action potential dynamics, $V_m(t)$, (top left), for the case of the direct current external stimulus, $I_{ext}(t)$, (bottom left). On the right, the histogram of ISIs, visually resembling an exponential distribution is depicted. Entropy amounts to $H = 2.3218$ for the number of bins determined via the Freedman–Diaconis rule [56].

To adapt experimental procedures, most often medical treatments, to different individuals in various environmental conditions, it is necessary to investigate how the conductivity of each ion channel impacts the action potential dynamics. Therefore, in the stochastic part of this study, ion channel conductivities are modeled as RVs, thus forming a three-dimensional stochastic problem. According to [57], their respective nominal values are set as follows: $\hat{g}_K = 36.0$ mS/cm^2, $\hat{g}_{Na} = 120.0$ mS/cm^2, and $\hat{g}_L = 0.3$ mS/cm^2. The parameters are uniformly distributed with the coefficient of variation $CV = 5\%$, 10%, 20%, and 50%, respectively.

Stochastic collocation with Lagrange interpolation and Gauss–Legendre quadrature with 3, 5, 7, and 9 points in each dimension are used for stochastic analysis, thus forming a set of 27, 125, 343, and 729 simulations for three-dimensional case. The convergence of the stochastic collocation method in computation of stochastic mean and variance of the mean ISI is accomplished with 125 simulations, while 343 simulations are needed for computation of skewness and kurtosis, as depicted in Figs. 12.25–12.28. It is worth noting that the increase of CV deteriorates the SC convergence, especially for the lowest level of collocation points.

The confidence interval for mean ISI over range of $k \in [0, 2]$ in simulation duration of $\Delta t_{sim} = 300$ ms using 125 deterministic simulations for $CV = 5\%$, 10%, 20%, and 50% are shown in Fig. 12.29.

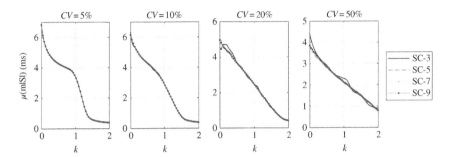

Figure 12.25 The convergence of the mean value for ISI for four levels of accuracy: 3, 5, 7, and 9 collocation points, respectively. Each column depicts simulations for a different coefficient of variation, *CV* = 5%, 10%, 20%, and 50%.

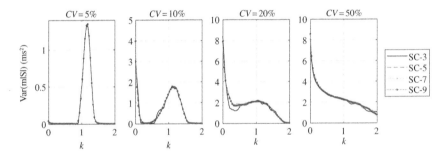

Figure 12.26 The convergence of the mean value for ISI for four levels of accuracy: 3, 5, 7, and 9 collocation points, respectively. Each column depicts simulations for a different coefficient of variation, *CV* = 5%, 10%, 20%, and 50%.

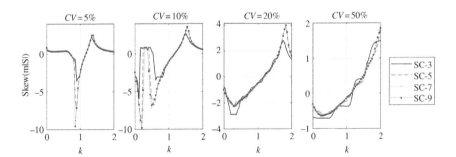

Figure 12.27 The convergence of the skewness value for ISI for four levels of accuracy: 3, 5, 7, and 9 collocation points, respectively. Each column depicts simulations for a different coefficient of variation, *CV* = 5%, 10%, 20%, and 50%.

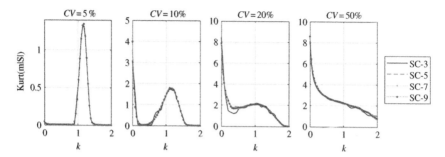

Figure 12.28 The convergence of the kurtosis value for ISI for four levels of accuracy: 3, 5, 7, and 9 collocation points, respectively. Each column depicts simulations for a different coefficient of variation, CV = 5%, 10%, 20%, and 50%.

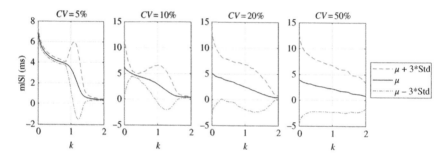

Figure 12.29 Confidence interval for mean ISI computed as $\mu \pm 3^*$Std. Each column depicts simulations for a different coefficient of variation, CV = 5%, 10%, 20%, and 50%.

The behavior of a neuron exposed to the effect of EM induction is the same, no matter the value of CV. An increase in the induction coefficient leads to a decrease in neuronal activity. The greatest influence of the uncertainty of ion channel conductivity is detectable for induction coefficient in range from [1, 1.5] after spiking state transits to a quiescent state.

Furthermore, in Fig. 12.30, uncertainty quantification of spike frequency, i.e. firing rate per 1 s, is depicted. Here the bifurcation parameter range is set to [0, 5] and simulation duration $\Delta t_{sim} = 1000$ ms. Results indicate that the change in electrical activity of a neuron occurs for the same values of k as in the previous simulation, shown in Fig. 12.29, with no change in uncertainty for this critical region.

Finally, ANOVA sensitivity analysis is presented in Figs. 12.31 and 12.32. Total effect, first- and higher-order sensitivity indices are depicted.

Interestingly, for CV = 5% and CV = 10%, the mean ISI hard drop in the critical range of k values for the tonic spiking electrical mode is affected by the variability in the interaction of the sodium and potassium ion channels, and the leakage

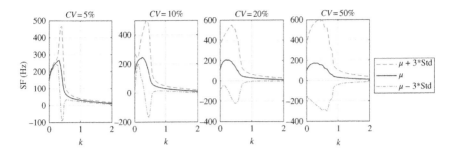

Figure 12.30 Confidence interval for firing rate (SF) computed as $\mu \pm 3^*$Std. Each column depicts simulations for a different coefficient of variation, CV = 5%, 10%, 20%, and 50%.

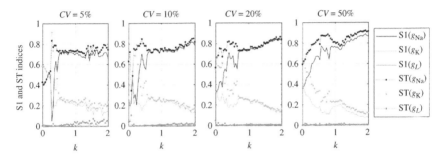

Figure 12.31 The first-order and total-effect sensitivity indices for mean ISI distribution over k, where external neuronal stimulus is constant DC current. Each column depicts simulations for a different coefficient of variation, CV = 5%, 10%, 20%, and 50%.

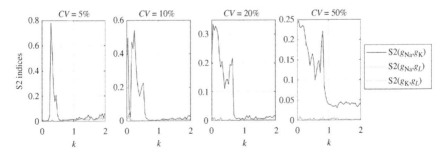

Figure 12.32 The second-order sensitivity indices for mean ISI distribution over k, where external neuronal stimulus is constant DC current. Each column depicts simulations for a different coefficient of variation, CV = 5%, 10%, 20%, and 50%.

channel variability exerts practically no impact. For $CV = 20\%$, the output, that is mean ISI over k in range [0, 2], is most sensitive w.r.t. the variability in sodium ion channel from the starting point of the critical induction coefficient range, $k \sim 0.75$.

For all CVs, the variability of the sodium channel beginning at $k \sim 1$ onward has the largest impact on the mean ISI, for the tonic bursting electrical mode. For $k < 1$, the first-order and the total-effect sensitivity indices revolve around similar values and the significance of the impact is difficult to discern. Both sodium and potassium ion channels, as well as the respective second-order combinations, achieve quite similar behavior and influence on the output.

12.4 Radiation Efficiency of Implantable Antennas

Applications of implantable antennas can be found in medicine, professional sports, occupational health, defence, etc. [58, 59]. The system consists of a sensor node or a medical device implanted in the body and the external on- or off-body access point. The antenna inside the implanted capsule is often designed as monopole, loop, helical, patch, or PIFA. The communication between the implanted capsule and access point is carried out via radiofrequency (RF) signal, usually in either Medical Device Radio Communication Service (MedRadio) or the Industrial, Scientific, and Medical (ISM) bands.

The design of the implanted capsule system faces two main challenges: (i) the capsule size is limited, and (ii) it is a rather tedious task to ensure robust link between the capsule and the external equipment. The reasons for the latter are manifold. Namely, body tissues are lossy media with uncertain EM properties. Additionally, to comply with the limits on maximal Specific Absorption Rate (SAR) and EM field levels, the battery lifetime and health security issues limit the transmit power of implanted capsule systems. Although rigorous numerical approaches are crucial for the design of implantable capsule systems, analytically solved canonical models still provide useful information in terms of optimal frequency choice for a specific application and fundamental limitations for available power density and link budget [14].

In this study the output of interest is the achievable radiation efficiency of implantable antennas, i.e. the upper limit of radiation efficiency for an implanted antenna of certain type [13, 14]. The loss mechanisms were analyzed in detail in [60] showing three different types of loss contributions:

- the losses due to the coupling of the near field (NF) and the lossy biological host tissue which are located close to the implant,
- the losses due to the field propagating through the body,
- the reflection at the body – free space (FS) interface.

The true values of tissues' permittivities and conductivities are known only within a certain uncertainty range. Therefore, the influence of their variation on the quality of in-body biotelemetry communication links is investigated. This will be a key parameter defining the link budget margins necessary for a specific application. Hence, the stochastic collocation method is used as a wrapper around the deterministic code in order to examine which of the loss mechanisms mostly depend on uncertainty of body parameters and highlight the potential impact of this uncertainty on the link budget.

The body is modeled as a multilayered sphere where each spherical layer represents a different body tissue. Four different models are depicted in Fig. 12.33, each with different combination of number of spherical layers and implant placements. It is worth noting that the total radiated power depends on the actual shape of the body. However, this is not the case with the power density inside and just outside the body in the vicinity of the implanted antenna [60, 61]. Therefore, the spherical model is a rather good approximation.

Each of four different implanted systems consists of electric dipole inserted in air bubble with radius $r_{impl} = 1$ mm. Geometry parameters for all four models are given in Table 12.3. Nominal values for the relative electric permittivity and conductivity of each tissue are taken from [1]. More detailed analysis, including the use of the magnetic dipole can be found in [60].

Deterministic DM fields inside and outside the body are obtained from spherical mode expansion while the final power density reaching the FS is expressed as [14]:

$$W_{\substack{reaching\ the \\ free\ space}} = W_{\substack{entering \\ the\ body}} \cdot \frac{r_{impl}^2}{\Delta^2} \cdot e_{\substack{losses\ in\ the \\ reactive \\ near\ field}} \cdot e_{\substack{propagating \\ field \\ absorption \\ losses}} \cdot e_{\substack{losses\ due\ to \\ reflection}} \tag{12.6}$$

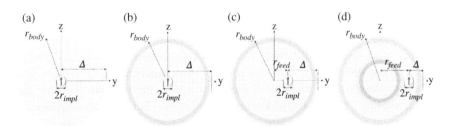

Figure 12.33 Four multilayered sphere models representing the body with an inserted implant. Different combinations of implant placements and number of layers are depicted (a–d).

Table 12.3 The tissue type and outer radius (r_t) for multilayered human body models. The distance between the implant and the sphere center (r_{feed}) is given for each model (Fig. 12.33).

One-layer model			Three-layer model			Seven-layer model	
r_t (cm)	Tissue		r_t (cm)	Tissue		r_t (cm)	Tissue
1	9.0	Muscle	1	9.0	Skin	1	10.0	Skin (dry)
	(a) $r_{feed} = 0$ cm		2	8.8	Fat	2	9.78	Fat
			3	8.3	Muscle	3	9.30	Muscle
			(b) $r_{feed} = 0$ cm			4	9.06	Cortical bone
			(c) $r_{feed} = 77$ mm			5	8.56	Dura matter
						6	8.51	CSF
						7	8.31	Brain
						(d) $r_{feed} = 77$ mm	

where $\Delta = r_{body} - r_{feed}$ (Fig. 12.33) and e stands for the efficiency due to different sources of losses.

For the sake of completeness, Fig. 12.34 depicts the impact of different sources of losses on the radiated power for the case of one-layer head model. The muscle relative permittivity and conductivity are fixed to nominal values.

The real part of the radial component of the power density evaluated at different points on the y-axis is depicted in the left panel of Fig. 12.34 for two operating frequencies, the MedRadio and ISM bands. Although there is a large difference in used frequencies, the final power density reaching FS is comparable [60].

Normalized power density just outside the spherical phantom and corresponding efficiencies for an implanted antenna placed in the center of a homogeneous spherical phantom are depicted in the right panel of Fig. 12.34. The contribution of each efficiency factor is treated separately, and the total losses are obtained by multiplying all three contributions. The largest power density at the surface of the body is obtained at around 1.2 GHz. This maximum is quite flat, i.e. there is no strong variation of power density for frequencies 800 MHz–2 GHz. The frequency variation of different efficiencies reveals that the efficiency due to reactive near-field losses is dominant for lower frequencies, as well as the reflection losses (note that for lower frequencies, the reactive NF and the corresponding value of the characteristic spherical characteristic impedances Z is present at the body surface, thus enlarging the reflection coefficient). On the other side, the propagation losses are much higher for higher frequencies since the body is larger in terms of wavelength.

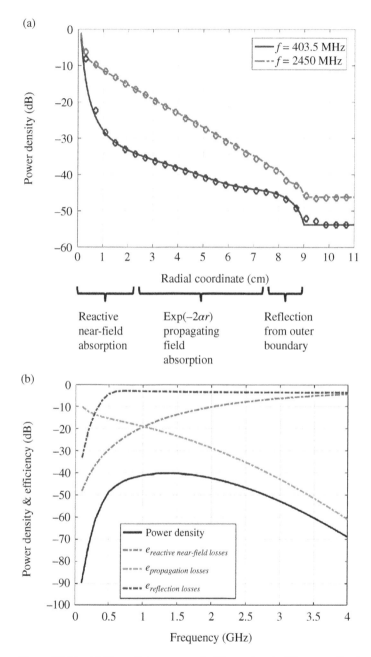

Figure 12.34 (a) The radial component of the real part of the normalized power density at different distances from the implanted antenna. Continuous lines are obtained using the developed spherical wave expansion method and diamond dots are obtained using CST Suite. (b) Normalized power density just outside the spherical phantom and corresponding efficiencies. Both plots are obtained for an implanted antenna placed in the center of a homogeneous spherical phantom.

In the stochastic part, relative electric permittivity and conductivity of tissues are modeled as RVs uniformly distributed around their respective nominal values with coefficient of variation of 20%. Stochastic collocation method is used for uncertainty quantification and the convergence is rather satisfactory [14]. The Lagrange interpolation and sparse-grid SC with CC integration rule are used. The goal is to estimate the uncertainty of the achievable power density outside the human body, i.e. of the link budget between the sensor and the access point of the in-body sensor network. This way it is possible to obtain fundamental insight that can be transposed into more realistic body phantoms and is thus very valuable to antenna designers.

The following figures, Figs. 12.35–12.38, depict the confidence intervals for the power density reaching the FS and the first-order sensitivity indices for four models of interest depicted in Fig. 12.33a–d. Confidence intervals are computed as mean ± one standard deviation and sensitivity indices are obtained via ANOVA sensitivity analysis.

Figure 12.35 exhibits confidence intervals and sensitivity indices for one-layer body model with antenna implanted in the sphere center and muscle tissue electric properties.

The right panel in Fig. 12.35 exhibits confidence interval with respect to frequency. CI width increases with the frequency. At frequencies up to 0.5 GHz the CI is very narrow, almost negligible. The left panel in Fig. 12.35 exhibits SA results. Propagating field (PF) absorption efficiency is approx. four times more sensitive to conductivity variation than on permittivity variation. On the other hand, other two efficiencies are more sensitive to permittivity variation. The sensitivity indices of power density just outside the body indicate the higher impact of conductivity, which demonstrates once again that the propagation losses are most dominant in the considered scenario in which the implanted antenna is not close to the surface of the body. The impact of the electric permittivity is between 50% and 20% in the considered frequency range.

In the second step, the antenna is kept in the center while fat and skin layers are added, thus forming a three-layered body model. Confidence intervals and sensitivity of the obtainable power density on the uncertainty of permittivity and conductivity of different layers of the phantom are depicted in Fig. 12.36.

The predominant layer (and therefore the largest influence of the uncertainty of permittivity and conductivity of that layer) is the layer where the implanted antenna is positioned, i.e. the muscle layer. The difference in the indices' values is drastic: The values are smaller by more than one order of magnitude for the skin tissue layer and by more than two orders of magnitude for the fat tissue layer. This is an additional proof that the propagation losses dominate in the scenarios where the source is far from the body surface.

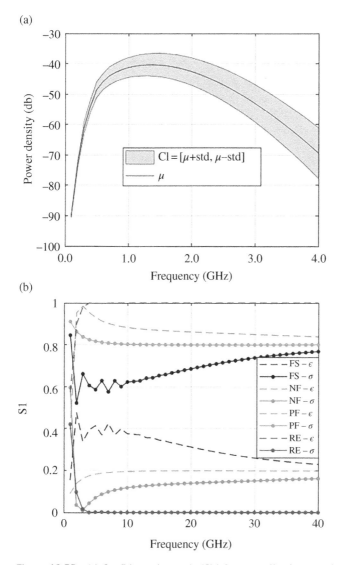

Figure 12.35 (a) Confidence intervals (CIs) for normalized power density ($W_{reaching\ the\ free\ space}$) just outside the one-layer spherical phantom with implanted antenna placed in the center. μ and std are mean and standard deviation. (b) The first-order sensitivity indices of muscle permittivity and conductivity w.r.t. four outputs of interest: power density reaching the free space (FS) and the efficiency factors (e) for losses in near field (NF), propagating field (PF) and reflection losses (RF).

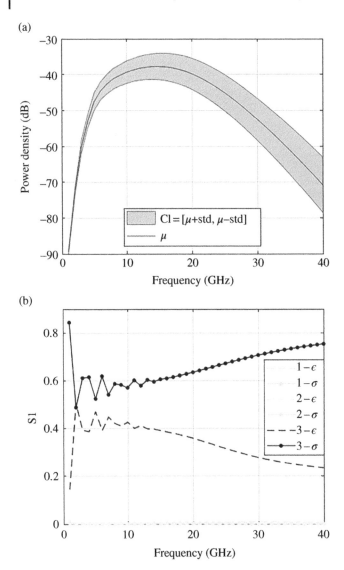

Figure 12.36 (a) Confidence intervals (CIs) for normalized power density ($W_{reaching\ the\ free\ space}$) just outside the three-layer spherical phantom with implanted antenna placed in the center. μ and std are mean and standard deviation. (b) The first-order sensitivity indices of input permittivities and conductivities w.r.t. power density reaching the free space (FS). Numbers 1, 2, and 3 denote skin, fat, and muscle tissues, respectively.

(a)

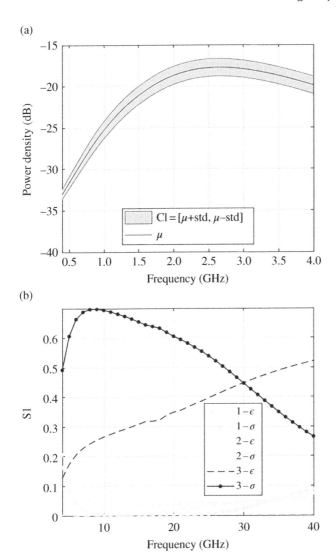

(b)

Figure 12.37 (a) Confidence intervals (CIs) for normalized power density ($W_{reaching\ the\ free\ space}$) just outside the three-layer spherical phantom with implanted antenna placed at distance r_{feed} from the center. μ and std are mean and standard deviation. (b) The first-order sensitivity indices of input permittivities and conductivities w.r.t. power density reaching the free space (FS). Numbers 1, 2, and 3 denote skin, fat, and muscle tissues, respectively.

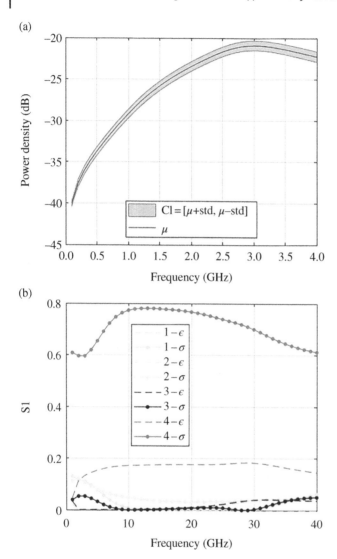

Figure 12.38 (a) Confidence intervals (CIs) for normalized power density ($W_{reaching\ the\ free\ space}$) just outside the seven-layer spherical phantom with implanted antenna placed at distance r_{feed} from the center. μ and std are mean and standard deviation. (b) The first-order sensitivity indices of input permittivities and conductivities w.r.t. power density reaching the free space (FS). Numbers 1, 2, 3, and 4 denote skin, fat, muscle, and bone tissues, respectively.

Confidence interval and sensitivity analysis for the case where the implanted antenna is close to the human phantom surface (the position of the implanted antenna is $r_{feed} = 77$ mm, i.e. at 6 mm distance from the fat tissue layer) is depicted in Fig. 12.37.

Again, the dominant influence is the uncertainty of permittivity and conductivity for the layer where the implanted antenna is positioned, i.e. the muscle layer. However, difference between sensitivity indices is not so large anymore. Furthermore, the frequency dependency of sensitivity indices is not "regular" anymore since the impedance matching of propagation through the multilayer tissue varies with frequency.

The last example is related to a design of a cranial implant; see [62] for details. The used multilayer spherical phantom consists of seven layers (dimensions of the phantom are given in Table 12.3). The cranial implant is located in the fourth layer, i.e. in the bone tissue which has a complex permittivity that is substantially different from the one of muscle tissue. Here an electric dipole excitation is considered. The size of the implanted antenna is $r_{impl} = 1$ mm. Confidence interval and sensitivity analysis are given in Fig. 12.38.

Again, since the source is close to the body surface, the variance of the power density is small due to small propagation losses. The dominant influence of uncertainty of permittivity and conductivity stems from the layer where the implanted antenna is positioned, i.e. the bone layer (although the permittivity and conductivity of the bone tissue are much smaller compared to the muscle tissue).

Comparing the results of the confidence intervals, the four cases can be grouped with respect to the capsule placement. Namely, the expected power density reaching the FS is lower when capsule is closer to the center. On the other hand, the variance, and therefore the CIs, are smaller when capsule is moved away from the center since the propagation losses are drastically reduced (as the waves travel a shorter distance through the tissues). The CI width tends to increase with frequency which is more pronounced for cases when capsule is in the center. As capsule is moved from the center, the CI width tends to become more homogeneous, i.e. it tends to be independent of frequency.

To sum up, the analysis is done using a spherical multilayer body model, and it is shown that the largest uncertainty influence comes from variations of parameters of the tissue in which the implanted antenna is located. Furthermore, for antennas positioned deeply in the human body, the PF absorption losses are the predominant effect with much larger dependency on the conductivity uncertainty. For implanted antennas close to the body surface, the variation of the predicted available power density is smaller. In this case, it was shown there is no dominant uncertainty effect and neither the three causes of EM field losses nor the tissue permittivity and conductivity cause a significant variation.

References

1 C. Gabriel, "Compilation of the dielectric properties of body tissues at RF and microwave frequencies," Report N.AL/OE-TR-1996-0037, Occupational and environmental health directorate, Radiofrequency Radiation Division, Brooks Air Force Base, Texas (USA), 1996.

2 C. Gabriel and A. Peyman, "Dielectric measurement: error analysis and assessment of uncertainty," *Physics in Medicine & Biology*, vol. 51, no. 23, p. 6033, 2006.

3 E. Chiaramello, M. M. Parazzini, S. Fiocchi, P. Ravazzani and J. Wiart, "Stochastic dosimetry based on low rank tensor approximations for the assessment of children exposure to WLAN source," *IEEE Journal of Electromagnetics, RF and Microwaves in Medicine and Biology*, vol. 2, no. 2, pp. 131–137, 2018.

4 O. Jawad, D. Lautru, J. M. Dricot, F. Horlin, A. Bernabi-Delai and P. De Doncker, "Estimation of specific absorption rate with kriging method," in *USNC-URSI Radio Science Meeting*, Lake Buena Vista, FL, USA, 2013.

5 I. Ahmed, H. Khammari, A. Shahid, A. Musa, K. Soon Kim, E. De Poorter and I. Moerman, "A survey on hybrid beamforming techniques in 5G: architecture and system model perspectives," *IEEE Communications Surveys & Tutorials*, vol. 20, no. 4, pp. 3060–3097, 2018.

6 M. Cvetković, A. Šušnjara, D. Poljak, S. Lallechere and K. El Khamlichi Drissi, "Stochastic collocation method applied to transcranial magnetic stimulation analysis," in *The Joint Annual Meeting of The Bioelectromagnetics Society and the European BioElectromagnetics Association - BioEM 2016*, Ghent, Belgium, 2016.

7 M. Cvetković, A. Šušnjara and D. Poljak, "Deterministic-Stochastic modeling of transcranial magnetic stimulation featuring the use of method of moments and stochastic collocation," *Engineering Analysis with Boundary Elements*, vol. 150, pp. 662–671, 2022.

8 A. Šušnjara, J. Ravnik, O. Verhnjak, D. Poljak and M. Cvetković, "Stochastic-deterministic boundary integral method for transcranial electric stimulation: a cylindrical head representation," in *27th International Conference on Software, Telecommunications and Computer Networks, SoftCOM 2019*, Split, Croatia, pp. 1–7, 2019.

9 J. Ravnik, A. Šušnjara, O. Verhnjak, D. Poljak and M. Cvetković, "Coupled boundary element: stochastic collocation approach for the uncertainty estimation of simulations of transcranial electric stimulation," in *WIT Transactions on Engineering Sciences, Boundary Elements and other Mesh Reduction Methods XLIV*, Online, 2021. https://www.witpress.com/elibrary/wit-transactions-on-engineering-sciences/131/38013.

10 A. Šušnjara, O. Verhnjak, D. Poljak, M. Cvetković and J. Ravnik, "Stochastic-deterministic boundary element modelling of transcranial electric stimulation using a three layer head model," *Engineering Analysis with Boundary Elements*, vol. 123, no. 2021, pp. 70–83, 2021.

11 A. Šušnjara, O. Verhnjak, D. Poljak, M. Cvetković and J. Ravnik, "Uncertainty quantification and sensitivity analysis of transcranial electric stimulation for 9-subdomain human head model," *Engineering Analysis with Boundary Elements*, vol. 135, no. 2022, pp. 1–11, 2022.

12 A. Lojić Kapetanović, A. Šušnjara and D. Poljak, "Stochastic analysis of the electromagnetic induction effect on a neuron's action potential dynamics," *Nonlinear dynamics*, vol. 105, no. 2021, p. 3585–3602, 2021.

13 Z. Šipuš, A. Šušnjara, A. Skrivervik, D. Poljak and M. Bosiljevac, "Influence of uncertainty of body permittivity on achievable radiation efficiency of implantable antennas – stochastic analysis," *IEEE Transactions on Antennas and Propagation*, vol. 69, no. 2021, p. 10, 2021.

14 Z. Šipuš, A. Šušnjara, A. Skrivervik, D. Poljak and M. Bosiljevac, "Uncertainty estimation of achievable radiation efficiency of implantable antennas," in *Proceedings of the 15th European Conference on Antennas and Propagation (EuCAP)*, Dusseldorf, Germany, 2021.

15 A. Pascual-Leone, N. J. Davey, J. Rothwell, E. M. Wassermann and B. K. Puri, *Handbook of Transcranial Magnetic Stimulation*, First Edition, A Hodder Arnold Publication, London, UK, 2002.

16 E. Wassermann, C. Epstein and U. Ziemann, *Oxford Handbookof Transcranial Stimulation*, First Edition, New York, NY, USA: Oxford University Press, 2008.

17 M. Hallett, "Transcranial magnetic stimulation and the human brain," *Nature*, vol. 406, no. 6792, pp. 147–150, 2000.

18 M. T. Rubens and T. P. Zanto, "Parameterization of transcranial magnetic stimulation," *Journal of Neurophysiology*, vol. 107, no. 5, p. 1257–1259, 2012.

19 Z. Turi, M. Lenz, W. Paulus, M. Mittner and A. Vlachos, "Selecting stimulation intensity in repetitive transcranial magnetic stimulation studies, a systematic review between 1991 and 2020," *European Journal of Neuroscience*, vol. 53, no. 10, p. 3404–3415, 2021.

20 M. Cvetković and D. Poljak, "An efficient integral equation based dosimetry model of the human brain," in *Proceedings of 2014 675 International Symposium on Electromagnetic Compatibility (EMC EUROPE) 2014*, Gothenburg, Sweden, 2014.

21 I. Laakso, T. Uusitupa and S. Ilvonen, "Comparison of SAR calculation algorithms for the finite-difference time-domain method," *Physics in Medicine & Biology*, vol. 55, no. 15, pp. N421–N431, 2010.

22 T. Rajapakse and A. Kirton, "Non-invasive brain stimulation in children: applications and future directions," *Translational Neuroscience*, vol. 4, no. 2, pp. 1–29, 2013.

23 A.J. Woods, A. Antal, M. Bikson, P.S. Boggio, A.R. Brunoni, P. Celnik, L.G. Cohen, F. Fregni, C.S. Herrmann, E.S. Kappenman and H. Knotkova, "A technical guide to tDCS, and related non-invasive brain stimulation tools," *Clinical Neurophysiology*, vol. 127, no. 2, pp. 1031–1048, 2016.

24 M. D. Johnson, H. H. Lim, T. I. Netoff, A. T. Connolly, N. Johnson, A. Roy, A. Holt, K. O. Lim, J. R. Carey, J. L. Vitek and B. He, "Neuromodulation for brain disorders: challenges and opportunities," *IEEE Transactions on Biomedical Engineering*, vol. 60, no. 3, pp. 610–624, 2013.

25 E. R. C. Kadosh, *The Stimulated Brain*, Elsevier, 2014.

26 S. Bai, C. Loo and S. Dokos, "A review of computational models of transcranial electrical stimulation," *Critical Reviews in Biomedical Engineering*, vol. 41, no. 1, pp. 21–35, 2013.

27 G. M. Noetscher, J. Yanamadala, S. N. Makarov and A. Pascual-Leone, "Comparison of cephalic and extracephalic montages for transcranial direct current stimulation – a numerical study," *IEEE Transactions on Biomedical Engineering*, vol. 61, no. 9, pp. 2488–2498, 2014.

28 O. I. Kwon, S. Z. Sajib, I. Sersa, T. I. Oh, W. Jeong, H. J. Kim and E. J. Woo, "Current density imaging during transcranial direct current stimulation using DT-MRI and MREIT: algorithm development and numerical simulations," *IEEE Transactions on Biomedical Engineering*, vol. 63, no. 1, pp. 168–175, 2016.

29 C. Lee, E. Kim and C. K. Im, "Techniques for efficient computation of electric fields generated by transcranial direct-current stimulation," *IEEE Transactions on Magnetics*, vol. 54, no. 5, pp. 1–5, 2018.

30 J. C. Horvath, O. Carter and J. D. Forte, "Transcranial direct current stimulation: five important issues we aren't discussing (but probably should be)," *Frontiers in Systems Neuroscience*, vol. 8, no. 2, pp. 1–8, 2014.

31 M. Parazzini, S. Fiocchi, A. Cancelli, C. Cottone, I. Liorni, P. Ravazzani and F. Tecchio, "A Computational model of the electric field distribution due to regional personalized or nonpersonalized electrodes to select transcranial electric stimulation target," *IEEE Transactions on Biomedical Engineering*, vol. 64, no. 1, pp. 184–195, 2017.

32 I. Laakso, S. Tanaka, S. Koyama, V. De Santis and A. Hirata, "Inter-subject variability in electric fields of motor cortical tDCS," *Brain Stimulation*, vol. 8, no. 5, pp. 906–913, 2015.

33 D. Poljak and M. Cvetković, *Human Interaction with Electromagnetic Fields: Computational Models in Dosimetry*, St. Louis, MI: Elsevier, 2019.

34 Electrode Position Nomenclature Committee, "Guideline thirteen: Guidelines for standard electrode position nomenclature," *Journal of Clinical Neurophysiology*, vol. 11, no. 1, p. 111–113, 1994.

35 ITIS Foundation, "Tissue properties: Dielectric Properties," [Online]. Available: https://itis.swiss/virtual-population/tissue-properties/database/dielectric-properties/ [Accessed 2 September 2021].

36 M. J. Ackerman, "The visible human project," *Proceedings of the IEEE*, vol. 86, no. 3, pp. 504–511, March 1998.

37 U.S. National Library of Medicine. "The Visible Human Project," [Online]. Available: https://www.nlm.nih.gov/research/visible/visible_human.html. [Accessed 2 September 2019].

38 J. M. Elloian, G. M. Noetscher, S. N. Makarov and A. Pascual-Leone, "Continuous wave simulations on the propagation of electromagnetic fields through the human head," *IEEE Transactions on Biomedical Engineering*, vol. 61, no. 6, pp. 1676–1683, 2014.

39 S. N. Makarov, G. M. Noetscher and A. Nazarian, *Low-Frequency Electromagnetic Modeling for Electrical and Biological Systems Using MATLAB*, Hoboken, NJ: John Wiley & Sons, Inc, 2016.

40 G. B. Saturnino, A. Thielscher, K. H. Madsen, T. R. Knösche and K. Weise, "A principled approach to conductivity uncertainty analysis in electric field calculations," *NeuroImage*, vol. 188, pp. 821–834, 2019.

41 A. Šušnjara, O. Verhnjak, D. Poljak, M. Cvetković and J. Ravnik, "Stochastic-deterministic boundary element modelling of transcranial electric stimulation using a three layer head model," *Engineering Analysis with Boundary Element*, vol. 123, no. 2020, pp. 70–83, 2021.

42 C. Schmidt, S. Wagner, M. Burger, U. Rienen and H. W. Carsten, "Impact of uncertain head tissue conductivity in the optimization of transcranial direct current stimulation for an auditory target," *Journal of Neural Engineering*, vol. 12, no. 4, pp. 1–12, 2015.

43 W. Gerstner, W. M. Kistler, R. Naud and L. Paninski, "Generalized integrate-and-fire neurons," in *Neuronal Dynamics: From Single Neurons to Networks and Models of Cognition (pp. 115-118)*, Cambridge: Cambridge University Press, 2014, pp. 115–118.

44 G. B. Ermentrout and N. Kopell, "Parabolic bursting in an excitable system coupled with a slow oscillation," *SIAM Journal on Applied Mathematics*, vol. 46, no. 2, pp. 233–253, 1986.

45 A. L. Hodgkin and A. F. Huxley, "A quantitative description of membrane current and its application to conduction and excitation in nerve," *Bulletin of Mathematical Biology*, vol. 52, no. 1–2, pp. 25–71, 1990.

46 R. FitzHugh, "Impulses and physiological states in theoretical models of nerve membrane," *Biophysical Journal*, vol. 1, no. 6, pp. 445–466, 1961.

47 C. Morris and H. Lecar, "Voltage oscillations in the barnacle giant muscle fiber," *Biophysical Journal*, vol. 35, no. 1, pp. 193–213, 1981.

48 E. M. Izhikevich, "Simple model of spiking neurons," *IEEE Transactions on Neural Networks*, vol. 14, no. 6, pp. 1569–1572, 2003.

49 H. R. Wilson, "Simplified dynamics of human and mammalian neocortical neurons," *Journal of Theoretical Biology*, vol. 200, no. 4, pp. 375–388, 1999.

50 E. M. Izhikevich, "Which model to use for cortical spiking neurons?," *IEEE Transactions on Neural Networks*, vol. 15, no. 5, pp. 1063–1070, 2004.

51 L. Lu, J. B. Kirunda, Y. Xu, W. Kang, R. Ye, X. Zhan and Y. Jia, "Effects of temperature and electromagnetic induction on action potential of Hodgkin–Huxley model," *European Physical Journal Special Topics*, vol. 227, pp. 767–776, 2018.

52 V. Walsh and A. Pascual-Leone, *Transcranial Magnetic Stimulation: A Neurochronometrics of Mind*, Cambridge: MIT Press, 2003.

53 L. O. Chua, "Memristor: the missing circuit element," *IEEE Transactions on Circuit Theory*, vol. 18, no. 5, pp. 507–519, 1971.

54 A. Dorval, "Probability distributions of the logarithm of inter-spike intervals yield accurate entropy estimates for small datasets," *Journal of Neuroscience Methods*, vol. 173, no. 1, pp. 129–139, 2008.

55 L. F. Shampine and M. W. Reichelt, "The matlab ode suite.," *SIAM Journal on Scientific Computing*, vol. 18, no. 1, pp. 1–22, 1997.

56 D. Freedman and P. Diaconis, "On the histogram as a density estimator: L2 theory," *Zeitschrift für Wahrscheinlichkeitstheorie und Verwandte Gebiete*, vol. 57, no. 2, pp. 453–476, 1981.

57 J. R. Clay, "Excitability of the squid giant axon revisited," *Journal of Neurophysiology*, vol. 80, no. 2, pp. 903–913, 1998.

58 E. Katz, *Implantable Bioelectronics*, Weinheim, Germany: Wiley, 2014.

59 A. Kiourti and K. S. Nikita, "A review of in-body biotelemetry devices: implantables, ingestibles, and injectables," *IEEE Transactions on Biomedical Engineering*, vol. 64, no. 7, pp. 1422–1430, 2017.

60 A. Skrivervik, M. Bosiljevac and Z. Šipuš, "Fundamental limits for implanted antennas: maximum power density reaching free space," *IEEE Transactions on Antennas and Propagation*, vol. 67, pp. 4978–4988, 2019.

61 J. Kim and Y. Rahmat-Samii, "Implanted antenna inside a human body: simulations, design, and characterization," *IEEE Transactions on Antennas and Propagation*, vol. AP-52, no. 8, pp. 1934–1943, 2004.

62 A. J. Moreno Montes, I. V. Trivino, M. Bosiljevac, M. J. Veljovic, Z. Šipuš and A. Skrivervik, "Antenna for a cranial implant: simulation issues and design strategies," in *14th European Conference on Antennas and Propagation, EuCAP 2020*, Online, 2020.

13

Stochastic–Deterministic Modeling of Wire Configurations in Frequency and Time Domain

In this chapter, three different applications of linear wire antennas are observed as stochastic–deterministic problems. Deterministic modeling of wire antenna configurations is described in Part I of this book. Namely, frequency domain (FD) analysis of wire configurations is described in Chapter 5 while time domain (TD) analysis is described in Chapter 6. In any case, the deterministic part can be summed up as follows: first, antenna current along the wire is computed from Pocklington's or Hallen's equation in FD and TD, respectively, by using the Galerkin–Bubnov Indirect Boundary Element Method (GB-IBEM). Besides the direct approach from Hallen's equation, TD response can also be obtained by coupling the FD approach with inverse fast Fourier transform (IFFT). Then, the electromagnetic (EM) field radiated by antenna is computed by solving the field integrals numerically. The main outputs of interest are the current along the wire, antenna impedance, and the radiated field. Other outputs of interest depend on the application type. In the following applications, an antenna is either a horizontal dipole or an array of dipoles. As for the environment, antenna is either in a free space, above a lossy half-space, or buried inside the half-space.

The examples described in this chapter are grouped into three categories: ground-penetrating radar (GPR), grounding systems, and air traffic control. GPR and grounding systems are analyzed in TD, while the air traffic control system is analyzed in FD. Input parameters exhibiting random nature and output parameters of interest are chosen according to each application.

13.1 Ground-Penetrating Radar

The subsurface investigation is necessary for many areas such as civil engineering, archeology, forensics, and mine detection [1]. In contrast to some alternative techniques, the GPR is a noninvasive and nondestructive approach with relatively fast

Deterministic and Stochastic Modeling in Computational Electromagnetics: Integral and Differential Equation Approaches, First Edition. Dragan Poljak and Anna Šušnjara.
© 2024 The Institute of Electrical and Electronics Engineers, Inc.
Published 2024 by John Wiley & Sons, Inc.

data acquisition [2]. The fundamental principle lies in the movement of GPR antennas over the surface of the inspected structure while emitting and receiving EM waves. The design of GPR antennas depends appreciably on the application of interest which makes the simulation of a particular type of GPR procedure important. In particular, the simulation of the field transmitted into the subsurface enables a more realistic interpretation of the target reflected wave. GPR antennas are studied by using different techniques which are generally classified into two categories, the FD [3] and the TD techniques [4–8]. The formulation is generally simpler for FD techniques implying less demanding computational implementation. On the other hand, TD methods allow better physical insight and a single simulation is needed for a large-frequency spectrum. A comparison between different TD and FD techniques is given in [8].

However, to obtain as accurate extraction of information as possible from GPR response, it is necessary to establish prior knowledge about the environmental settings of GPR applications. Such settings include dielectric properties of the inspected subsurface, terrain configuration, antenna position, etc. [9]. Unavoidable, such properties have stochastic nature, hence their variations should be taken into account. For example, dielectric properties of soil depend on temperature, frequency, water content, and fractions of its individual constituents [9–12].

The following two subsections bring examples of stochastic analysis of GPR antenna current and the field transmitted to the ground [13] and [14], respectively. Some typical input parameters considered random variables are soil relative permittivity, soil conductivity, antenna height, and observation point depth. The geometry of interest in both examples features GPR antenna modeled as a dipole placed horizontally above the lossy half-space as depicted in Fig. 13.1. The presented geometry is simple but convenient for testing new approaches and benchmarking. Both FD and TD deterministic formulations of the problem are considered in the black box manner while stochastic collocation is used as a wrapper.

13.1.1 The Transient Current Induced Along the GPR Antenna

The geometry of interest is a straight, thin dipole with length $L = 1$ m and radius $a = 6.74$ mm placed parallel to the lossy half-space at distance h (Fig. 13.1). The wire is a perfect conductor, with a voltage source applied to the gap in the center of the antenna. The source has a Gaussian-shaped waveform given as [15]:

$$V(f) = V_0 \cdot \sqrt{\pi}\, t_w\, e^{-(\pi f t_w)^2}\, e^{-j2\pi f t_0} \tag{13.1}$$

where the amplitude, time delay, and half-width are set to $V_0 = 1$ V, $t_0 = 1.43$ ns, and $t_w = 2/3$ ns, respectively.

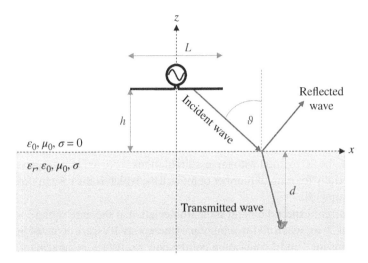

Figure 13.1 A GPR dipole antenna horizontally placed above a lossy half-space.

The current induced along the wire is governed by Pocklington's integro-differential equation in the FD [3]:

$$E_x^{exc} = j\omega \frac{\mu}{4\pi} \int_{-\frac{L}{2}}^{\frac{L}{2}} I(x')g(x,x')dx' - \frac{1}{j4\pi\omega\varepsilon_0} \frac{\partial}{\partial x} \int_{-\frac{L}{2}}^{\frac{L}{2}} \frac{\partial I(x')}{\partial x'} g(x,x')dx' \quad (13.2)$$

where E_x^{exc} is the excitation field, $g(x, x')$ is Green's function and $I(x')$ is the unknown current calculated via GB-IBEM. More details about the GB-IBEM solution of Pocklington's equation can be found elsewhere, e.g. in [3].

Since the output of interest is the transient current in the center of the wire, the solution is transformed to the TD in the following manner. First, the current is calculated for the frequency range 10 Hz–28.64 GHz, thus obtaining the transfer function of a system $H(f)$ in the FD. Then, $H(f)$ is multiplied by the spectrum of the Gaussian pulse, $V(f)$. Finally, the frequency response is transformed to the TD by using the IFFT using the Matlab function "IFFT" [16] and [17].

In the stochastic part, three input parameters are modeled as random variables with uniform distributions: the height of the antenna (i.e. antenna distance from the subsurface), $h \sim U(12, 18)$ cm, the soil permittivity $\varepsilon_r \sim U(14, 18)$, and soil conductivity $\sigma \sim U(0.1, 9.9)$ mS/m. The soil is modeled as an average soil type whose permittivity and conductivity are taken from [1].

Lagrange stochastic collocation with Gaussian quadrature is utilized to obtain the stochastic response for the antenna current. The experimental design is built with 3 and 5 collocation points in one dimension, resulting in $3^3 + 5^3 = 152$ deterministic simulations. All possible combinations of input RVs are considered, starting from three one-dimensional cases, then considering all two-dimensional cases, and finally a three-dimensional case. A variance-based approach is used to obtain the sensitivity analysis (SA) in the following manner: the variances from one-dimensional and two-dimensional cases are normalized with the total variance (three-dimensional case) and then compared. The execution time of a single deterministic simulation is 17.78 min on ASUS PC with i5-5200 CPU and 2.20 GHZ processor, which is rather impractical for traditional Monte Carlo simulations. On the other hand, the SC method with far less simulations is proven to have good convergence, especially for a small number of input RVs, which is the case in this computational setup [15].

The transient current expectation and standard deviation at the wire center for the time interval of 30 ns, when all three input parameters are RVs, are depicted in Fig. 13.2. The results for 3 and 5 collocation points show a satisfactory agreement.

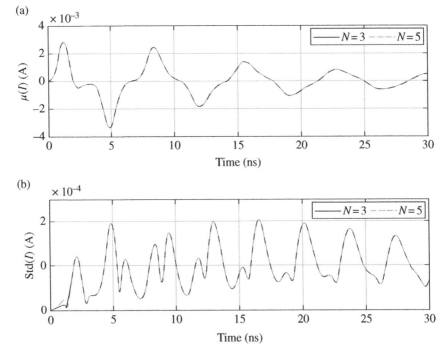

Figure 13.2 The mean (a) and standard deviation (b) of the transient current at the wire center obtained with 3 and 5 collocation points, i.e. with 27 and 125 deterministic simulations.

There is a certain discrepancy in the computation of standard deviation in the early time response. However, the value is too small, so this can be neglected. Hence, to access the first- and second-order statistics of the current in the middle of the wire, only 3 collocation points are required, i.e. 27 deterministic simulations. Interestingly, the computation showed that when only conductivity is random, 5 collocation points are necessary for convergence. However, as it is going to be demonstrated later, the influence of this variable on the current is very small.

The crude estimate of confidence intervals (CIs) given as the mean ± 3 standard deviations is depicted in Fig. 13.3. Throughout the first part of the whole-time interval, the CI is rather narrow and almost negligible for the early time response. At the beginning of the simulation, only the peaks exhibit a noticeable deviation from the mean trend. On the contrary, in the second part of the captured time interval, the deviation is larger and uniform.

Impact factors of input parameters, calculated by normalizing the variances from univariate and bivariate cases, are shown in Fig. 13.4. The influence of antenna height is dominant throughout the whole simulation interval. The influence of the soil relative permittivity cannot be ignored in the time instants where the current reaches its local minimum and maximum values. The overall influence of soil conductivity is evidently small.

The impact factor originating from the interactions between the input parameters is depicted in Fig. 13.5. Again, the antenna height domination is obvious while the interactions of other input parameters have negligible impact on the total variation of the current.

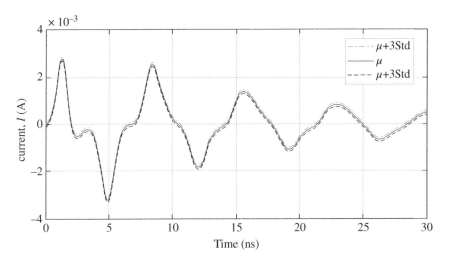

Figure 13.3 The confidence intervals of the transient current at the wire center obtained with 27 deterministic simulations.

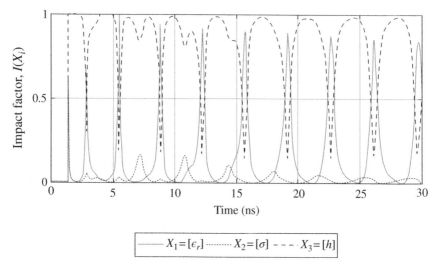

Figure 13.4 Impact factor of the three input parameters computed as $I(X_i) = V(X_i)/V(X)$ where $i = 1,2,3$, $V(\cdot)$ is the variance, X_i is the i-th input parameter, and $X = [\varepsilon_r, \sigma, h]$.

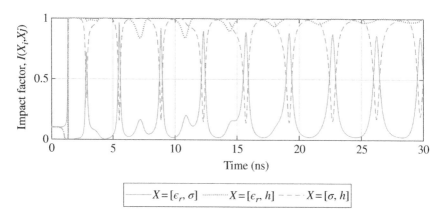

Figure 13.5 Impact factor of the input parameters' interactions computed as $I(X_i, X_j) = V(X_i, X_j)/V(X)$ where $i, j = 1,2,3$, $V(\cdot)$ is the variance, X_i or X_j is the ith or jth input parameters and $X = [\varepsilon_r, \sigma, h]$.

13.1.2 The Transient Field Transmitted into a Lossy Soil

This example differs from the previous one in the formulation. Namely, the TD formalism is used to obtain the current and hence the transmitted field. The geometry is the same as depicted in Fig. 13.1. The dipole antenna dimensions are $L = 1$ m and $r = 6.74$ mm for length and radius, respectively. The TD Gaussian wave pulse is as follows [14]:

$$V(t) = V_0 \cdot e^{-\left(\frac{t-t_0}{t_w}\right)} \tag{13.3}$$

where $V_0 = 1$ V, $t_0 = 1.43$ ns, and $t_w = 2/3$ ns.

The current along the antenna is governed by Hallen space-time integral equation [14]:

$$\int_{-L/2}^{L/2} \frac{I\left(x', t - \frac{R}{c}\right)}{4\pi R} dx' - \int_{-\infty}^{t} \int_{0}^{L} r(\vartheta, \tau) \frac{I\left(x', t - \frac{R_i}{c} - \tau\right)}{4\pi R_i} dx' d\tau$$

$$= F_0\left(t - \frac{R}{c}\right) + F_L\left(t - \frac{L-x}{c}\right) + \frac{1}{2Z_0} \int_{-L/2}^{L/2} E_x^{inc}\left(x', t - \frac{|x - x'|}{c} t\right) dx' \tag{13.4}$$

where $I(x',t)$ is the unknown space-time dependent axial current, c is the light velocity, and Z_0 is the free-space impedance. The distances between the observation point x and source point x' on actual wire and its image in the soil are R and R_i, respectively. The purpose of the functions F_0 and F_L is to account for the multiple reflections from the wire ends. Integral equation is solved numerically by means of the space-time version of GB-IBEM and more details can be found in [18].

The transient field is obtained from the following integral:

$$E_x^{tr}(r, t) = \frac{\mu_0}{4\pi} \int_{-\infty}^{t} \int_{-L/2}^{L/2} \Gamma_{tr}^{MIT}(\tau) \frac{\partial I\left(x', t - \frac{R''}{v} - \tau\right)}{\partial t} dx' d\tau \tag{13.5}$$

where $E_x^{tr}(r, t)$ is the horizontal component of the transmitted electric field, R'' is the distance between the dipole and the observation point in the soil, and v is the propagation velocity through the medium depending on its relative permittivity. The transmission coefficient is denoted with Γ_{tr}^{MIT} and it is computed according to MIT approach [7].

Stochastic part features four input parameters modeled as random variables: soil relative permittivity (ε_r), electrical conductivity (σ), antenna height (h), and the observation point position (d). The last two parameters account for possible surface roughness and the uncertainty in the depth of the buried object, respectively. Input parameters' uniform distributions are as follows: $\varepsilon_r \sim (7, 30)$, $\sigma \sim (0.1, 9.9)$ S/m, $h \sim (12, 18)$ cm, and $d \sim (0.93, 1.07)$ m. The distribution ranges for antenna height and observation depth are chosen intuitively according to the expected variations of these variables. On the other hand, distribution ranges for dielectric properties of soil are taken from the literature [1]. The observed type of soil is wet sand.

Stochastic analysis is carried out for one-dimensional cases and a four-dimensional case, by using 3, 5, 7, and 9 collocation points in each dimension. Lagrange SC with Gauss–Legendre quadrature is used. The number of collocation points chosen for four-dimensional cases is based on accomplished convergence in one-dimensional cases. The output parameters of interest are as follows: the electric field transmitted into the soil versus time (E_{tr} vs. t), time delay of a signal (t_{delay}) and the maximum value of the transmitted electric field in the observed time interval of 40 ns (E_{max}).

The mean of electric field transmitted into the soil computed in OAT manner is depicted in Fig. 13.6.

The convergence of the SC method, when electric permittivity is the only input parameter modeled, as RV is not achieved (Fig. 13.6a). Moreover, signal mean has a distorted shape. Namely, the relation between the relative permittivity and the propagation velocity is $v^2 = c^2/\varepsilon_r$. Therefore, it is expected for the relative permittivity to have the greatest impact on the time delay of a signal. However, increasing the number of collocation points results in adding more signals with different starting points to the calculation. Consequently, for the given range of relative permittivity, stochastic moments for the electric field versus time cannot be obtained. As far as the other input parameters are considered, the convergence of the SC method is accomplished even with 3 collocation points (Fig. 13.6b–d).

Likewise, the convergence in the computation of standard deviation cannot be accomplished for the case when only permittivity is RV. On the contrary, the convergence for the remaining three one-dimensional cases is satisfactory. Only 3 collocation points are needed when conductivity and observation point depth are RVs and 5 collocation points when antenna height is RV. Nevertheless, some conclusions about impact of input parameters on the transmitted field variation can be drawn.

Figure 13.7 depicts the CI of the transmitted field computed as mean $\pm$ 1 standard deviation in time interval (10, 40) ns.

CIs are the widest when relative permittivity is RV. Hence, relative permittivity is the most dominant input parameter (red color in Fig. 13.7). Soil conductivity affects the overall shape of the signal to the least extent; however, its influence dominates at the signal peaks (Fig. 13.7b). The observation point position has a strong impact on overall signal shape, and it is followed by the antenna height (Fig. 13.7c and Fig. 13.7d, respectively). Unlike soil conductivity, these two parameters have a considerable impact on the whole signal, not just its peaks.

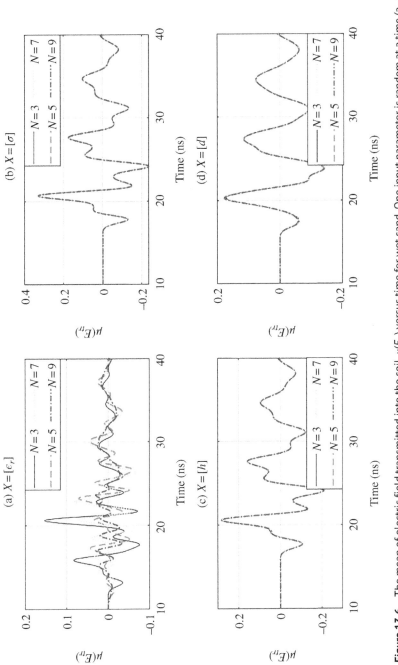

Figure 13.6 The mean of electric field transmitted into the soil, $\mu(E_{tr})$ versus time for wet sand. One input parameter is random at a time (a–d). N is the number of deterministic simulations.

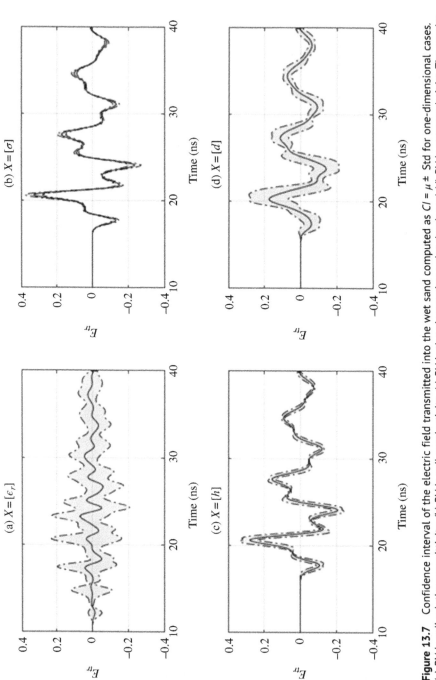

Figure 13.7 Confidence interval of the electric field transmitted into the wet sand computed as $CI = \mu \pm$ Std for one-dimensional cases. (a) RV is soil relative permittivity, (b) RV is soil conductivity, (c) RV is the observation point depth and (d) RV is antenna height. The results are obtained with five collocation points.

Since the SC method failed to converge in case of wet sand because of the large distribution range of the wet soil permittivity, another full tensor model with 625 deterministic simulations is built for the case of average type of soil whose permittivity has much narrower distribution range, $\varepsilon_r \sim (14, 18)$. Soil conductivity, antenna height, and observation point depth have the same distributions as for wet sand. CI and SA are exhibited in Fig. 13.8.

Impact factor obtained in OAT manner for average soil type is depicted in Fig. 13.9.

The observation point depth has the highest impact, followed by the relative permittivity. They are followed by antenna height, while soil conductivity is the least influential parameter. It is worth noting that complementary to [19], the results in Figs. 13.7 and 13.9 provide useful information about the time-domain sensitivity of the system regarding uncertain environment.

Furthermore, the time instant in which the signal is observed for the first time, i.e. when the signal value is greater than zero, is denoted as the signal time delay. The results of stochastic analysis for the time delay are depicted in Fig. 13.10 for wet sand and average soil.

Clearly, the most influential parameter for wet sand is relative permittivity while in case of average soil, the permittivity and target depth have the same impact. The influence of the antenna height is very small for both soil types. Time delay of a signal is not affected at all by the soil conductivity for both soil types. The CI for wet sand is [6.79, 25.63] ns and for average soil [12.82, 17.73] ns. It is to be noted that the range for time delay calculated by using the extreme values of the input

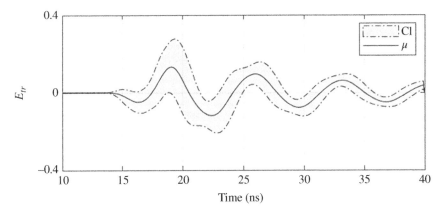

Figure 13.8 Confidence interval of the electric field transmitted into the average soil computed as $CI = \mu \pm$ Std for four-dimensional case. The results are obtained with 625 deterministic simulations.

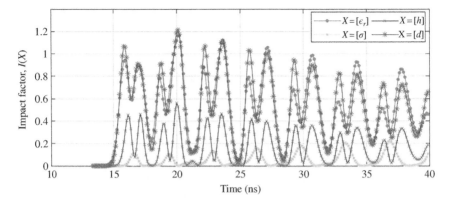

Figure 13.9 Impact factors of input parameters in case of average soil type. Impact factor is computed as $I(X_i) = V(X_i)/V(X)$, where $i = 1,2,3,4$ for ε_r, σ, h and d, respectively. V is the variance, while $X = [\varepsilon_r, \sigma, h, d]$.

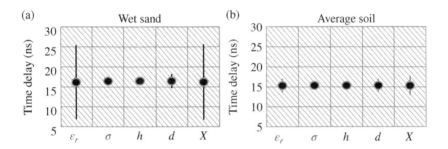

Figure 13.10 Confidence intervals (CIs) of the signal time delay for wet sand (a) and average soil (b) computed as mean ± 3 standard deviations. First four bars represent one-dimensional cases $X = [\varepsilon_r]$, $X = [\sigma]$, $X = [h]$, and $X = [d]$ respectively, while the fifth bar represents the four-dimensional case, $X = [\varepsilon_r, \sigma, h, d]$. Dots denote the time delay mean while bars denote the CI width.

parameters is [9.25, 22.80] ns and [13.12, 17.74] ns for wet sand and average soil, respectively.

Generally, the shape of the transmitted signal depends on the excitation pulse and the signal amplitude decays over time. Therefore, the first peak of the signal is the maximum value of the transmitted field. The mean value and the standard deviation of the maximum field transmitted into the ground are depicted in Fig. 13.11.

The most influential parameter is soil conductivity which is expected since greater conductivity implies bigger attenuation of the transmitted signal. There

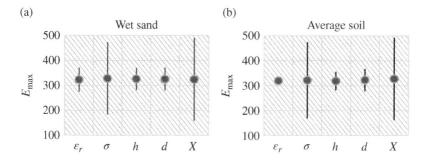

Figure 13.11 Confidence intervals (CI) of the maximum value of the electric field transmitted into the wet sand (a) and average soil (b) computed as mean ± 3 standard deviations. First four bars represent one-dimensional cases $X = [\varepsilon_r]$, $X = [\sigma]$, $X = [h]$, and $X = [d]$, respectively, while the fifth bar represents the four-dimensional case, $X = [\varepsilon_r, \sigma, h, d]$. Dots denote the maximum field mean while bars denote the CI width.

is a noticeable difference in the rank of the least significant input parameter. Namely, permittivity, height, and depth have a very similar influence on maximum field value for the wet sand. However, if the range of relative permittivity is reduced, the influence of relative permittivity is much smaller than the influence of the other two variables. Nevertheless, the multivariate simulation for both soil types has similar average value and standard deviation since the major origin of variations is the soil conductivity. CI of maximum field for wet sand is [158.51, 490.31] mV/m and for average soil is [162.33, 492.45] mV/m. The maximum field calculated by using the extreme values of the input variables is inside the interval [237.10, 475.40] mV/m and [247.30, 476.70] mV/m for wet sand and average soil, respectively.

13.2 Grounding Systems

Grounding system is the main component of the lightning protection system (LPS) [20, 21] and the crucial parameter in the design process is its transient impedance. The transient impedance is defined as the ratio of potential rise instantaneous values at the current injection point to the current injected [22]. The usual computational approach is the Transmission Line Model (TLM) [22, 23]. However, a more accurate and thus rigorous approach is the full wave model which accounts for the radiation effects. Full wave model is based on the corresponding integral equation in either frequency or TD arising from the antenna theory and the thin-wire approximation [24]. The computation is carried out in two steps: the first results in electrode current distribution, while the second features the calculation of other parameters of interest such as scattered voltage, transient impedance, and radiated field [25].

The significance of input parameters' random nature when dealing with lightning sources and LPS has been emphasized in many studies. The precision of LPS models may be highly modified when modeling random nature of key parameters such as grounding resistance and/or stray capacitance of protection elements as reported in [26]. Furthermore, lightning strike TD pattern including time-to-half, front time, and maximum slope of lightning current was considered in stochastic manner in [27]. Even scientific reports such as CIGRE brochures recognize uncertainty quantification as an important part of modeling [28]. Stochastic nature is reflected in different input parameters such as material and soil electrical properties and system geometry [29]. These types of uncertainties can be categorized as epistemic uncertainties since they can be reduced to a certain extent (e.g. through a more accurate measurement of material or soil characteristics). On the other hand, natural phenomena such as lighting strikes introduce aleatory uncertainties to the EM environment. More precisely, the variability of lighting pulse time parameters cannot be reduced but only better characterized. The difference between the epistemic and aleatory uncertainties is described in Chapter 9.

This chapter outlines three examples of deterministic-stochastic modeling of the transient impedance with both environmental and source properties modeled as random variables. The geometry of interest features a horizontal, perfectly conducting (PEC) grounding electrode of length L and radius a, embedded in a lossy medium at depth d and excited by an equivalent current source of magnitude I_s, as shown in Fig. 13.12. The soil is a lossy half-space characterized by electric permittivity ε and conductivity σ. Dimensions of the electrode are assumed to satisfy the thin-wire approximation $(L \gg a)$ [30].

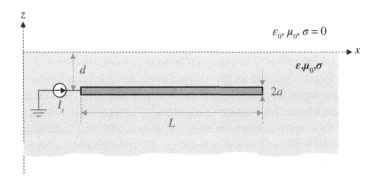

Figure 13.12 Grounding electrode as horizontal perfectly conducting thin wire.

Current induced along the grounding electrode is governed by Pocklington integro-differential equation in the FD [24].

$$j\omega \frac{\mu_0}{4\pi} \int_0^L I(x')g(x,x')dx' - \frac{1}{4\pi\omega\varepsilon_{eff}} \frac{\partial}{\partial x} \int_0^L \frac{\partial I(x')}{\partial x'} g(x,x')dx' = 0 \tag{13.6}$$

where $I(x')$ is the unknown current distribution, ε_{eff} stands for the complex permittivity of the medium, and $g(x,x')$ is the Green's function:

$$g(x,x') = \frac{e^{-\gamma R_1}}{R_1} - \Gamma_{ref} \frac{e^{-\gamma R_2}}{R_2} \tag{13.7}$$

Here γ denotes the propagation constant of the medium computed as $\gamma^2 = j\omega\mu\sigma - \omega^2\mu\varepsilon$, while R_1 and R_2 correspond to distances from the source and the image to the observation point, respectively. The earth-air interface is considered via the reflection coefficient Γ_{ref} arising from the Modified Image Theory (MIT) [31].

The analytical solution of the Pocklington equation yields [24]:

$$I(x,s) = I_g \frac{\sinh[\gamma(L-x)]}{\sinh(\gamma L)} \tag{13.8}$$

According to the Generalized Telegraphers Equation, the scattered voltage along the electrode is given as [32]:

$$V_{sct}(x) = -\frac{1}{4\pi\omega\varepsilon_{eff}} \int_0^L \frac{\partial I(x')}{\partial x'} g(x,x')dx' \tag{13.9}$$

Finally, the scattered voltage is:

$$V_{sct}(x) = -\frac{I_g}{4\pi\omega\varepsilon_{eff} \sinh(\gamma L)} \int_0^L \cosh[\gamma(L-x)]g(x,x')dx' \tag{13.10}$$

The transient impedance of the grounding electrode $z(t)$ is derived from the TD value of the scattered voltage (obtained by IFFT algorithm) as follows [32]:

$$z(0,t) = \frac{v(0,t)}{i(0,t)} \tag{13.11}$$

where $v(0,t)$ stands for the time values of voltage at the beginning of the electrode (i.e. at $x=0$), and $i(0,t)$ is the current at the equivalent source owing to lightning strike current.

Input parameters for the three test cases in the stochastic part are given in Table 13.1 Two test cases are three-dimensional, and one is six-dimensional

Table 13.1 Input parameters for the three test cases.

		Test case #1 Three-dimensional	Test case #2 Three-dimensional	Test case #3 Six-dimensional
Electrode parameters:				
Length	L (m)	10	$\sim U\,(7.5, 12.5)$	$\sim U\,(8, 12)$
Radius	a (mm)	5	5	5
Depth	d (cm)	50	50	$\sim U\,(20, 80)$
Soil parameters:				
Relative permittivity	ε_r	10	$\sim U\,(5, 15)$	$\sim U\,(5, 15)$
Conductivity	σ (mS/m)	$\sim U\,(1, 10)$	$\sim U\,(0.5, 1.5)$	$\sim U\,(0.5, 1.5)$
Lighting pulse parameters:				
Front time	τ_1 (µs)	$\sim U\,(0.4, 4)$	1	$\sim U\,(0.4, 4)$
Time-to-half	τ_2 (µs)	$\sim U\,(50, 70)$	10	$\sim U\,(50, 70)$

Uniform distribution of random variables is denoted with $\sim U$.

stochastic problem. Stochastic collocation with Lagrange interpolation and Gauss–Legendre quadrature is used. Higher dimensions are considered with full tensor approach. The results from the third test case validated with the Monte Carlo method.

13.2.1 Test Case #1: Soil And Lighting Pulse Parameters are Random Variables

The first test case is a three-dimensional stochastic problem with vector of input parameters $X = [\sigma, \tau_1, \tau_2]$ [33]. The goal is to investigate the influence of variability of soil conductivity, lightning front time, and lightning time-to-half on the transient impedance CI width. Figure 13.13 depicts the CI and impact factors.

Figure 13.13a depicts rather wide CI of the transient impedance for the observed time interval, especially after the early time response. For most of the time, the CI width is constant.

The impact of the three input parameters exhibiting the random nature is obtained in OAT manner (Fig. 13.13b). The most dominant parameter is soil conductivity at all time instants. In the early time response, the lighting pulse front time has some noticeable impact.

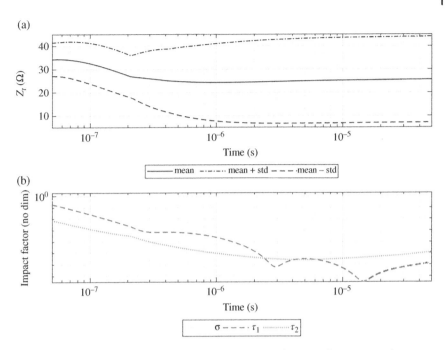

Figure 13.13 (a) Transient impedance confidence intervals versus time computed as mean ± 1 standard deviation. (b) Impact factors computed in OAT manner. Soil conductivity, lighting pulse front time, and time to half are random variables.

13.2.2 Test Case #2: Soil and Electrode Parameters are Random Variables

The second test case is a three-dimensional stochastic problem with vector of input parameters $X = [L, \sigma, \varepsilon_r]$. Hence, the aim is to investigate the influence of variability of electrode length, soil conductivity, and relative permittivity on the transient impedance CI width [34]. The $1/10\,\mu s$ lighting pulse is taken into consideration. Figure 13.14 depicts the CI and impact factors.

A complete simulation time can be divided into two segments; before $0.1\,\mu s$ (early time behavior) and after $0.1\,\mu s$ (steady state). The CI width from Fig. 13.14a is again rather wide, i.e. $\pm 30\,\Omega$ in a steady state, thus implying significant impact of applied input parameters' ranges on the value of impedance in any given time instant. The SA from Fig. 13.4b is computed by following ANOVA approach. The first-order sensitivity indices are almost the same as the total effect sensitivity indices for most of time instants, meaning that mutual interactions have minimal effect on the transient impedance total variance. In the first half-time, the relative permittivity has the highest impact while in the second half-time, the soil

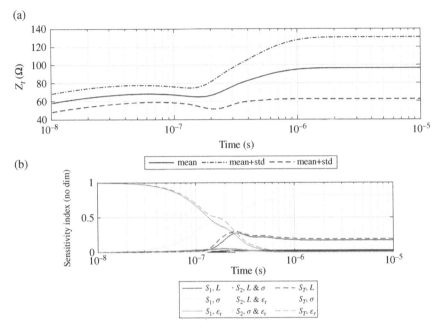

Figure 13.14 (a) Transient impedance confidence intervals versus time computed as mean ± 1 standard deviation. (b) Sensitivity indices computed in ANOVA manner; S_1, S_2, and S_T stand for first-order, second-order and total effect indices, respectively. Electrode length, soil conductivity, and relative permittivity are random variables.

conductivity is the most significant parameter. The electrode length is ranked as the third important parameter. Although the soil conductivity prevails in the second half time, the impact of the electrode length is still visible. The second-order sensitivity indices are small, once again showing that the input parameter-combined impact is rather small.

13.2.3 Test Case #3: Soil, Electrode, and Lighting Pulse Parameters are Random Variables

Finally, the third test case is a six-dimensional stochastic problem with vector of input parameters $X = [\varepsilon_r, d, \tau_1, \tau_2, L, \sigma]$. Hence, the aim is to investigate the influence of variability of soil permittivity, electrode depth, lighting pulse front time, lighting pulse time-to-half, electrode length, and soil conductivity, respectively, on the transient impedance CI width [35]. The input parameters have similar values as in second test case, with the difference in treating the lighting pulse time parameters and electrode depth as random. Figure 13.15 depicts the transient impedance stochastic mean obtained with SC and MC methods as well as the relative error of SC method with respect to MC method.

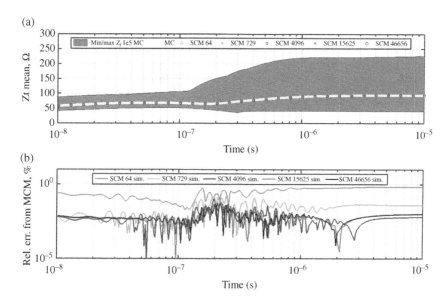

Figure 13.15 The results of stochastic analysis for the six-dimensional test case.
(a) Transient impedance mean obtained with MC and SC methods. (b) The relative error of SC
method w.r.t. MC method computed as $Err^{(u)} = (\mu(Z_t)^{(SC-u)} - \mu(Z_t)^{(MC)})/\mu(Z_t)^{(MC)}$ at each time
instant. $\mu(Z_t)$ is the mean transient impedance and u is the SC level.

The transient impedance mean is obtained from several accuracy levels of SC
method, i.e. with 64, 729, 4.096, 15.625, and 46.656 deterministic simulations. Note
that extreme values of transient impedance and mean transient impedance
obtained with MC method are also depicted. Mean and the two extremes are
obtained with three different sets of 100 000 MC simulations. The resulting mean
transient impedance exhibits the expected behavior. The difference between the
mean impedance in the early time period and in a steady state is 30 Ω. The SC
method convergence is good, which is visible from Fig. 13.15a. In addition,
Fig. 13.15b depicts the relative error of SC method with respect to MC. Excellent
agreement between SCM and MC is evident since highest relative error levels are
below 1%.

The first-order sensitivity indices based on ANOVA SA are depicted in Fig. 13.16.
The influence of various parameters changes over time and complete simulation
time can be divided into two segments: the early time behavior up to 0.1 µs and
steady state after 0.1 µs.

The convergence of SC method in computing the first-order sensitivity indices is
tested by computing the gap between the sensitivity indices obtained with different
SC levels. The results are depicted in Fig. 13.17.

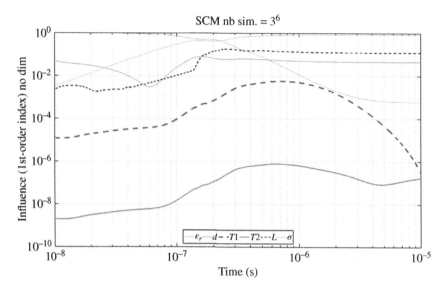

Figure 13.16 The first-order sensitivity indices as influence criteria for the six-dimensional case ($X = [\varepsilon_r, d, \tau_1, \tau_2, L, \sigma]$).

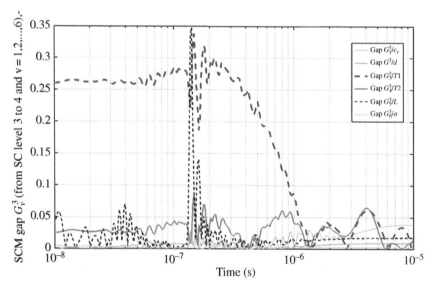

Figure 13.17 The gap between the two consecutive SC levels computed as $G^{(u)} = |S_1^{(u)} - S_1^{(u+1)}|/|S_1^{(u+1)}|$ for each input parameter and at each time instant. S_1 denotes the first-order index and u denotes different SC level. The depicted results are for the consecutive levels $u = 3$ and $u + 1 = 4$.

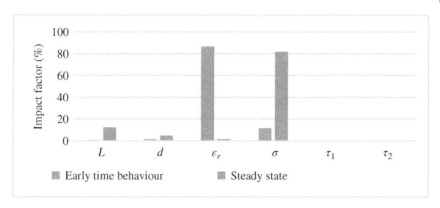

Figure 13.18 The integrated influence of input parameters for the early time behavior and steady state.

The comparison of data from SCM levels 3 and 4 shows very weak discrepancy since majority of results are below 10% level. The results given for lightning pulse front time (dashed blue) electrode length (dotted black) vary up to 35%. However, during that simulation period (early time behavior), their influence is shown to be relatively weak.

Furthermore, the integrated influence of input parameters is depicted in Fig. 13.18.

For a steady state period, soil permittivity is the most influential parameter followed by electrode length just as in test case #2. The third and the fourth important parameters are the electrode depth and soil permittivity. The lighting pulse front time and time-to-half are insignificant during the whole-time duration. For the early-time behavior, the ranking of parameters with noticeable impact is as follows: soil permittivity, soil conductivity, and electrode depth. The impact of electrode length is barely noticeable. This is also in accordance with the results in the second test case.

13.3 Air Traffic Control Systems

Poor weather conditions may deteriorate the visibility and make it challenging for a pilot to land an aircraft smoothly and safely on the ground. In such conditions, the reliability of landing on runway is enabled by the instrument landing system (ILS). The ILS provides the lateral and vertical guidance for the pilot with the localizer and the glide slope, respectively [36]. The ILS, thus, creates the glide path for safe aircraft landing whereas the angle between the glide path and the ground, i.e.

the glide path angle is standardized by the Federal Administration Aviation (FAA) Flight Procedures [37].

The signal transmitted by the ILS system and the one received by the pilot are not the same due to the reflecting properties of the runway. Namely, the total signal consists of the direct signal and its ground-reflected image. However, terrain configuration and environmental conditions can change the ground reflection coefficient, thus altering the main lobe direction and consequently the glide path angle. Some typical situations feature grass or snow on the runway [36]. Since the EM properties of vegetation or snow cannot be exactly determined, it is important to investigate to which extent their variation impacts the variation of the glide slope angle as presented in [38] and [39]. In other words, it is important to quantify the CIs of the output angle. Generally, no matter how many layers are introduced in the runaway model, the EM properties of each of them can be considered RVs.

In this study, the runaway is considered a lossy two-layered half-space as depicted in Fig. 13.19. The first layer is modeled as either snow cover or vegetation of finite thickness d, while the second layer is the half-space with the properties of some average soil type. Typically, the glide path antenna is located 120 m from the runway centerline and approximately 300 m from the runway threshold, i.e. the ILS reference point where downward extended glide path intersects the runway centerline.

Furthermore, in Fig. 13.20, the glide path antenna system is depicted with the related main lobe and the glide path angle, θ.

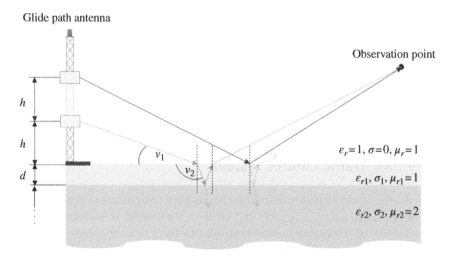

Figure 13.19 Runway modeled as a two-layered lossy half-space with glide path antenna. Both direct and reflected rays are depicted for the observation point of interest.

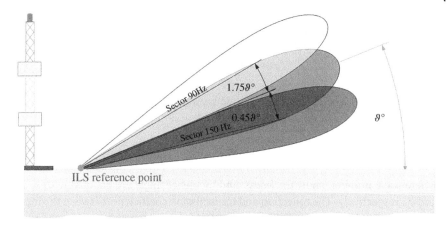

Figure 13.20 The system of a glide path antenna with the corresponding main lobe and the glide path angle, θ.

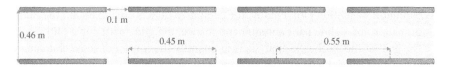

Figure 13.21 The panel antenna model of the glide path antenna system.

The glide path antenna is an array consisting of eight half-wave dipoles and the metal reflector as depicted in Fig. 13.21. The operating frequency is 328 MHz. The physical dimensions of the antenna are as follows: dipoles are placed at the distance 0.15λ from the reflector, wire diameter is 6 mm, and the antennas' heights are 3 and 6 m ($h = 3$ m in Fig. 13.19), respectively.

The currents induced along the M dipoles are governed by the set of M-coupled Pocklington integro-differential equations [40]:

$$E_x^{exc} = -\frac{1}{j4\pi\omega\varepsilon_0} \sum_{n=1}^{M} \int_{-L_n/2}^{L_n/2} \left[\frac{\partial^2}{\partial x^2} + k_1^2\right] g_{mn}^{tot}(x,x')I_n(x')dx', m = 1, 2, \ldots M$$

(13.12)

where the unknown current distribution induced along the n-th wire due to known excitation is denoted with $I_n(x')$. The function $g^{tot}(x,x')$ is the total Green's function consisting of a free space Green's function and the Green's function for the wire image multiplied by the reflection coefficient that takes into account the presence of a two-layered half-space [40]:

$$g_{mn}^{tot}(x, x') = \frac{e^{-jk_1 R_{1mn}}}{R_{1mn}} - \Gamma_{ref}^{tot} \cdot \frac{e^{-jk_1 R_{2mn}}}{R_{2mn}} \tag{13.13}$$

The reflection coefficient for the two-layered half-space has the following form [40]:

$$\Gamma_{ref}^{tot} = r_{12} + \frac{\tau_{12}\tau_{21}r_{23}e^{-2jk_2 d \sin \vartheta_2}}{1 - r_{21}r_{23}e^{-2jk_2 d \sin \vartheta_2}} \tag{13.14}$$

where wave number, snow thickness, and incidence angle are denoted with k, d, and θ_2, respectively. The reflection and transmission coefficients between the incident (n) and the reflecting/transmitting medium (m) are denoted with r_{mn} and τ_{mn}, respectively. The coefficients can be computed by using any type of reflection/coefficient approximation between the two media such as Fresnel's approximation or the one stemming from the MIT [41].

The set of coupled equations is solved by means of the IBEM. Once the current is known, the field is computed from the field integrals that have the same form as Pocklington's equation [40]. Finally, the computed field is used to obtain the radiation pattern of a glide path antenna and, hence, the glide path angle [40].

The goal of the stochastic analysis is to quantify the impact of some characteristic parameters of the first layer, such as snow or vegetation, on the variation of the maximal radiation angle, β [38] and [39]. The stochastic collocation with Lagrange interpolation and Gauss–Legendre quadrature is used while the higher dimensions are treated via the tensor grid approach. In the following examples, the CIs are computed as mean $\pm$ 3 standard deviations and probability density functions are given.

13.3.1 Runway Covered with Snow

Two types of snow are considered separately, dry and wet snow since they differ greatly in water content [38]. Namely, dry snow has 0% water content. Snow density $(\rho_d \text{ (kg/m}^3))$ and thickness $(d \text{ (cm)})$ are modeled as uniformly distributed random variables for both dry and wet snow, with water content $(W \text{ (%)})$ as third random input parameter for wet snow. Snow dielectric parameters are computed from empirical relations given in [42]. Uniform distributions for input parameters are: (10, 275) kg/m³, (0, 0.1)%, and (0, 5) cm, for snow density, water content, and snow depth, respectively.

Figure 13.22 depicts the results of stochastic analysis for the maximum angle β in case of dry snow. CIs are shown in Fig. 13.22a when only snow density is RV with snow depth as deterministic parameter in intervals 0–13 cm. The possible values are spread widely around the expected value. The minimal value of angle β is 3.6°

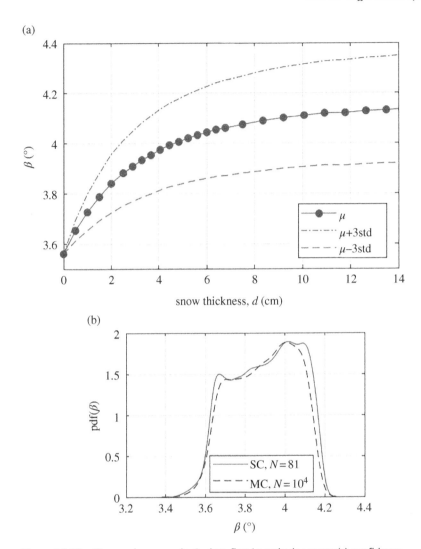

Figure 13.22 The maximum angle β when first layer is dry snow: (a) confidence intervals when only snow density is RV; μ and Std denote the mean and standard deviation of β; (b) probability density function, pdf when both density and thickness are RVs; N is number of deterministic simulations.

for $d = 0$ cm, i.e. when there is no snow cover on the runway, while the maximal possible value is 4.35° at $d = 13$ cm with confidence level of 99.7%.

If snow thickness is also modeled as RV, thus creating the two-dimensional stochastic problem, the CI is $CI = [3.32, 4.39]°$. Impact factors obtained in OAT manner are 0.53 and 0.66 for snow density and snow depth, respectively. The snow depth has slightly higher impact compared to snow density, but none of the factors can be neglected. Figure 13.22b depicts the pdf for the angle β for two-dimensional case. The pdf is obtained by SC and MC methods, with 81 and 10^4 simulations, respectively. Two peaks are observed, one at 3.66° and the other at 4.03°.

CIs of maximum angle for wet snow are depicted in Fig. 13.23a. Snow density and water content are RVs while snow depth is deterministic in range of 0–13 cm.

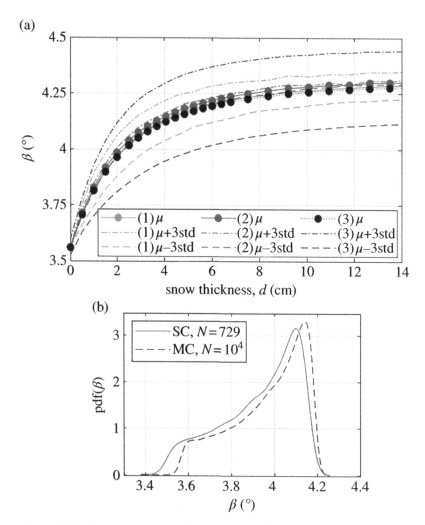

Figure 13.23 The maximum angle β when first layer is dry snow: (a) confidence intervals when only snow density is RV; μ and Std denote the mean and standard deviation of β; (b) probability density function, pdf when water content, density, and thickness are RVs; N is number of deterministic simulations.

Comparing the width of CIs from the two one-dimensional cases, it is obvious that the water content has bigger impact on the angle variability. However, the joint influence of both variables is even more significant.

When all three input parameters are RVs, the confidence interval is $CI = [3.45, 4.49]°$. Impact factors computed in OAT manner for a three-dimensional case are 0.03, 0, and 0.97 for density, water content, and thickness, respectively, thus showing that the major influence pertains to snow depth. Figure 13.23b exhibits the pdf of maximum angle for three-dimensional case. The pdf of the angle β is obtained by 729 SC and 10^4 MC simulations. The distribution is asymmetrical, with the highest peak on the right-hand side, at around 4.01°.

To summarize, when analyzing the snow characteristics for fixed snow thickness, the water content is the most influential input parameter. However, in case of dry snow, whose water content is 0%, the snow density remains the only significant parameter. Moreover, the joint impact of the two parameters is very strong in case of wet snow. On the other hand, if snow thickness is also considered RV, this parameter has the highest influence. When the snow depth increases, the value of the maximum angle rises until its own maximum, which leads to a raised ground plane effect [36]. Dry snow may cause angles lower than nominal value.

Finally, the shape of the pdf is different for the dry and wet snow, respectively, and the presence of the water content is shown to have a high impact on the skewness and kurtosis of the angle distribution.

13.3.2 Runway Covered with Vegetation

The second example assumes that the first layer is a vegetation whose electric permittivity and conductivity are random variables, thus creating the two-dimensional stochastic model [39]. Both input parameters are uniformly distributed, with conductivity in the range of $\sigma = (0.001, 0.1)$ S/m and relative permittivity in the range of $\varepsilon_r = (12, 30)$.

Figure 13.24a–c shows results for maximum angle when vegetation layer thickness is in the range of 0–13 cm. Comparing the one-dimensional cases $X = [\sigma]$ and $X = [\varepsilon]$ and two-dimensional case $X = [\sigma, \varepsilon]$, it is obvious that the vegetation permittivity has a higher impact on the angle variation.

Probability density function for angle β obtained by 729 SC simulations is depicted in Fig. 13.24d. As it can be noticed, the distribution is concentrated around value of 2.99° which is quite close to mean value of 3.013°. Standard deviation is 0.0483° which is 1.6% of the expected value. The corresponding CI is $CI = [2.9050, 3.0610]°$.

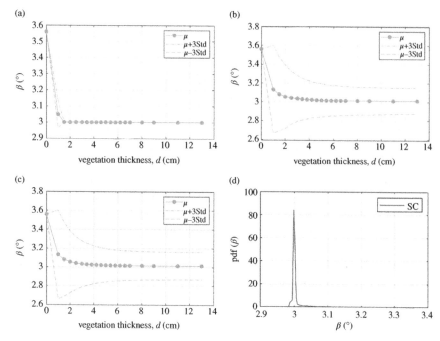

Figure 13.24 The maximum angle β when first layer is vegetation: (a) confidence intervals when only vegetation conductivity is RV; (b) confidence intervals when only vegetation permittivity is RV; (c) confidence intervals when both vegetation conductivity and permittivity are RVs; μ and Std denote the mean and standard deviation of β; (d) probability density function, pdf when conductivity and permittivity are RVs.

References

1 H. M. Jol, Ground Penetrating Radar Theory and Applications, Amsterdam, Netherlands: Elsevier Science, 2009.

2 A. Benedetto and L. Pajewski, Civil Engineering Applications of Ground Penetrating Radar, ebook: Springer Cham, 2015.

3 D. Poljak and V. Doric, "Transmitted field in the lossy ground from ground penetrating radar (GPR) dipole antenna," *Computational Methods and Experimental Measurements XVII 3*, vol. 59, pp. 3–11, 2015.

4 A. Giannopoulos, "Modelling ground penetrating radar by gprmax," *Construction and Building Materials*, vol. 19, no. 10, pp. 755–762, 2005.

5 L. Gurel and U. Oguz, "Three-dimensional fdtd modelling of a ground penetrating radar," *IEEE Transactions on Geoscience and Remote sensing*, vol. 38, no. 4, pp. 1513–1521, 2000.

6 T. Weiland, "A discretization model for the solution of Maxwell's equations for six-component fields," *Archiv fuer Elektronik und Uebertragungstechnik*, vol. 31, pp. 116–120, 1977.

7 D. Poljak, S. Šesnić, A. Šušnjara, D. Parić and K. El Khamlichi Drissi, "Direct time domain evaluation of the transient field transmitted into a lossy ground due to GPR antenna radiation," *Engineering Analysis with Boundary Elements*, vol. 82, no. 2017, pp. 27–31, 2017.

8 C. Warren, S. Šesnić, A. Ventura, L. Pajewski, D. Poljak and A. Giannopoulos, "Comparison of time-domain finite-difference, finite-integration, and integral-equation methods for dipole radiation in half-space environments," *Progress In Electromagnetics Research M*, vol. 57, pp. 175–183, 2017.

9 I. Giannakis, A. Giannopoulos and C. Warren, "A realistic FDTD numerical modelling framework of ground penetrating radar for landmine detection," *IEEE Journal of Selected Topics in Applied Earth Observations and Remote Sensing*, vol. 9, no. 1, pp. 37–51, 2016.

10 W. G. Fanno and V. Trainotti, "Dielectric properties of soil," in *Annual Report Conference on Electrical Insulation and Dielectric Phenomena*, Kitchener, Ontario, 2001.

11 D. Čavka, N. Mora and F. Rachidi, "A comparison of frequency-dependent soil models: application to the analysis of grounding systems," *IEEE Transactions on Electromagnetic Compatibility*, vol. 56, no. 1, pp. 177–187, 2014.

12 Z. Khakiev, K. Kislitsa and V. Yavna, "Efficiency evaluation of ground-penetrating radar by the results of measurment of dielectric properties of soils," *Journal of Applied Physics*, vol. 112, no. 12, pp. 124909(1–4), 2012.

13 A. Šušnjara, D. Poljak, V. Dorić, S. Lallechere, K. El Khamlichi Drissi, P. Bonnet and F. Paladian, "Frequency domain deterministic-stochastic analysis of the transient current induced along a ground penetrating radar dipole antenna over a lossy half-space," *Ground Penetrating Radar*, vol. 1, no. 2, pp. 37–51, 2018.

14 A. Šušnjara, D. Poljak, S. Šesnić, K. El Khamlichi Drissi, P. Bonnet, F. Paladian and S. Lallechere, "Stochastic-deterministic and sensitivity analysis of the transient field generated by GPR dipole antenna and transmitted into a lossy ground," in *2017 IEEE International Symposium on Electromagnetic Compatibility & Signal/Power Integrity (EMCSI)*, Washington, DC, USA, 2017.

15 S. Lalléchère, S. Antonijevic, K. El Khamlichi Drissi and D. Poljak, "Optimized numerical models of thin wire above an imperfect and lossy ground for GPR statistics," in *Proc. International Conference on Electromagnetics in Advanced Applications (ICEAA)*, Torino, Italy, 2015.

16 A. Šušnjara, D. Poljak, S. Šesnić and V. Dorić, "Time domain and frequency domain integral equation method for the analysis of ground penetrating (GPR) antenna," in *Proc. 24th International Conference on Software, Telecommunications and Computer Networks (SoftCOM 2016)*, Split, Croatia, 2016.

17 The MathWorks, Inc. (2010). Matlab Version 7.11.0.584 (R2010b). Available: https://www.mathworks.com.

18 D. Poljak, S. Antonijević, S. Šesnić, S. Lallechere and K. El Khamlichi Drissi, "On deterministic-stochastic time domain study of dipole antenna for GPR applications," *Engineering Analysis with Boundary Elements*, vol. 73, pp. 14–20, 2016.

19 S. Lalléchère, S. Šesnić, P. Bonnet, K. El Khamlichi Drissi, F. Paladian and D. Poljak, "Sensitivity analysis of the time transient currents induced along thin wires buried in lossy and uncertain environments," in *EuCAP*, Paris, France, 2017.

20 S. Visacro, "A comprehensive approach to the grounding response to lightning currents," *IEEE Transactions on Power Delivery*, vol. 22, no. 1, pp. 381–386, 2007.

21 L. Grcev and F. Dawalibi, "An electromagnetic model for transients in grounding systems," *IEEE Transactions on Power Delivery*, vol. 5, no. 4, pp. 1773–1781, 1990.

22 R. Velazquez and D. Mukhedkar, "Analytical modelling of grounding electrodes transient behavior," *IEEE Transactions on Power Apparatus and Systems*, vol. 103, no. 6, pp. 1314–1322, 1984.

23 F. M. Tesche, M. Ianoz and T. Karlsson, EMC Analysis Methods and Computational Models, New York: John Wiley & Sons, Inc., 1997.

24 S. Šesnić and D. Poljak, "Antenna model of the horizontal grounding electrode for transient impedance calculation: analytical versus boundary element method," *Engineering Analysis with Boundary Elements*, vol. 37, no. 6, pp. 909–913, 2013.

25 S. Šesnić, D. Poljak and S. V. Tkachenko, "Analytical modeling of a transient current flowing along the horizontal grounding electrode," *IEEE Transactions on Electromagnetic Compatibility*, vol. 55, no. 6, pp. 1132–1139, 2013.

26 A. Xémard, A. Pagnetti, M. Martinez, P. D. Moreau, F. Paladian, P. Bonnet and C. A. Nucci, "Effect of uncertainties on the precision of lightning studies," in *Intternational Colloquium on Lightning and Power Sytems*, Lyon, France, 2014.

27 B. Jurisic, A. Xémard and I. Uglesic, "Evaluation of transmitted over-voltages through a power transformer taking into account uncertainties on lightning parameters," in *33rd International Conference on Lightning Protection*, Estoril, Portugal, 2016.

28 CIGRE, "Guide to Procedures for Estimating the Lightning Performance of Transmission Lines," CIGRE brochure 63, 1991.

29 G. Spadacini, F. Grassi and S. Pignari, "Statistical estimation of the electromagentic noise induced by field-to-wire coupling in random bundles of twisted-wire pairs," in *2015 Asia-Pacific Symposium on Electromagnetic Compatibility*, Taipei, Taiwan, 2015.

30 S. Šesnić and D. Poljak, "Direct time domain analytical solution for the transient impedance of the horizontal grounding electrode," in *Proceedings of the 2nd URSI Atlantic Radio Science Meeting*, Gran Canaria, 2018.

31 T. Takashima, T. Nakae and R. Ishibashi, "Calculation of complex fields in conducting media," *IEEE Transactions on Electrical Insulation*, vol. 15, no. 1, pp. 1–7, 1980.

32 D. Poljak, V. Doric, F. Rachidi, K. El Khamlichi Drissi, K. Kerroum, S. V. Tkachenko and S. Šesnić, "Generalized form of telegraphers equations for the electromagnetic field coupling to buried wires of finite length," *IEEE Transactions on Electromagnetic Compatibility*, vol. 51, no. 2, pp. 331–337, 2009.

33 S. Šesnić, A. Šušnjara, S. Lalléchère, D. Poljak, K. El Khamlichi Drissi, P. Bonnet and F. Paladian, "Advanced analysis of the transient impedance of the horizontal grounding electrode: from statistics to sensitivity indices," in *2017 XXXIInd General Assembly and Scientific Symposium of the International Union of Radio Science (URSI GASS)*, Montreal, Canada, 2017.

34 S. Šesnić, D. Poljak, A. Šušnjara, S. Lalléchère and K. El Khamlichi Drissi, "Transient impedance of the horizontal grounding electrode: sensitivity analysis of the direct time domain analytical solution," in *2019 URSI International Symposium on Electromagnetic Theory (EMTS)*, San Diego, CA, USA, 2019.

35 S. Šesnić, S. Lalléchère, D. Poljak, A. Šušnjara and K. El Khamlichi Drissi, "Sensitivity analysis of the direct time domain analytical solution for transient impedance of the horizontal grounding electrode using ANOVA approach," *Electric Power Systems Research*, vol. 190, p. 106861, 2021.

36 F. Marcum, *Design of an Image Radiation Monitor for ILS Glide Slope in the Presence of Snow*, Ohio University: Doctoral disertation, 1995.

37 United States Department of Transportation, "Federal Aviation Administration," 2023. [Online]. Available: https://www.faa.gov/ [Accessed 10 January 2023].

38 A. Šušnjara, V. Dorić, S. Lalléchère, D. Poljak, M. Birkić, P. Bonnet and F. Paladian, "Sensitivity analysis of the main lobe direction for glide slope antenna due to snow cover on runway," in *UMEMA 2017 Uncertainty Modeling for Engineering Applications*, Torino, Italy, 2017.

39 D. Poljak, V. Dorić, A. Šušnjara, M. Birkić, S. Lallechere and K. El Khamlichi Drissi, "Deterministic-stochastic modeling of a glide path antenna system above a multilayer," in *2020 International Symposium on Electromagnetic Compatibility - EMC EUROPE*, Rome, Italy, 2020.

40 D. Poljak, V. Dorić and M. Birkić, "Analysis of LPDA radiation above a multilayer," in *2018 26th International Conference on Software, Telecommunications and Computer Networks (SoftCOM)*, Split, Croatia, 2018.

41 A. Šušnjara, V. Dorić and D. Poljak, "Electric field radiated by a dipole antenna and transmitted into a two-layered lossy half space: comparison of plane wave approximation with the modified image theory approach," in *3rd International Conference on Smart and Sustainable Technologies 2018: SpliTech2018*, Split, Hrvatska, 2018.

42 J. H. Bradford, J. T. Harper and J. Brown, "Complex dielectric measurments from ground-penetrating radar data to estimate snow liquid water content in the pendular regime," *Water Resources Research*, vol. 45, no. W0840, p. 12, 2009.

14

A Note on Stochastic Modeling of Plasma Physics Phenomena

Predictive numerical simulations of plasma behavior in tokamaks are of crucial importance in studying international thermonuclear experimental reactor (ITER) operation and could be useful for the interpretation of future ITER experiments. These simulations include magnetohydrodynamics (MHD) equilibrium governed by GSE and transport phenomena described by six transport equations as described in Chapter 8 of this book. A direct extension is the hybrid deterministic–stochastic modeling of the GSE and transport equations featuring the use of FEM (Finite Element Method), SCM (Stochastic Collocation Method), and ANOVA (Analysis of Variance) approaches. Fusion community can benefit from newly developed deterministic-stochastic codes that can predict the confidence interval of the output of interest given the uncertainties of the input variables. Moreover, a sensitivity analysis provides information on the impact of a given random variable at the input to the output of interest. The hybrid stochastic-deterministic approach is new in this field of research and there aren't many publications so far on this topic. In this chapter, one such example is given which serves as an opener to the subject.

14.1   Tokamak Current Diffusion Equation

The current diffusion equation (CDE) governs the current diffusion through the conductive plasma inside a tokamak [1, 2]. The axisymmetric tokamak geometry is depicted in Fig. 14.1.

The evolution of the magnetic field configuration in axisymmetric tokamak devices is represented by a coupled system of the CDE and the Grad–Shafranov equation [3]. The two equations are iterated until they yield consistent results

Deterministic and Stochastic Modeling in Computational Electromagnetics: Integral and Differential Equation Approaches, First Edition. Dragan Poljak and Anna Šušnjara.
© 2024 The Institute of Electrical and Electronics Engineers, Inc.
Published 2024 by John Wiley & Sons, Inc.

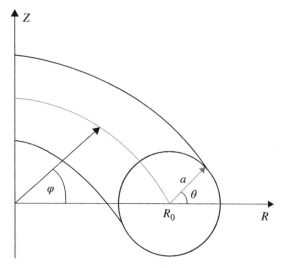

Figure 14.1 The axisymmetric tokamak geometry.

in terms of the magnetic flux and plasma current profiles (ion density, electron density, ion temperatures, electron energy transport, and rotation transport).

The CDE can be written in the following form [1]:

$$\frac{\partial \psi}{\partial t} - \frac{g_2}{\mu_0 \sigma_\| \rho_1^2 g_1} \frac{\partial^2 \psi}{\partial t^2} - \left[\frac{g_2}{\mu_0 \sigma_\| \rho_1^2 g_1} \frac{\partial}{\partial t} \ln\left(\frac{V' g_2}{F}\right) + \frac{x}{\rho_1} \frac{d\rho}{dt} + \frac{x}{2B_0} \frac{dB_0}{dt} \right]$$

(14.1)

where ψ is poloidal magnetic flux, Φ is the toroidal magnetic field flux in a given flux surface, B_0 the vacuum magnetic field at a given R_0, $\sigma_\|$ the parallel conductivity, V' the plasma volume enclosed by a flux surface, and j_{ni} is the noninductive current density. The coordinate ρ is the flux coordinate from the expression $x = \rho/\rho_1$ where ρ_1 is ρ value at the plasma boundary (the last closed flux surface). g_1 and g_2 are computed as $g_1 = \langle 1/R^2 \rangle$ and $g_2 = \langle |grad\ \rho|^2 \rangle$.

The CDE is solved via FEM combined with an implicit time domain backward Euler scheme. More details on deterministic solution can be found in [1].

The stochastic CDE is six-dimensional with six input parameters varied by 10% from their nominal values [4]. Input parameters exhibiting random nature are $g_1(x)$, $g_2(x)$, $V'(x)$, $\sigma_\|(x)$, $F(x)$, and $j_{ni}(x)$. Output parameters of interest are the flux ψ, the flux first derivative $d\psi/dx$ denoted with $d\psi$ and the flux second derivative $d^2\psi/dx^2$ denoted with $d^2\psi$ at time instant of 905 s. The Lagrange stochastic collocation method is applied with Gauss-Legendre collocation points to compute the stochastic mean and variance. Additionally, ANOVA-based sensitivity analysis is carried out to identify the parameters with the highest impact with respect to output flux and its first two derivatives.

Figs. 14.2–14.4 depict the convergence of the stochastic collocation method in computation of mean and variance of poloidal flux and its first two derivatives.

The convergence is tested by consecutively increasing the number of collocation points in one dimension, i.e. 3, 5, and 7, thus resulting in total of 729, 15 625, and 117 649 deterministic simulations. The convergence of SC method in all cases is quite satisfactory; 729 deterministic simulations are needed for computation of mean and variance.

Next, confidence intervals are depicted in Fig. 14.5 for all three outputs of interest. CIs are computed as mean ± 3 standard deviations.

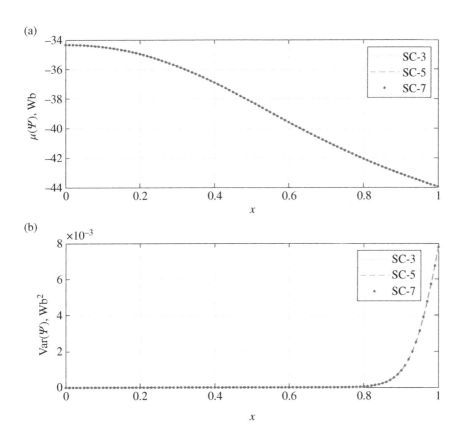

Figure 14.2 The convergence of the poloidal flux (a) mean and (b) variance with three levels of accuracy, SC-3, SC-5, and SC-7, i.e. total of 729, 15 625, and 117 649 deterministic simulations.

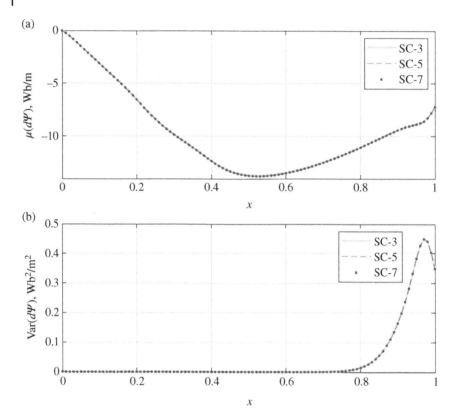

Figure 14.3 The convergence of the poloidal flux first derivative (a) mean and (b) variance with three levels of accuracy, SC-3, SC-5, and SC-7, i.e. total of 729, 15 625, and 117 649 deterministic simulations.

The dispersion around the mean is present at the plasma boundary which greatly affects the boundary conditions for the next time step and hence the plasma profiles for transport equations and their stability. Besides that, the CIs are quite narrow, almost negligible. The dispersion is more pronounced for the derivatives (Neumann boundary conditions) than for the flux. The maximal dispersion for flux is 0.6023%, while for the first derivative is 24.61%.

Furthermore, ANOVA sensitivity analysis is depicted in Figs. 14.6–14.8. The total effect and the first-order sensitivity indices are computed for flux, flux first derivative, and flux second derivative. Total effect and first-order sensitivity indices in all three cases have the same values which means that interactions between the parameters have a negligible impact on variations of the output parameters. Only the total effect indices are depicted.

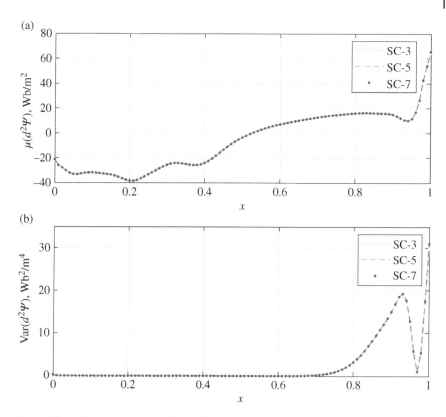

Figure 14.4 The convergence of the poloidal flux second derivative (a) mean and (b) variance with three levels of accuracy, SC-3, SC-5, and SC-7, i.e. total of 729, 15 625, and 117 649 deterministic simulations.

Fig. 14.6 depicts the total effect sensitivity indices for the poloidal flux.

Up to $x = 0.68$, the impact of g_{m2} is dominant. The ranking of other input parameters is as follows: the second place pertains to parameters f and j_{ni}, parameters g_{m1} and σ are at third place, although their impact is rather weak while V' has no impact at all. Between $x = 0.68$ and $x = 0.79$, the impact of σ, f, and j_{ni} decreases while the impact of V' increases. By the end of the interval, V' has the strongest influence on flux variance while the impact of g_{m1} and σ reaches zero. For the rest of the domain, i.e. $x > 0.79$, the impact of f and j_{ni} decreases to zero, while σ and V' have a significant impact, V' being the most influential parameter.

Up to $x = 0.68$, the impact of g_{m2} is dominant. The ranking of other input parameters is as follows: the second place pertains to parameters f and j_{ni}, parameters g_{m1} and σ are at third place, although their impact is rather weak while V' has no

(a) Confidence intervals of poloidal flux versus normalized flux coordinate

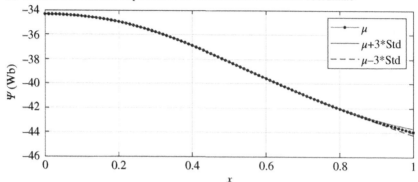

(b) Confidence intervals of poloidal flux first derivative versus normalized flux coordinate

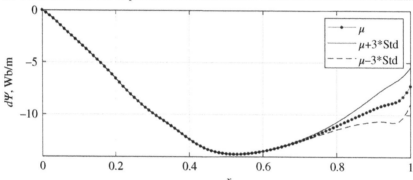

(c) Confidence intervals of poloidal flux second derivative versus normalized flux coordinate

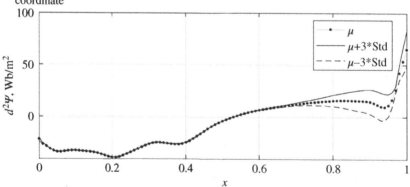

Figure 14.5 Confidence intervals of the (a) poloidal flux, (b) its first and (c) second derivative computed as mean ± 3 standard deviations.

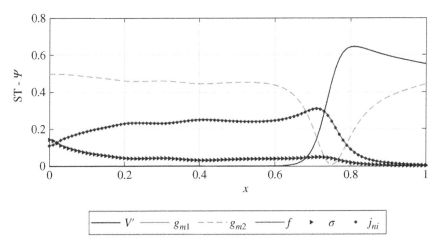

Figure 14.6 Total effect sensitivity indices from ANOVA-based approach for poloidal flux.

impact at all. Between $x = 0.68$ and $x = 0.79$, the impact of σ, f and j_{ni} decreases while the impact of V' increases. By the end of the interval, V' has the strongest influence on flux variance while the impact of g_{m1} and σ reaches zero. For the rest of the domain, i.e. $x > 0.79$, the impact of f and j_{ni} decreases to zero, while σ and V' have a significant impact, V' being the most influential parameter.

Fig. 14.7 depicts the total effect sensitivity indices for the poloidal flux first derivative.

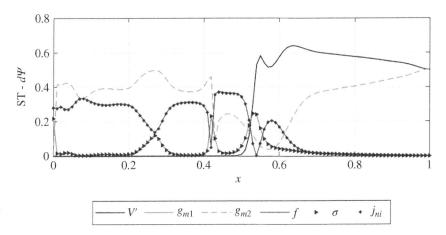

Figure 14.7 Total effect sensitivity indices from ANOVA-based approach for poloidal flux first derivative.

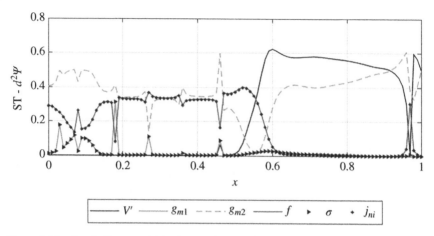

Figure 14.8 Total effect sensitivity indices from ANOVA-based approach for poloidal flux second derivative.

Up to $x = 0.42$, the impact of g_{m2} is dominant. The second most influential parameters are f and j_{ni}, while the third place belongs to parameters g_{m1} and σ. Parameter V' has no impact at all. Between $x = 0.42$ and $x = 0.55$, the impact of f and j_{ni} is the highest, followed by σ. Parameters f and j_{ni} are at the third place while the impact of V' is negligible, although, by the end of the interval, it slowly increases. For the rest of the domain, i.e. $x > 0.55$, the impact of f and j_{ni} decreases to zero, while σ and V' have strong impact, V' being the most influential parameter. At the boundary, $x = 1$, σ, and V' have the same impact.

Finally, Fig. 14.8 depicts the total effect sensitivity indices for the poloidal flux second derivative.

Up to $x = 0.19$, the impact of g_{m2} is dominant. The ranking of other input parameters is as follows: the second place pertains to parameters f and j_{ni}, while the third place belongs to parameters g_{m1} and σ. Parameter V' has no impact at all. Between $x = 0.19$ and $x = 0.42$, the impact of f, j_{ni}, and σ is the strongest, while f and j_{ni} are almost negligible. The impact of V' is completely negligible. In the interval $x = [0.43, 0.55]$, the impact of σ decreases while the impact of V' increases. The impact of f and j_{ni} is the strongest. For the rest of the domain, i.e. $x > 0.56$, the impact of f and j_{ni} decreases to zero, except for small number of observation points. Parameters σ and V' have a significant impact, V' being the most influential parameter for most of the observation points. At the boundary, $x = 1$, σ and V' have the same impact.

References

1 S. Šesnić, V. Dorić, D. Poljak, A. Šušnjara, J. F. Artaud and J. Urban, "A finite element versus analytical approach to the solution of the current diffusion equation in tokamaks," *IEEE Transactions on Plasma Science*, vol. 46, no. 4, pp. 1027–1034, 2018.

2 (EUROfusion MST1 Team) Meyer, H., Eich, T., Beurskens, M., Coda, S., Hakola, A.; Martin, P., ..., Dorić, V., ..., Poljak, D. et al., "Overview of progress in European medium sized tokamaks towards an integrated plasma-edge/wall solution," *Nuclear Fusion*, vol. 57, no. 10, p. 102014, 2017.

3 D. Poljak, S. Šesnić, A. Rubić and E. Maze, "A note on the use of analytical and domain discretisation methods for the analysis of some phenomena in engineering physics," *International Journal for Engineering Modelling*, vol. 31, no. 1–2, pp. 1–2, 2018.

4 A. Šušnjara, D. Poljak, M. Cvetković, H. Dodig, S. Lallechere and K. El Khamlichi Drissi, "A Note on Stochastic Collocation Applied in Dosimetry, Magneto-Hydrodynamics and Ground Penetrating Radar Simulations," UMEMA 2018 Abstracts Collection, Split, Croatia, pp. 38–39, 2018.

Index

IEEE PRESS SERIES ON ELECTROMAGNETIC WAVE THEORY

Time-Harmonic Electromagnetic Fields
Robert F. Harrington

Antenna Theory & Design, Revised Edition
Robert S. Elliott

Differential Forms in Electromagnetics
Ismo V. Lindell

Conformal Array Antenna Theory and Design
Lars Josefsson, Patrik Persson

Multigrid Finite Element Methods for Electromagnetic Field Modeling
Yu Zhu, Andreas C. Cangellaris

Electromagnetic Theory
Julius Adams Stratton

Electromagnetic Fields, Second Edition
Jean G. Van Bladel

Electromagnetic Fields in Cavities: Deterministic and Statistical Theories
David A. Hill

Discontinuities in the Electromagnetic Field
M. Mithat Idemen

Understanding Geometric Algebra for Electromagnetic Theory
John W. Arthur

The Power and Beauty of Electromagnetic Theory
Frederic R. Morgenthaler

Electromagnetic Modeling and Simulation
Levent Sevgi

Multiforms, Dyadics, and Electromagnetic Media
Ismo V. Lindell

Low-Profile Natural and Metamaterial Antennas: Analysis Methods and Applications
Hisamatsu Nakano

From ER to E.T.: How Electromagnetic Technologies Are Changing Our Lives
Rajeev Bansal

Electromagnetic Wave Propagation, Radiation, and Scattering: From Fundamentals to Applications, Second Edition
Akira Ishimaru

Time-Domain Electromagnetic Reciprocity in Antenna Modeling
Martin Štumpf

Boundary Conditions in Electromagnetics
Ismo V. Lindell, Ari Sihvola

Substrate-Integrated Millimeter-Wave Antennas for Next-Generation Communication and Radar Systems
Zhi Ning Chen, Xianming Qing

Electromagnetic Radiation, Scattering, and Diffraction
Prabhakar H. Pathak, Robert J. Burkholder

Electromagnetic Vortices: Wave Phenomena and Engineering Applications
Zhi Hao Jiang, Douglas H. Werner

Advances in Time-Domain Computational Electromagnetic Methods
Qiang Ren, Su Yan, Atef Z. Elsherbeni

Foundations of Antenna Radiation Theory: Eigenmode Analysis
Wen Geyi

Advances in Electromagnetics Empowered by Artificial Intelligence and Deep Learning
Sawyer D. Campbell, Douglas H. Werner

Deterministic and Stochastic Modeling in Computational Electromagnetics: Integral and Differential Equation Approaches
Dragan Poljak, Anna Šušnjara

Printed and bound by CPI Group (UK) Ltd, Croydon, CR0 4YY

16/04/2025

14658418-0005